MIGRATION AND CLIMATE CHANGE

Migration and Climate Change provides the first authoritative overview of the relationship between climate change and migration, bringing together both case studies and syntheses from different parts of the world. It discusses policy responses, normative issues and critical perspectives from the point of view of human rights, international law, political science, and ethics, and addresses the concepts, notions and methods most suited to confronting this complex issue. The book constitutes a unique and thorough introduction to one of the most discussed but least understood consequences of climate change and brings together experts from different disciplines, including anthropology, climatology, demography, geography, law, political science and sociology, providing a valuable synthesis of research and debate.

ETIENNE PIGUET is professor of geography at the University of Neuchâtel (Switzerland) and nominated expert for the Intergovernmental Panel on Climate Change 5th report (working group II on the impacts, adaptation and vulnerability to climate change). He specializes in migration and refugee studies and is a member of the steering committee of the population geography commission/International Geographic Union. He is an internationally recognized expert on the links between human migration and environmental degradation. Author of several books on migration policy and immigrant minorities in Switzerland, as well as of a 2008 report on climate change and migration for the Office of the High Commissioner for Refugees.

ANTOINE PÉCOUD has been with UNESCO's Section on International Migration and Multicultural Policies since 2003 and is a research associate at the Unité de Recherche Migrations et Société, University of Paris VII, and at Migrations Internationales, Espaces et Sociétés, University of Poitiers (France). He holds a B.A. from the University of Lausanne and a D.Phil. in social and cultural anthropology from the University of Oxford. His research has focused on migration management, immigrant entrepreneurship in Germany and the human rights implications of international migration. Co-edited books include *Migration Without Borders. Essays on the Free Movement of People* (2007); *Migration and Human Rights. The United Nations Convention on Migrant Workers' Rights* (2009); and *The Politics of International Migration Management* (2010).

PAUL DE GUCHTENEIRE is head of the Programme on International Migration and Multicultural Policies at UNESCO and director of the journal *Diversities*. He has worked as an epidemiologist at the Netherlands Cancer Research Foundation and is a past director of the Steinmetz Institute of the Royal Netherlands Academy of Arts and Sciences and a former president of the International Federation of Data Organizations. His current research focuses on the human rights dimension of international migration and the development of policies for migration management at the international level. His publications include *Democracy and Human Rights in Multicultural Societies* (2007), *Migration without Borders: Essays on the Free Movement of People* (2007), *Migration and Human Rights: The United Nations Convention on Migrant Workers' Rights* (2009), as well as several works on data collection and analysis in the social sciences.

MIGRATION AND CLIMATE CHANGE

Edited by

ETIENNE PIGUET
(University of Neuchâtel)

ANTOINE PÉCOUD
(UNESCO)

and

PAUL DE GUCHTENEIRE
(UNESCO)

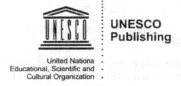

UNESCO Publishing

United Nations
Educational, Scientific and
Cultural Organization

CAMBRIDGE
UNIVERSITY PRESS

CAMBRIDGE
UNIVERSITY PRESS

University Printing House, Cambridge CB2 8BS, United Kingdom

One Liberty Plaza, 20th Floor, New York, NY 10006, USA

477 Williamstown Road, Port Melbourne, VIC 3207, Australia

314-321, 3rd Floor, Plot 3, Splendor Forum, Jasola District Centre, New Delhi - 110025, India

79 Anson Road, #06-04/06, Singapore 079906

Cambridge University Press is part of the University of Cambridge.

It furthers the University's mission by disseminating knowledge in the pursuit of education, learning and research at the highest international levels of excellence.

www.cambridge.org
Information on this title: www.cambridge.org/9781107662254

First published 2011

A catalogue record for this publication is available from the British Library

Library of Congress Cataloging in Publication data
Migration and climate change / edited by Etienne Piguet, Antoine Pécoud, Paul de Guchteneire.
p.   cm.
ISBN 978-1-107-01485-5 (hardback)
1. Climatic changes – Social aspects.   2. Global environmental change – Social aspects.   3. Migration, Internal – Environmental aspects.   4. Emigration and immigration – Environmental aspects.   I. Piguet, Etienne.   II. Pécoud, Antoine.   III. Guchteneire, P. F. A. de (Paul F. A. de)   IV. Title.
QC903.M54   2011
304.8–dc22
2011011605

ISBN   978-1-107-01485-5   Hardback
ISBN   978-1-107-66225-4   Paperback

CONTENTS

FIGURES

TABLES

NOTES ON CONTRIBUTORS

SUSANA B. ADAMO is an associate research scientist at the Center for International Earth Science Information Network (CIESIN), part of the Earth Institute of Columbia University (United States). She is also adjunct assistant professor at SIPA (School of International and Public Affairs), and co-coordinator of the Population and Environment Research Network (PERN). A social demographer with a background in geography, her research interests include the dynamics of internal migration in developing countries, migration–environment interactions, and rural/urban demography. Originally from Argentina, prior to joining CIESIN she was a postdoctoral scholar at the Carolina Population Center, and a professor at the National University of Córdoba and the Facultad Latinoamericana de Ciencias Sociales (FLACSO)-Buenos Aires. She holds an M.A. in population studies from FLACSO-México and a Ph.D. in demography/sociology from the University of Texas at Austin.

TAMER AFIFI is Associate Academic Officer in the Environmental Migration, Social Vulnerability and Adaptation (EMSVA) section of United Nations University Institute for Environmental and Human Security (Germany). He mainly works in the area of Environmental Change and Forced Migration. His work focuses on economic aspects and modelling techniques in association with environmental migration, with a focus on North Africa and the Middle East (MENA), Sub-Saharan Africa and Latin America. He holds a Ph.D. in economics from the University of Erlangen-Nürnberg (Germany), an M.A. in economics of development from the Institute of Social Studies, The Hague (Netherlands), and a B.Sc. in economics from the Faculty of Economics and Political Science, Cairo University, Egypt. Tamer has worked for five years as a lecturer of economics at the Faculty of Commerce, Sohag University, Egypt. Prior to his work at UNU-EHS, Tamer joined the Bonn International Graduate School of the Center for Development Research, University of Bonn.

ALISSON FLÁVIO BARBIERI holds a B.Sc. in economics and a Master's degree in demography from the Universidade Federal de Minas Gerais (UFMG, Brazil), and a Ph.D. in city and regional planning at the University of North Carolina at Chapel Hill (United States). His research focuses on the relationship between population and economic change and environmental dynamics, population mobility and urbanization, human dimensions of global environmental changes, and design and analysis of planning and public policies in developing countries. He has conducted or participated in several research projects on Latin America, particularly in the Amazon and large metropolitan areas of Brazil. He is currently associate professor of demography at the UFMG, researcher at the Center for Regional Development and Planning (CEDEPLAR/ UFMG), and associate faculty at the Department of International Health, Johns Hopkins University (United States).

PRATIKSHYA BOHRA-MISHRA is a Ph.D. candidate at the Woodrow Wilson School and Office of Population Research at Princeton University (United States). After completing high school in Nepal, she came to the US to study for an undergraduate degree. She worked as an economic and financial consultant at LECG (Law and Economic Consulting Group) in New York City before starting a Ph.D. in public policy and demography and has focused her research on the determinants of migration and remittance; the impact of remittance; the relationship between migration and violence as well as environmental degradation; and immigrant assimilation. She has authored and co-authored several papers in journals such as *Demography, International Migration Review*, and *Journal of Ethnic and Migration Studies*.

STEPHEN CASTLES is research professor of sociology at the University of Sydney (Australia) and Associate Director of the International Migration Institute (IMI), University of Oxford (United Kingdom). Until August 2009 he was professor of migration and refugee studies at the University of Oxford and Director of IMI. A sociologist and political economist, he works on international migration dynamics, global governance, migration and development, and regional migration trends in Africa, Asia and Europe. From 2001 to 2006, he was Director of the Refugee Studies Centre at Oxford University. He has been an adviser to the Australian and British governments, and has worked for the ILO, the IOM, the European Union and other international bodies. Recent publications include *The Age of Migration: International*

Population Movements in the Modern World (with Mark Miller, 4th edn, 2009); *Migration, Citizenship and the European Welfare State: A European Dilemma* (with Carl-Ulrik Schierup and Peo Hansen, 2006); and *Migration and Development: Perspectives from the South* (edited with Raúl Delgado Wise, 2008).

ULISSES E. C. CONFALONIERI holds degrees in medicine, veterinary medicine and biological sciences. He started his career doing laboratory and field research on infectious diseases in Brazil. In the early 1980s, he became involved with health care and health policy issues for rural and tribal communities in the Brazilian Amazon region. Following the Rio Earth Summit in 1992, he became interested in environmental issues. He has since contributed to IPCC reports both as coordinating lead author and review editor. He was also a convening lead author for the Millennium Ecosystem Assessment. His current research focuses on the influences of climate variability and change, as well as ecosystem, biodiversity and land-cover changes upon human population health, especially on tropical and infectious diseases. He is currently professor at the National School of Public Health of the Oswaldo Cruz Foundation, Brazilian Ministry of Health, and professor of public health at the Federal University in Rio de Janeiro. He is author or co-author of some eighty scientific articles or book chapters.

CHRISTEL COURNIL is lecturer in public law at the University of Paris XIII (France). Her research interests include migration law, human rights and environmental law. She has published *Le statut interne de l'étranger et les normes supranationales* (2005) and coedited *Les changements climatiques et les défis du droit* (with Catherine Colard Fabregoule, 2009).

EMMANUEL DAVID is assistant professor of sociology at Villanova University (United States). He received a Ph.D. in sociology from the University of Colorado at Boulder, where he taught in the Women and Gender Studies Program. His research and teaching interests include the sociology of gender, social movements, cultural memory, and the social dimensions of disaster. His dissertation was awarded Honorable Mention in the 2009 Samuel H. Prince Dissertation Award competition, which is sponsored by the International Sociological Association's Research Committee on Disasters. He has published several articles and book chapters on gender and Hurricane Katrina.

PAUL DE GUCHTENEIRE is head of the Programme on International Migration and Multicultural Policies at UNESCO and director of the journal *Diversities*. He has worked as an epidemiologist at the Netherlands Cancer Research Foundation and is a past director of the Steinmetz Institute of the Royal Netherlands Academy of Arts and Sciences and a former president of the International Federation of Data Organizations. His current research focuses on the human rights dimension of international migration and the development of policies for migration management at the international level. His publications include *Democracy and Human Rights in Multicultural Societies* (2007), *Migration without Borders: Essays on the Free Movement of People* (2007), *Migration and Human Rights: The United Nations Convention on Migrant Workers' Rights* (2009), as well as several works on data collection and analysis in the social sciences.

ALEX DE SHERBININ is a senior researcher at the Center for International Earth Science Information Network (CIESIN) of Columbia University's Earth Institute, deputy manager of the NASA Socioeconomic Data and applications Center, and co-coordinator of the Population-Environment Research Network. A geographer whose research focuses on the human aspects of environmental change at local, national and global scales, he has held prior positions at the Population Reference Bureau and the International Union for Conservation of Nature.

CHARLES EHRHART is currently coordinating CARE International's global response to climate change. In this role, he leads a Secretariat providing strategic direction and technical support to operations in around seventy of the world's poorest countries. Charles holds a Ph.D. in social anthropology from the University of Cambridge (United Kingdom), and has published and presented widely on poverty and environmental change.

ASTRID EPINEY is professor in European and public international law at the University of Fribourg (Switzerland) and Director at the Institute of European Law. Her teaching, research and publications lie mainly in the fields of European and international environmental law, European constitutional law, fundamental freedoms in the European Union, transport law and European migration law as well as the relationship between Switzerland and the European Union.

ALLAN FINDLAY is professor of geography at the University of Dundee (Scotland, United Kingdom). An editor of the international journal *Population, Space and Place*, he is chair of the IGU Commission on Population Geography (2008–12) and former chair of the Commission on Population and Vulnerability (2004–08). Over the last decade he has been privileged to learn from populations affected by sudden-onset environmental events in India, Bangladesh and the Philippines, as well as studying adaptation to climate change in Malawi.

ALISTAIR GEDDES is a lecturer in human geography and Research Councils UK academic fellow at the University of Dundee (Scotland, United Kingdom). His background is in applied quantitative analysis of spatially referenced data sets, including use of geographic information systems. Since moving to Dundee, his work has become increasingly focused on human mobility, including recently completed work on UK students abroad and current projects on the scope and experience of forced labour among migrant workers in the UK. He holds a Ph.D. in geography from Pennsylvania State University, and earlier degrees from the University of Edinburgh.

FRANÇOIS GEMENNE is a research fellow affiliated with the Institute for Sustainable Development and International Relations (IDDRI, Sciences Po Paris), as well as with the Centre for Ethnic and Migration Studies (CEDEM, University of Liège). He teaches at Sciences Po Paris, the University of Paris 13 and the Free University of Brussels. His research deals with populations displaced by environmental changes and the policies of adaptation to climate change. He has conducted field studies in New Orleans after Hurricane Katrina, in the archipelago of Tuvalu, as well as in China, Central Asia and Mauritius. Between 2007 and 2009, he supervised the research clusters on Asia-Pacific and Central Asia of the EU research project EACH-FOR. He has consulted for the International Organization of Migration (IOM) and the Asian Development Bank (ADB). He holds a joint doctorate in political science from Sciences Po Paris and the University of Liège, as well as a Master in Development, Environment and Societies from the University of Louvain and a Master of Research in Political Science from the London School of Economics. Publications include *Anticiper pour s'adapter* (with L. Tubiana and A. Magnan, 2010), *Géopolitique du changement climatique* (2009) and *Nations and their Histories: Constructions and Representations* (co-edited with Susana Carvalho, 2009).

GRAEME HUGO is a professor with the Department of Geographical and Environmental Studies and Director of the National Centre for Social Applications of Geographic Information Systems at the University of Adelaide (Australia). His research interests are in population issues in Australia and South-East Asia, especially migration. He is the author of over 300 books, articles in scholarly journals and chapters in books, as well as a large number of conference papers and reports. His recent research has focused on migration and development, environment and migration, and migration policy. In 2009 he was awarded an ARC Australian Professorial Fellowship over five years for his research project, Circular Migration in Asia, the Pacific and Australia: empirical, theoretical and policy dimensions. In 2009 with colleagues he completed a study on Climate Change and Migration in Asia and the Pacific for the Asian Development Bank.

LORI M. HUNTER received her Ph.D. in sociology and population studies from Brown University (United States). She is associate professor of sociology and environmental studies with the University of Colorado at Boulder, where she is also a faculty research associate with the Institute of Behavioral Science Research Programs on Population, Environment & Society. Her research and teaching focus on human/ environment interactions, with specific examinations of public perception of environmental issues and the associations between human population dynamics and environmental context. Her research has appeared in *Population Research and Policy Review, Social Science Quarterly,* and *Society and Natural Resources.* Currently editor-in-chief of the journal *Population and Environment,* she is a former chair of the Steering Committee of the Population-Environment Research Network, and an Associate Director of the University of Colorado's Population Center.

KHALID KOSER is Academic Dean, head of the New Issues in Security Programme, and director of the New Issues in Security Course (NISC) at the Geneva Centre for Security Policy (Switzerland). He is also non-resident Senior Fellow in foreign policy studies at the Brookings Institution and research associate at the Graduate Institute of International and Development Studies in Geneva. Chair of the United Kingdom's Independent Advisory Group on Country Information, he is co-editor of the *Journal of Refugee Studies* and on the editorial boards of *Ethnic and Racial Studies* and *Forced Migration Review.* With extensive field experience in Afghanistan, the Balkans, the Horn of Africa, Southern

Africa and Western Europe, he has published widely on international migration, asylum, refugees and internal displacement. His most recent book is *International Migration: A Very Short Introduction* (2008).

MICHELLE T. LEIGHTON is a US Fulbright Scholar in Kyrgyzstan and Munich Re Foundation Chair on Social Vulnerability for the United Nations University – Environment and Human Security Institute. She is affiliated with the University of San Francisco School of Law, formerly as Director of Human Rights Programs for the Center for Law and Global Justice, and is a member of the German Marshall Fund's transatlantic study team on climate-induced migration. As a specialist in international human rights law, desertification and migration, she has counselled NGOs, government and intergovernmental agencies, including the UN Environment Programme, CCD Secretariat, UN Food and Agriculture Organization, International Organization for Migration, and the US Congressional Commission on Immigration Reform. She served as the Acting Dean of John F. Kennedy University Law School and on the faculties of the Kyrgyz Academy of Law, Bishkek; East China University of Politics and Law, Shanghai; UC Berkeley School of Law, University of San Francisco; and Golden Gate University. She has published numerous articles in books and journals. Ms Leighton received her LL.M degree from the London School of Economics and Political Science, J.D. with honours from Golden Gate University, and BA from University of California at Davis. In 1989, she co-founded the Natural Heritage Institute in San Francisco.

DOUGLAS S. MASSEY is the Henry G. Bryant professor of sociology and public affairs at Princeton University and was formerly on the faculties of the University of Pennsylvania and the University of Chicago. He is past President of the American Sociological Association and the Population Association of America, and current President of the American Association for Political and Social Science. He is a member of the US National Academy of Sciences and the American Academy of Arts and Sciences. His most recent publication is *Brokered Boundaries: Constructing Immigrant Identity in Anti-Immigrant Times* (co-authored with Magaly Sanchez, 2010).

JANE MCADAM holds B.A. and LL.B. degrees from the University of Sydney (Australia) and a D.Phil. from the University of Oxford (United Kingdom). She is an associate professor in the Faculty of Law at the

University of New South Wales (Australia). She is also director of research in the School of Law; director of International Law Programs in the Faculty of Law; and director of the International Refugee and Migration Law project at the Gilbert + Tobin Centre of Public Law. A research associate at the University of Oxford's Refugee Studies Centre and director of its International Summer School in Forced Migration in 2008, she previously taught in Sydney's Faculty of Law and at Oxford's Lincoln College. She is the author of *Complementary Protection in International Refugee Law* (2007); *The Refugee in International Law* (with G. S. Goodwin-Gill, 3rd edn, 2007); and the editor of *Forced Migration, Human Rights and Security* (2008). Associate rapporteur of the Convention on Refugee Status and Subsidiary Protection Working Party for the International Association of Refugee Law Judges and adviser to the United Nations High Commissioner for Refugees on the legal aspects of climate-related displacement; she has also been a consultant to the Australian and British governments on migration issues.

ANTHONY OLIVER-SMITH is emeritus professor of anthropology at the University of Florida (United States). He is affiliated with the Center for Latin American Studies and the School of Natural Resources and Environment at the same university. He also held the Munich Re Foundation Chair on Social Vulnerability at the United Nations University Institute on Environment and Human Security in Bonn from 2005 to 2009. He has done anthropological research and consultation on disasters and involuntary resettlement in Peru, Honduras, India, Brazil, Jamaica, Mexico, Japan and the United States. His work on disasters has focused on issues of post-disaster aid and reconstruction, vulnerability analysis and social organization, including class/race/ethnicity/gender-based patterns of differential aid distribution, social consensus and conflict, and social mobilization of community-based reconstruction efforts. His work on involuntary resettlement has focused on the impacts of displacement, place attachment, resistance movements, and resettlement project analysis. He has authored, edited or co-edited eight books and over sixty journal articles and book chapters.

ANTOINE PÉCOUD has been with UNESCO's Section on International Migration and Multicultural Policies since 2003 and is a research associate at the Unité de Recherche Migrations et Société, University of Paris VII, and at Migrations Internationales, Espaces et Sociétés, University of Poitiers (France). He holds a B.A. from the University of Lausanne and a D.Phil. in social and cultural anthropology from the University of Oxford. His research has focused on migration management, immigrant entrepreneurship in Germany and the human rights implications of international migration. Co-edited books include *Migration Without Borders. Essays on the Free Movement of People* (2007); *Migration and Human Rights. The United Nations Convention on Migrant Workers' Rights* (2009); and *The Politics of International Migration Management* (2010).

ETIENNE PIGUET, who received his Ph.D. from the University of Lausanne in 1998, is professor of geography at the University of Neuchâtel (Switzerland), where he holds the Geography of Mobilities chair. He is a member of the population geography commission steering committee of the International Geographic Union. His specialization is migration studies and his publications and research fields include immigration policies, labour market issues, ethnic and minority business, asylum-seekers and refugees, and urban segregation. His most recent research interests focus on the links between migration and climate change/environmental degradation.

MARTINE REBETEZ is a climatologist, specializing in climate change research. She is a senior scientist at the Swiss Federal Institute WSL for Forest, Snow and Landscape Research, and an associate professor of physical geography at the University of Neuchatel, Switzerland. She has published numerous research papers on climate change, mainly focused on changes in climate parameters occurring in the Alps and in Europe, as well as on their consequences. She was also a reviewer of the 2007 IPCC report.

KOKO WARNER is the Head of the Environmental Migration, Social Vulnerability, and Adaptation Section at UNU-EHS. Warner is a Lead Author for IPCC's 5th Assessment Report, Working Group 2 on Adaptation (chapter 20). Warner researches risk management strategies of the poor in adapting to changing environmental and climatic conditions. Warner serves on the science and management boards of the European Commission, such as EACH-FOR (a first-time global survey of environmentally induced migration in 23 countries) and CLICO (Climate Change, Hydro-conflicts and Human Resources). She is Co-Chair of the German Marshall Fund project on Climate Change and

Migration with Susan Martin. She helped found and is on the Steering Committee of the Climate Change, Environment, and Migration Alliance (CCEMA) and works extensively in the context of the UNFCCC climate negotiations on adaptation (particularly in risk management and migration). She oversees the work of the Munich Re Foundation Chair on Social Vulnerability project at UNU-EHS, a network of seven endowed professors and a community of scholars working on related topics. Koko is the UNU focal point to the UNFCCC, and focal point for climate adaptation and the Nairobi Work Programme. She is a member of the UN's Interagency Standing Committee, Task Force on Climate Change, Migration and Displacement. Warner was a Fulbright Fellow, and holds a master's degree in international economic development and environmental economics from George Washington University and earned a Ph.D. in environmental economics from the University of Vienna's Department of Economics. She was a researcher at IIASA, ETH Zürich's Swiss Federal Institute for Snow and Avalanche Research, prior to joining UNU-EHS. She has published in *Nature*, *Climate Policy*, *Global Environmental Change*, *Disasters*, *Environmental Hazards*, *Natural Hazards*, the *Journal of Population and Environment* and the *Geneva Papers on Risk and Insurance – Issues and Practice*, and has written numerous book chapters. She serves on the editorial board of the *International Journal of Global Warming*.

Introduction: migration and climate change

ETIENNE PIGUET, ANTOINE PÉCOUD AND PAUL DE
GUCHTENEIRE

Climate change has become a major concern for the international community. Among its consequences, the impact on migration is increasingly
attracting the attention of policy-makers and researchers. Yet knowledge in
this field remains limited and fragmented: there are uncertainties surrounding the actual mechanisms at stake, the number of persons affected and the
geographical zones concerned; there are debates between those who stress
the direct impact of the environment on population flows and those who
rather insist on the social, economic and political contexts in which such
flows occur; different disciplines make their respective inputs to the literature, ranging from empirical case studies to analytical discussions.
Moreover, the available information is heterogeneous, as research outcomes
coexist with numerous 'grey' publications, such as policy reports (Barnett
and Webber, 2009; WBGU, 2008; IPCC, 2007; Stern, 2007), advocacy
brochures by IGOs and NGOs (Jakobeit and Methmann, 2007; Christian
Aid, 2007; CARE et al., 2009) and conference proceedings (IOM/UNFPA,
2008; IOM, 2009; Afifi and Jäger, 2010).

This volume therefore provides a comprehensive overview of the
climate change–migration nexus. It presents empirical insights on the
links between climate change, the environment and migration, while
bringing together case studies and synthesis from disciplines such as
anthropology, climatology, demography, geography, law, political science and sociology. It investigates the key issues raised by the climate
change–migration nexus, including the social and political context in
which the topic emerged; states' policy responses and the views of different institutional actors; critical perspectives on the actual relationship
between the environment and (forced) migration; the concepts most
adequate to address this relationship; gender and human rights implications; as well as international law and policy orientations.

Two major interconnected arguments arise in the contributions. The first concerns the weight of environmental and climatic factors in migration and their relationship to other push or pull factors, whether of a social, political or economic nature. Understanding the role of the environment in migration dynamics implies analysing how and why people are vulnerable to climate change, as well as examining the different strategies they develop to cope with (or adapt to) environmental stress – migration being one among other such strategies. The second argument is about the political framework in which such migration flows should take place and how to treat the people who move in connection with environmental factors. This implies a discussion of the possible protection to be granted to those in situations of vulnerability and the responsibilities of states and of the international community in providing such protection. The two issues are deeply intertwined, as the extent to which the environment determines migration is intimately connected to the status associated with the people concerned.

This introduction first provides a short historical overview of the debate, then discusses the impact on migration of three major environmental factors linked to climate change (tropical cyclones, heavy rains and floods; drought and desertification; and sea level rise). The following sections explore the core issues that run through the volume: the plurality of factors that shape migratory dynamics, the social determinants of people's vulnerability to climate change, the diversity in the migration patterns associated with climate change, and issues of data collection and methodology. The different concepts used by researchers in the field, along with their analytical and political implications, are reviewed, which leads to a discussion of the legal implications of environmental migration and the responsibilities of states. The last section explores the possible policy orientations to address the climate change–migration nexus.

A short history of the debate

Environmental migration is an issue that is commonly presented as 'new' or as part of 'future trends'. Yet, as several chapters recall, it is a long-standing phenomenon: for example, Michelle Leighton (Chapter 13) provides evidence that desertification and droughts have always been closely associated with the movement of people (see also Jane McAdam, Chapter 5, and Anthony Oliver-Smith, Chapter 7).

Environmental factors ranked highly in the first systematic theories of migration. In 1889, Ravenstein (1889, p. 286) mentioned 'unattractive climate' as 'having produced and still producing currents of migration' (along with 'bad or oppressive laws, heavy taxation, uncongenial social surroundings and compulsion' and, most importantly in his view, economic motivations). The American geographer Ellen Churchill Semple later wrote that 'the search for better land, milder climate and easier conditions of living starts many a movement of people which, in view of their purpose, necessarily leads them into an environment sharply contrasted to their original habitat' (1911, p. 143). However, despite these early historical insights, references to the environment as an explanatory factor were to progressively disappear from the migration literature over the course of the twentieth century. Indeed, core publications such as J. W. Gregory (1928), Donald R. Taft (1936) or Julius Isaac (1947) do not mention environmental factors. The same applies to Zelinsky's hypothesis on 'mobility transition' (1971) and to Stouffer's 'intervening opportunities' approach (1940; 1960). The environment is also absent from neoclassical economic theory (Harris and Todaro, 1970), as well as from the so-called 'ecological models' (Sly and Tayman, 1977).[1] Since the late 1980s, there have been numerous theoretical publications on migration, but without any mention of environmental factors.[2]

Four main trends explain this decreasing interest in natural or environmental factors. First, according to a powerful Western-centric idea, technological progress would decrease the influence of nature on human life; Petersen (1958) thus views environmental migration as a 'primitive' form of migration bound to decline as human beings gradually increase their control over their environment. Second, environment-based explanations of migration were progressively rejected for their supposedly deterministic nature, to the benefit of socio-cultural approaches or Marxist/economic perspectives. A third reason is the rise of the economic paradigm in migration theory: while already present in Ravenstein's work, economic factors were given the most central role, whether in Marxism-inspired or neoclassical research (Harris and Todaro, 1970; Castles and Kosack, 1973).[3]

---

[1] When the term 'environment' is used in this context, it has nothing to do with natural variables but refers to population factors such as the density of habitation, the ethnic composition of neighbourhoods, etc.

[2] See notably Salt (1987); Portes and Böröcz (1996); Zolberg et al. (1989); Massey et al. (1993); Massey et al. (1998); Arango (2000); Geyer (2002); Ghatak et al. (1996); Cohen (1995); Hammar et al. (1997); Brettell and Hollifield (2007). One notable exception is Richmond (1994) (see François Gemenne, Chapter 9).

[3] Note nevertheless that environmental factors are implicit in the new economics of migration; households' collective risk strategies in rural societies include, for example,

Finally, forced migration studies, while they could have included environmentally induced displacements, rather developed upon a strong political premise according to which 'states make refugees' (Marx, 1990).

It is in this intellectual context that 'environmental migrants' came back into the picture, as one of the pressing issues raised by climate change (see François Gemenne, Chapter 9). In the 1980s and early 1990s, a few landmark publications raised the issue and provided alarmist estimates of the number of people foreseen to move; Norman Myers (1993) argued for example that up to 150 million environmental refugees were to be expected by the end of the twenty-first century (see also El-Hinnawi, 1985; Jacobson, 1988). In 1990, the first UN intergovernmental report on climate change stated that 'the gravest effects of climate change may be those on human migration as millions will be displaced' (IPCC-1, 1990, p. 20). And in 1994, para. 10.7 of the *Programme of Action of the International Conference on Population and Development* (held in Cairo and widely understood as the first major occurrence of migration issues in international debates) stated that 'Governments are encouraged to consider requests for migration from countries whose existence, according to available scientific evidence, is imminently threatened by global warming and climate change' (ICPD, 1994).

As Gemenne argues, these early research and policy discussions were heavily embedded in a climate change agenda, characterized by a strategy to raise awareness surrounding the potential impact of climate change on migration – and on security at large. In this approach, 'environmental migrants' were portrayed as forced to leave their country and as moving exclusively for climate change-related reasons, while the tone of the debate was future-oriented – hence favouring usually alarmist predictions rather than empirical analysis of already existing flows. This clearly clashed with most migration researchers' convictions and led to a long-standing divide between natural and social scientists: while the former took for granted the interrelation between environmental deterioration and migration and stressed the very high number of people concerned, the latter considered the environment as, at most, one driver of migration among many others and were very cautious regarding the estimates put forward (Black, 2001; Castles, 2002). As Stephen Castles adds (Chapter 16), alarmist predictions that aimed at sensitizing

droughts or other environmental factors (thus motivating the emigration of part of the household, see Stark and Bloom, 1985).

governments and public opinions rather contributed to further stigmatize migrants from low-income countries, while migration researchers reacted in a very defensive way that did little to favour a sound debate between disciplines.

Today it would seem that, although the debate still goes on, the disciplinary divide is gradually being overcome: environmental scientists tend to be more cautious while migration specialists do recognize the role of the natural environment in migration dynamics.[4] On the whole, most scholars now dismiss the apocalyptic predictions that used to influence debates; there is also a consensus on the fact that available evidence regarding the processes at stake is still far from satisfactory.[5] Yet, in a context in which climate change has become an overarching priority for a wide range of actors worldwide, the vision of 'climate refugees' escaping environmental disasters remains a powerful way to catch the imagination of the public – hence the numerous initiatives taken by politicians, environmental activists, international organizations and, to a certain extent, by lawyers, climatologists or social scientists (CARE et al., 2009; Biermann and Boas, 2010; Collectif Argos, 2010). Alarmist future predictions thus remain popular; as Nicholas Stern wrote in his 2007 report on the economic consequences of global warming: 'Greater resource scarcity, desertification, risks of droughts and floods, and rising sea levels could drive many millions of people to migrate' (Stern, 2007, p. 20).

In sum, there are at least three lessons to be learnt from this history of the debate. First, the controversy between natural and social scientists is deeply rooted in intellectual history and the weight given to environmental factors in migration dynamics is therefore both a matter of 'hard facts' and of intellectual traditions; thus a single historical migratory event can be initially understood in environmental terms, and be later reframed in economic or political terms.[6] In this respect, the current focus on environmental migration appears less as a 'new' research issue

---

[4] It is even among social (rather than natural) scientists that some of the most doomsday-like predictions can be found; e.g. Reuveny (2008) writes that rich countries 'may ultimately lose control over incoming migration' because of environmental degradations.

[5] For recent studies and synthesis illustrating these trends, see Hugo (2008), Kniveton et al. (2008), Piguet (2008), Barnett and Webber (2009), Jäger et al. (2009), Morrissey (2009), Tacoli (2009), Renaud et al. (2007), Boano et al. (2008), Brown (2008), Perch-Nielsen et al. (2008), Jonsson (2010).

[6] Examples of this paradigmatic shift include the Irish famine exodus of the mid-nineteenth century and the 1930s droughts in the American Dust Bowl, which are nowadays reinterpreted as complex socio-political processes rather than 'simple' environmental disasters (Scally, 1995; King, 2007; McLeman et al., 2008).

than as an expression of another paradigmatic shift. Second, this field of study is inherently political, which means that research and statements regarding the climate change–migration nexus are very hard to dissociate from the highly politicized debate on climate change itself. Third, as a result of this specific history, this field of study is contested while poor in empirical evidence. The bibliometric study provided in Allan Findlay and Alistair Geddes (Chapter 6) shows how terms such as 'environmental migrants' have been increasingly used over the last two decades, but with a surprisingly low number of in-depth studies; it would seem that many people use the term, but that very few actually do research.

Before proceeding to examination of the core issues raised by the contributions to this volume, the next section reviews the available knowledge on three main environmental factors that are predicted to grow in significance due to climate change in the years to come (see Martine Rebetez, Chapter 2) and that are held to have an impact on migration: (1) the increase in strength and frequency of tropical cyclones, heavy rains and floods; (2) droughts and desertification; (3) sea level rise.

Tropical cyclones, heavy rains and floods

Tropical cyclones,[7] storms and floods are typical examples of rapid-onset phenomena impacting on population displacement. The approximate estimates of the number of persons already affected yearly by flooding (99 million between 2000 and 2008[8]) and by tropical cyclones and storms (39 million) give an idea of the amplitude of the threat (Rodriguez et al., 2009), but the number of people who would be affected by a climate change-induced increase of such disasters is very difficult to estimate. No climate model is indeed able to accurately predict the exact localization and timing of such disasters and there is therefore no certainty as to whether or not the affected zones will be densely populated.

According to a number of detailed studies,[9] rapid-onset phenomena lead overwhelmingly to short-term internal displacements rather than

---

[7] We use the generic term 'tropical cyclone' to include hurricanes (western Atlantic/eastern Pacific), typhoons (western Pacific), cyclones (southern Pacific/Indian Ocean), tropical storm, etc.

[8] We use the classification of natural disasters from International Disaster Database EM-DAT http://www.emdat.be/classification (Rodriguez et al., 2009). Floods are classified as hydrological disasters whereas hurricanes are labelled as meteorological disasters.

[9] See in particular Lonergan (1998), Hunter et al. (2003), Kliot (2004), Paul (2005), Pais and Elliott (2008), Poncelet (2008).

long-term or long-distance migration. This is linked to the fact that victims, who live mainly in poor countries, lack the resources to move. They tend to stay where they live or to move only within a short distance. Moreover, many return and reconstruct their homes in the disaster zone. A synthesis of results on the fate of victims of natural disasters displaced in eighteen sites showed (already twenty years ago) that there are few exceptions to the strong propensity to return and to the weak potential of long-term migration (Burton et al., 1993). Paradoxically, extreme events may even act as *pull* rather than *push* factors: in the case of the Indian Ocean tsunami in 2004, relatives moved to the area to find out whether their family had been affected and to offer support; in addition, reconstruction projects increased the demand for labour and attracted migrant workers from other areas; finally, new economic opportunities arose from the presence of numerous aid-providing institutions (Paul, 2005; Naik et al., 2007). This being said, macro-level investigations that compare rates of emigration with local exposure to disasters lead to more contrasting results. Several studies demonstrate that a high frequency of disasters (including floods, storms, hurricanes, drought and frost) encourages people to move away from their town or country (see Saldaña-Zorrilla and Sandberg (2009) for Mexico, Naudé (2008) for sub-Saharan Africa, Reuveny and Moore (2009) for developing countries and Afifi and Warner (2008) for a sample of 172 countries around the world).[10]

Overall, the potential of tropical cyclones, floods and torrential rains to provoke long-term and long-distance migration, while ascertained, remains limited. As pointed out by Kniveton et al. (2008), the level of vulnerability can be tremendously different from one region to another and it is only if the affected society is highly dependent on the environment for livelihood and if social factors exacerbate the impact of the disaster – as was typically the case with Hurricane Katrina (Reuveny, 2008) – that significant migration takes place.[11]

---

[10] At a more micro-level, Carvajal and Pereira (2008) show that, in Nicaragua, a household highly exposed to Hurricane Mitch had a higher probability of sending a member abroad than a household with similar adaptive capacity but in a non-exposed area. On the contrary, Neumayer (2005) found no correlation between emigration and natural disasters in the zones of departure, but a significant link to the political situation in his study on asylum-seekers towards Europe.

[11] Even in this often-cited case of long-term displacements, estimates of the number of returnees are still difficult to establish due to a lack of reliable data (Hernandez, 2009).

Drought and desertification

In the recent past, the number of persons affected by climatic disasters such as extreme temperatures, droughts or wildfire is estimated at around 83 million each year (between 2000 and 2008; Rodriguez et al., 2009). The IPCC foresees that 74 million to 250 million people will be affected, in 2020, by increased water shortages in Africa and Asia; it also states that 'freshwater availability in Central, South, East and Southeast Asia, particularly in large river basins, is projected to decrease due to climate change which, along with population growth and increasing demand arising from higher standards of living, could adversely affect more than a billion people by the 2050s' (IPCC, 2007, p. 10).

Compared with cyclones and flooding, a lack of drinking and irrigation water usually generates much less sudden impacts, and thus leads to more progressive patterns of mobility. Empirical evidence is mixed. On the one hand, there are many well-known cases of mass population movements attributed to droughts in Africa (Sahel, Ethiopia), South America (Argentina, Brazil), the Middle East (Syrian Arab Republic, Islamic Republic of Iran), and Central and Southern Asia (Black and Robinson, 1993). The impact of droughts on migration is also documented in the Malian Gourma region by a historical overview over the twentieth century (Pedersen, 1995). In South America, Leighton notes that 'the periodic drought and desertification plaguing Northeast Brazil contributed to factors causing 3.4 million people to emigrate between 1960 and 1980' (Leighton, 2006, p. 47). On the other hand, many researchers question the link between drought and emigration by emphasizing the multiplicity of causes determining migration and the other survival strategies available to affected populations (De Haan et al., 2002). According to Kniveton et al., 'drought seems to cause an increase in the number of people who engage in short-term rural to rural type migration. On the other hand, it does not affect, or even decreases international, long-distance moves' (2008, p. 34). In the absence of a consensus, three broad kinds of results can be identified in the literature (see also Leighton, Chapter 13).

The first confirms the link between drought and emigration. Barrios et al. (2006) use a cross-country data set of seventy-eight countries over a thirty-year period and observe that shortages in rainfall increased rural exodus in the sub-Saharan African continent (but not elsewhere in the developing world) and thus contributed significantly to urbanization in Africa. In the Americas, Munshi (2003) establishes a correlation between emigration to the United States and low rainfall in the region of origin in Mexico (see also Leighton Schwartz and Notini, 1994). Van der Geest et al. (2010) use geographical analysis to evaluate the relation between internal migration, rainfall and vegetation dynamics in Ghana. They conclude that migration propensities are higher in environmentally less-endowed districts and that the lack of rainfall is a predictor of migration, but this result does not hold for the region of Accra and signals the necessity to differentiate migration sub-systems. Finally, Afifi and Warner, in their above-mentioned study of 172 countries, find that indexes of desertification, water scarcity, soil salinization and deforestation are all correlated with emigration (Afifi and Warner, 2008).

A second group of case studies, on the contrary, concludes that droughts have minimal impact on migration. The most often cited relies on two surveys (1982 and 1989) conducted in rural Mali with over 7,000 individuals and 300 households before and after a series of droughts affecting the country; a reduction (and not an increase) in international emigration was observed due to the lack of available means to finance the journey, even if short-term internal migration of women and children did rise (Findley, 1994). Smith (2001) also found a limited impact on emigration during the 1994 droughts in Bangladesh, as less than 1% of households had to resort to emigration. This result is coherent with the analysis on interprovincial migrations in Burkina Faso by Henry et al. (2003), where environmental variables and droughts contributed only marginally to the explanation of migrations; the authors conclude that, in this country, even if migration is influenced by biophysical changes in the environment, claims that environmental change alone is causing massive displacements are not supported by the data. Kniveton et al. find a similar result in their analysis of the relationship between climate variability and migration to the United States in the drought-prone Mexican regions of Zacatecas and Durango between 1951 and 1991 (2008, pp. 42–47): they find no significant correlation in Zacatecas whereas, in Durango, more rainfall generates more emigration and not the contrary. In the same way, Naudé finds no correlation between emigration and water scarcity (proxied by the surface of land under irrigation) across forty-five sub-Saharan African countries (Naudé, 2008).

Finally, several studies show contrasting patterns according to the type of migration concerned (long-term versus short-term and long-distance versus short-distance). In another study on Burkina Faso, Henry et al. (2004) collected individual migration histories among 3,911 individuals and environmental data at community level in about 600 places of origin mentioned by migrants; the environmental indicator consists of rainfall data covering the 1960–1998 period and the dependent variable is the risk of the first village departure. Findings suggest that people from the drier regions are more likely to engage in both temporary and permanent migrations to other rural areas and that short-term rainfall deficits increase long-term migration to rural areas but decrease short-term moves to distant destinations. The research presented by Pratikshya Bohra-Mishra and Douglas Massey (Chapter 4) does not directly address the question of drought but brings comparable results. It shows that if the quality of drinking water has no impact on population displacements, perceived defor-estation, population pressure and agricultural decline do produce elevated rates of local population mobility, but no significant increases in internal or international migration. These results partly contradict a previous study using the same method in the same area, but with a smaller sample and a shorter time span (Shrestha and Bhandari, 2007). The evidence that scarcity of water and desertifica-tion do have an impact on migration patterns, but that they mainly generate short-distance moves and that their impact is mediated by numerous other variables, is also confirmed by local case studies, among others in the context of the EACH-FOR project (see Warner et al., Chapter 8; Hamza et al., 2008; also Meze-Hausken, 2004).

Again, a link may be assumed to exist between rain deficits and migration, but it remains highly contextual – so that projections of increased migrations linked to drought-related phenomena are haz-ardous. Just as for rapid-onset phenomena, it would be difficult to provide an estimate of the magnitude of populations at risk and of the potential migration flows arising from droughts induced by global warming.

Sea level rise

In contrast to the two environmental factors discussed so far (tropical cyclones-heavy rains-floods and drought-desertification), the link between sea level rise and migration appears much more straightforward (see Oliver-Smith, Chapter 7). Unlike most other hazards, sea level rise is virtually irreversible and manifests itself in a more or less linear way over a long period. In the absence of new infrastructures such as dykes, this would make definitive out-migration the only possible solution, while allowing for progressive and planned departures. Sea level rise is also at the heart of some of the most dramatic and publicized manifestations of climate change, including the possible disappearance of island states (see McAdam, Chapter 5).

Compared with other climatic events, sea level rise is a rather new phenomenon and the number of available studies remains limited. Historical evidence nevertheless exists; for example, the Chesapeake Bay islands on the Atlantic coast of the United States have experienced sea level rise since the mid nineteenth century at rates of about 0.35 cm/year, which contributed, beside other factors, to the abandonment of most of the islands by their resident populations in the early twentieth century (Arenstam Gibbons and Nicholls, 2006). The consequences of sea level rise can be quite reliably predicted and localized, because the configuration of coastlines, their altitude and their population are simple to integrate into geographic information systems (GIS) that permit simulations and projections. It is therefore possible to calculate – on a global scale – the number of persons living in low-elevation coastal zones and threatened by rising water levels, higher tides, further-reaching waves, salinization or coastal erosion.

McGranahan et al. (2007) define 'low elevation coastal zones' as being situated at an altitude of less than 10 m. Even though these zones only account for 2.2% of dry land on Earth, they are currently home to 10.5% of the world population – i.e. around 602 million people, of which 438 million live in Asia and 246 million in the poorest countries of the world. Anthoff provides a slightly lower figure, at 397 million people, which is nevertheless considerable (Anthoff et al., 2006). Yet it would be premature to conclude that these people will all be forced to evacuate their homes in the near future. The IPCC report evokes a 7 m rise in sea level (consecutive to the possible melting of the Greenland ice cover), but this would occur over several centuries or even millenaries. Of more concern is the scenario of future CO_2 emission based on continuing economic growth with a moderation of fossil fuel use (scenario A1B of the IPCC), which predicts an increase of 0.3 m to 0.8 m of sea level by 2300 (IPCC, 2007). More recent estimations (quoted by Rebetez, Chapter 2, and Oliver-Smith, Chapter 7) show that this process might go significantly faster than previously thought. On this basis, it seems reasonable to consider that populations living at an altitude of less than 1 m above sea level are directly vulnerable – and within a few decades. According to Anthoff et al. (2006), 146 million people would be concerned here, 75% of which in the major river deltas and estuaries in South Asia (Indus, Ganges-Brahmaputra, etc.) and East Asia (Mekong, Yangtze, Pearl

River, etc.). Although far less populated, certain islands (such as Tuvalu or Maldives) are the most threatened in the short term, as they are situated only centimetres above sea level.

In sum, sea level rise probably constitutes the aspect of climate change that represents the clearest threat in terms of long-term forced migration. But as recalled by Oliver-Smith (Chapter 7), reaction to sea level rise is more complex than the mere abandonment of lands. Migration can indeed happen long before an area really becomes uninhabitable, and, symmetrically, concerned populations can elaborate strategies of adaptation and mitigation that may significantly postpone the necessity to leave. The recent decision by the Netherlands Government to improve its dyke protection system illustrates that financial resources constitute a key factor in this respect (Kabat et al., 2009).[12]

Multiple determinants of migration

The studies reviewed above, along with most of the contributions to this volume, highlight the complexity of the relationship between environmental factors and migration and the fact that climate change is only one factor among several others in explaining migration dynamics. In its simplest form, this refers to the fact that any migratory movement is the product of several converging factors and that environmental stress is always mixed with other causes, which may include economic constraints or opportunities, social networks, political context, etc.

Several contributions further show how factors fostering mobility are not only numerous but intertwined. For example, Alisson Flávio Barbieri and Ulisses Eugênio Confalonieri (Chapter 3) document how environmental change can generate health problems, while Leighton (Chapter 13) demonstrates the interconnections between climate change and food security. In such cases, identifying the 'primary' cause of migration is probably impossible, as all causes may reinforce each other. Findlay and Geddes (Chapter 6) contribute to this debate by arguing that environmental factors play a role only if they emerge in a context already characterized by political, demographic, economic or social tensions; climate change would thus be an *additional* burden, which can have a multiplier effect. In other words, climate change is unlikely to trigger migration in wealthy and democratic societies, which echoes Amartya Sen's well-known work on famines, according to which

[12]   See also http://www.deltacommissie.com/.

these are due less to environmental factors than to ill-founded political choices and lack of democracy (Sen, 1981). Findlay and Geddes also show how environmental and non-environmental factors can interact in a step-by-step manner: if people have already moved for predominantly economic reasons, they would be more likely to move again because of climate change.

Discussing multicausality therefore implies acknowledging the non-direct relationship between climate change and migration, and the factors that mediate between the two (see Warner et al., Chapter 8). As Rebetez makes clear (Chapter 2), climate change is a complex environmental process that does not have uniform consequences everywhere; and as Oliver-Smith notes (Chapter 7), societies have always had to adapt to changing environmental contexts – a multifaceted process of technological, organizational, institutional, socio-economic and cultural nature that is likely to be just as complex as climate change itself. The number of variables is therefore important, leading to high uncertainty and local variability.

Policy-wise, multicausality implies that high-income countries are unlikely to suddenly witness the arrival of 'environmental migrants', as policy-makers sometimes seem to believe. To a large extent, future migration flows will resemble current ones – at least from the perspective of receiving states. This is not to say that climate change has no impact, but rather that its impact will be difficult to identify at first sight.

Social dimension of vulnerability

The mediating function of social factors in the relationship between climate change and migration points to the fact that people do not have access to the same resources when it comes to reacting or adapting to environmental change. The idea that vulnerability is shaped by such social factors runs throughout this volume and several contributors refer to the wide range of social variables that determine people's exposure to climate change. From a social sciences perspective, this would seem to go without saying; yet, as Findlay and Geddes argue (Chapter 6), studies on the climate change–migration nexus have long privileged top-down approaches in which so-called 'hot spots' are identified and mechanically understood as places where migration will occur – regardless of 'from below' considerations on the ways in which people will react and adapt. This is manifest in many of the available maps on the topic, which show the geographical zones likely to be affected by climate change – but say nothing of the social context.

Lori Hunter and Emmanuel David (Chapter 12) contribute to this discussion with a careful analysis of the neglected role of gender. They show that changes in livelihood patterns affect men and women differently, because of their different social positions. Gender is also known to influence the perception of risk (which is a crucial variable in migration strategies), as well as the way people experience displacement. As Hunter and David argue, such gender relations are obscured by the focus on groups and communities that dominate the scholarship on migration. The importance of gender differences (which are also power relations) is also documented by Bohra-Mishra and Massey (Chapter 4) and Leighton (Chapter 13).

Another core variable in the construction of vulnerability is of course class resources and wealth. As Bohra-Mishra and Massey further note, climate change affects disproportionally poor agrarian communities, precisely those that have the least resources to leave their home (a point also made by Graeme Hugo, Chapter 10). In Bangladesh, Findlay and Geddes (Chapter 6) report that migration is too expensive and risky to be considered by people affected by floods. Oliver-Smith (Chapter 7) quotes the findings of studies according to which the migratory impact of sea level rise would depend as much on social and human factors as on sea level rise itself. In other words, the consequences of climate change vary according to the context, as the same environmental factor will have different impacts according to the characteristics of the people it affects. It follows that environmental degradation does not mechanically lead to displacement and that we should resist the 'tendency to equate populations at risk with population displacement' (Hugo, 2008, p. 31).

Diversity in migration and mobility patterns

To understand the impact of climate change on migration, it is necessary to disentangle the different kinds of mobility that may be connected to environmental factors. Indeed, notions such as 'displacement', 'mobility' or 'migration' (and the associated predicted numbers of people concerned) refer to situations that range from a few hours spent in a temporary shelter in fear of a hurricane to the relocation of whole communities whose land has disappeared following sea level rise.

There are at least three variables to take into account. First, migration may be short or long term. Discussions could gain in clarity if, for example, the UN-inspired distinction between temporary displacements (less than three months), short-term migration (three months to one

year) and long-term migration (more than one year) was more systematically used (UN, 1998). Most authors argue that, at present, temporary
and short-term patterns of migration are predominantly associated with
environmental change. The temporality of migration also has to do with
the nature of environmental processes: as Hugo recalls (Chapter 10),
slow-onset phenomena such as desertification or sea level rise are likely
to be associated with long-term migration, whereas sudden disasters
such as tropical cyclones will generate temporary displacement. But
this typology is far from systematic. Leighton (Chapter 13) shows how
droughts have long fuelled seasonal migration dynamics, which also
points to the differences between permanent departures and back-and-
forth types of mobility.

A second key distinction is between short- and long-distance migration, or between internal and international moves. Debates on the
climate change–migration nexus often seem to focus overwhelmingly
on international migration, and particularly on flows from 'South' to
'North'. But as Khalid Koser makes clear (Chapter 11), this bias tells
more on Western fears than on actual trends, as there is evidence that
most migration triggered by environmental factors concerns internal
migration (see also Chapters 4 and 13).

The third distinction is between forced and voluntary migration. The
often-used notion of 'environmental refugee' conveys the idea that people
are forced to leave their home because of the natural environment. But the
more or less constrained nature of migration is open to debate. It is indeed
extremely difficult to capture the decision-making process among potential
migrants and to understand why, how and when people decide to leave.
This also points to the above-mentioned social dimension of vulnerability,
as people's strategies depend upon their resources and opportunities.
Finally, the possible interventions of governments in moving people (documented in Chapter 10) further contribute to challenge the distinction
between forced and voluntary movement.

This echoes the long-standing debate on the extent to which migration
stems from a failure to adapt. The dominant view is that people who
move because of environmental factors are in fact unable to adapt – and
thus have no option but to leave. In this view, migration is the worst
scenario and the option to avoid, and policies should strive to enable
people to stay. But several contributions argue that migration is not only
a reactive, but also a proactive strategy; rather than being a last-resort
option, it represents a coping mechanism and a way of adapting, for
example through seasonal migration patterns or by arranging for one

member of the family to leave (and thus enabling the other members to stay).

This being said, the·distinctions between various forms of migration are not always neat. For example, Hugo argues that temporary migration may eventually turn out to be permanent, as people wish to return but cannot for various reasons (Chapter 10). Short-term mobility may also make people more prone to envisage international migration at a later stage. The distinction between forced and voluntary migration may also be quite fine, as people develop strategies in reaction to external constraints.

Methodology

As Castles argues (Chapter 16), discussions on the relationship between climate change and migration have long been marked by a methodological divide and, despite recent attempts towards improvements (Kniveton et al., 2009; Bilsborrow, 2009; Piguet, 2010b), it is widely recognized that a lack of rigour and clarity characterizes research on the climate–migration nexus. As Leighton emphasizes (Chapter 13), data pertaining to the environmental and migratory dynamics rarely come from the same sources and are therefore difficult to combine. Moreover, researchers from different disciplinary backgrounds and empirical traditions have different methodological orientations and have not always managed to work together.

The contributions to this volume illustrate the variety of methods that can be used to investigate the impact of environmental factors on migration, as well as the possible complementarities between them. A first methodological approach is mainly descriptive and prospective. It focuses on the identification of the main regions and populations threatened by environmental degradation (the so-called 'hot spots') and on integrated assessments of the vulnerability and resilience of their inhabitants, which provide insights into possible future migrations. Oliver-Smith thus investigates the possible consequences of sea level rise (Chapter 7), whereas Hunter and David use a 'sustainable livelihoods' framework to differentiate the migratory consequences of climate change according to gender (Chapter 12). Barbieri and Confalonieri incorporate, in a fairly innovative and explorative way, environmental degradation scenarios in economic models, which are then used to forecast migration (Chapter 3).

The second research strategy is more analytical and attempts to disentangle the specific environmental component among other drivers of migration. The purpose is to question the role and weight of environmental factors in already occurring phenomena. Bohra-Mishra and Massey (Chapter 4) make use of a 108-month panel study conducted between 1997 and 2006 in 151 Nepalese neighbourhoods. Other contributors use less quantitative and more ethnographic methods to report project findings from the Nile delta, the Sahel, Vietnam (Chapter 8) and more generally Africa (Chapter 13), Bangladesh (Chapter 6) and Tuvalu (Chapter 5). One contribution makes use of historical analogues allowing insights from forced resettlement (Chapter 11) for the future consequences of climate change.

An important feature emerging from all studies is that, although fruitful results can emerge from either quantitative, qualitative or mixed methodologies, it is of paramount importance to take into account not only the objective characteristics of the environmental degradations but also people's perceptions and representations of these evolutions and of their potential migration consequences. The measure of the impact of environmental factors on displacement should be complemented by an examination of the socio-cultural perceptions and representations of these threats among concerned populations (Mortreux and Barnett, 2009), a turn recently advocated in relation to climate change studies in general (Hulme, 2008).

Conceptual issues

Conceptual issues are a major source of confusion in the debate on the climate change–migration nexus. There are persistent disagreements over how to refer to people migrating because of environmental factors; while popular, terms such as 'environmental migrants' or 'climate refugees' have raised controversies that are both scientific/academic and political (see Gemenne, Chapter 9, and Findlay and Geddes, Chapter 6).

From a research perspective, the juxtaposition of the terms 'environment' or 'climate' with 'migrants' or 'refugees' has been criticized for implying a monocausal relationship between environmental factors and human mobility, and thus for negating the multicausality discussed above. As noted by Castles, 'the term environmental refugee is simplistic, one-sided and misleading. It implies a monocausality which very rarely exists in practice ... [Environmental and natural factors] are part of a complex pattern of multiple causality, in which [they] are closely linked to economic, social and political ones' (Castles, 2002, p. 5). In this sense, there will never be any 'environmental migrant' (or 'climate refugee') because it will never be possible to identify a group of people who migrate *only* because of environmental variables.

Although quite widely accepted, the definition of 'environmental migrants' provided by the International Organization for Migration suffers from the same shortcoming:

> Persons or groups of persons who, for compelling reasons of sudden or progressive change in the environment as a result of climate change that adversely affect their lives or living conditions, are obliged to leave their habitual homes, or choose to do so, either temporarily or permanently, and who move either within their country or abroad.[13]

The term 'environmentally induced population movements' (EIPM) might constitute a more neutral solution, but it is vague and not very appealing to the general public. Another option is the term 'environmentally displaced persons' (EDPs), which was for example used in the EACH-FOR research project (Warner et al., Chapter 8). It encompasses three subcategories: environmental migrants (people who choose to move voluntarily from their place of residence primarily for environmental reasons); environmental displacees (people who are forced to leave their place of residence because their livelihoods are threatened as a result of adverse environmental processes and events); and development displacements (people who are intentionally relocated or resettled due to a planned land-use change). The boundaries between these three subgroups nevertheless remain blurred.

Politically, conceptual discussions have focused on the use of the 'refugee' notion. Legally, this refers to the status recognized by the UN 1951 Geneva Convention and to its definition of 'refugee' as a person leaving his/her country of residence for 'well-founded fear of being persecuted for reasons of race, religion, nationality, membership of a particular social group or political opinion'. Environmental reasons are absent from this definition, which can lead to two opposite positions: we may either advocate for an extension of this definition to include environmental factors (and hence for a modification of the Geneva Convention or for a new treaty specifically addressing the case of 'environmental refugees'); or we may reject the very reference to 'refugees' in the case of climate change, mostly for fears of diluting a specific legal

---

[13] This definition was put forward in a 2007 background paper (MC/INF/288) at the 94th IOM Council (http://www.iom.int/jahia/webdav/site/myjahiasite/shared/shared/mainsite/microsites/IDM/workshops/evolving_global_economy_2728112007/MC_INF_288_EN.pdf). It also appears in IOM (2008, p. 399) and various other publications.

category into a broader and ill-defined category. This led the UNHCR to voice 'serious reservations with respect to the terminology and notion of environmental refugees or climate refugees', noting that 'these terms have no basis in international refugee law and the majority of those who are commonly described as environmental refugees have not crossed an international border. Use of this terminology could potentially under-mine the international legal regime for the protection of refugees and create confusion regarding the link between climate change, environ-mental degradation and migration' (UNHCR, 2009, p. 7).

Indeed, in a context in which respect for the Geneva Convention is already under threat, incorporating environmental factors in refugee debates could eventually jeopardize the protection afforded to recog-nized refugees. The reasons given are threefold. First, this could strengthen the already widespread fears surrounding uncontrollable waves of poor refugees to high-income countries, thereby fuelling xen-ophobic reactions or serving as a justification for increasingly restrictive asylum policies. It could also further blur the already fragile distinction between voluntary (i.e. economic) and forced (i.e. political) migration – thus undermining the very foundations of the asylum principle. And finally, in a more fundamental manner, it could introduce a sort of 'natural' connotation to asylum issues, which would be incompatible with the political nature of the persecutions considered by the Geneva Convention: 'In so far as the term environmental refugee conflates the idea of disaster victim and refugee, its use brings with it the danger that the key features of refugee protection could be undermined and the lowest common denominator adopted. Because environmental can imply a sphere outside politics, use of the term environmental refugee may encourage receiving states to treat the term in the same way as economic migrants to reduce their responsibility to protect and assist' (McGregor, 1993, p. 162). In other words, the danger here would be 'to abdicate political responsibility by overplaying the hand of nature' (Cambrézy, 2001, p. 48).

This 'hand of nature' argument could be challenged, on the ground that climate change (unlike tsunamis or earthquakes) is not a neutral or apolitical phenomenon, but to a large extent the product of world economic development. The 'world' would thus be responsible for the situation of climate 'refugees' (which is not the case with the many traditional refugees who leave local conflicts or dictatorships that may not be directly connected to world politics). As Zetter writes: 'The strength of the climate change argument lies in a common conception that specific moral burdens rest on global society. Such global burdens do not readily appear to exist for the other, more localized, categories of migrants such as refugees and IDP' (Zetter, 2009, p. 400). We could go one step further and argue that a small number of wealthy states are, in fact, at the origins of most climate change, and that past CO_2 emissions could consequently determine the respective share of responsibility. According to the IOM's 2008 *World Migration Report*, 'some analysts are beginning to argue that migration is both a necessary element of global redistributive justice and an important response to climate change; and that greenhouse gas emitters should accept an allocation of "climate migrants" in proportion to their historical greenhouse gas emissions' (IOM, 2008, p. 399). In this context, states and populations in the 'South' display resentment (and make claims) towards the 'North' on the basis of its responsibility in fuelling climate change – even if developed states have so far remained largely indifferent (see for example Chapter 6).

As has become clear, the conceptual discussion around the definition best suited to describe and analyse the link between migration and environmental change goes far beyond purely conceptual issues and raises the question of the protection and status to be granted to the people concerned, and of the responsibilities of the international community towards them. Given the far-reaching complexity of these debates, a consensus is unlikely to be reached in the near future, either among researchers, or in policy and public debates. As a consequence, differences in terms, notions and definitions are likely to persist. But as Walter Kälin (former Representative of the Secretary-General on the Human Rights of Internally Displaced Persons) has stated, 'we should not be distracted by semantic discussions with little practical meaning about whether to call affected persons climate change refugees, environmental migrants or something else. Instead, what is needed is a thorough analysis of the different contexts and forms natural disaster induced displacement can take' (Kälin, 2008). In other words, as long as participants in the debate share core concerns (including multi-causality and the recognition of the social construction of vulnerability), a variety of terminologies does not hamper the development of a coherent common approach on the issues at stake.

Protection of "environmental migrants" and states' responsibilities

As argued, the different terms referring to people who migrate in connection with environmental factors imply different representations of how states could or should treat these people and of the protection that they should receive. The starting point of this complex and sensitive issue is the current absence of standards in defining this protection; as Christel Cournil argues (Chapter 14), none of the concepts mentioned above has a legal definition – leading to an institutional and normative vacuum (see also Chapters 5, 13 and 15).

In the absence of specific norms, we could try to rely on existing instruments and explore how they relate to the issues relating to "environmental migrants"; this is the object of contributions on internal and international migration respectively. In the case of people moving within their own country (which, as argued above, is the most frequent case), Koser (Chapter 11) shows that existing soft law instruments, and notably the *Guiding Principles on Internal Displacement*, do recognize some environmental factors (e.g. disasters) as a cause for displacement. But he also warns that they suffer from implementation challenges, which are due to problems of definition and to the non-binding nature of the Principles. Concerning international migration, Astrid Epiney (Chapter 15) examines the largely neglected elements of existing international law that could be of relevance to environmental migration (such as the international responsibility for wrongful acts). She concludes that they address only part of the issues raised by environmental migration; moreover, they are difficult to implement, in particular because of the difficulty of identifying single responsible states in the case of environmental disasters or climate change.

If there is a consensus on the existence of legal loopholes, there are disagreements over the remedies to this situation. There have been numerous calls for the elaboration of new standards to define the responsibilities of states and the protection of the people concerned. Cournil (Chapter 14) provides an in-depth exploration of such 'legal fiction' and reviews the different proposals that have been put forward. These range from amending the Geneva Convention to the development of entirely new instruments, either at the bilateral, regional or international level. McAdam (Chapter 5) provides counter-arguments and warns that calls for new normative instruments will not only face a deep lack of political willingness, but also more structural obstacles. In particular, she argues that the categories of 'environmental migrants' may be too vague and ill-defined to justify a new treaty, which would risk being politically visible but legally useless. Moreover, the collective dimension of migration in the case of environmental change, along with the absence of a clearly defined persecutor, makes the analogy with refugees problematic.[14]

Indeed, the establishment of a new treaty faces several challenges. Not only will it be difficult to reach an international agreement on the definitions of the people concerned and the criteria to grant protection, but negotiations are likely to come up against highly sensitive issues surrounding the responsibilities of industrialized nations – an obstacle that has proven very prominent in international discussions pertaining to climate change. In addition, there is the risk of exercising a downward pressure on existing treaties such as the Geneva Convention. In this context, and regardless of the different perceptions that exist, it seems likely that environmental factors will increasingly fuel migration, but without a specific legal framework (at least at international level). Yet this does not prevent an examination of the policy orientations relevant to situations of environmental migration.

Possible policy orientations

What are the policies that have been elaborated to respond to environmentally induced migration? And what are the policy orientations that could be envisaged to address the challenges raised by the movement of people in a context of environmental change? Given the heterogeneity in the types of climate stress that can foster migration, it is worth distinguishing between different kinds of policy options.

First, there is the case of disasters and sudden climatic events. There have always been cyclones, floods or other natural catastrophes and most, if not all, regions of the world have experienced the challenge of addressing the situations of the people concerned. The problem lies in the efficiency of the already existing mechanisms, especially if it is assumed that climate change will increase the frequency and/or intensity of some kinds of disaster (see Chapter 2), thus putting humanitarian efforts under further stress. This calls for reinforcing rescue mechanisms and, in the case of less-developed countries, for greater international solidarity, not least in making the necessary funds available. This is one

---

[14] Another legal issue connected to climate change and migration regards statelessness. In the case of sinking island states, not only would inhabitants need to leave their home, but entire countries could disappear. Migrants from these states would then risk becoming stateless, which calls for innovative legal and policy approaches (see UNHCR, 2009, and Piguet, 2010a, for a discussion).

of the *raisons d'être*, at the international level, of the United Nations Disaster Assessment and Coordination (UNDAC) teams, managed by the UN Office for the Coordination of Humanitarian Affairs (OCHA). Overall, the main objective should therefore be to make a more extensive use of existing policy mechanisms and to adapt them to the specific challenges raised by climate change.

Yet, we should keep in mind that the impact of climate change on migration will also manifest itself through much less sudden events. As Hugo argues (Chapter 10), governments and policy-makers seem to react above all to disasters that force people to leave overnight; this applies to some of the most documented cases of environmental migration, such as the 2004 Asian tsunami and the 2005 Hurricane Katrina in New Orleans. By contrast, the 'silent crisis' fuelled by progressive environmental change, while affecting potentially very high numbers of people, is the object of much less policy attention. In some extreme cases, resettlement may constitute the appropriate policy, in order to enable large numbers of people to leave their home on a permanent basis. As Hugo further recalls, these are not new policies either, as resettlement has regularly been implemented in other contexts, especially in relation to large-scale infrastructure projects such as dams. Yet he also shows that this is a costly and complex process, which all too often has failed because of insufficient funding, mismanagement or neglect of socio-cultural factors. Again therefore, the relevant policy approach would be to improve existing policy options, through increased funding and international cooperation.

That being said, resettlement is not an option for all those people concerned by progressive manifestations of climate change. There is therefore a need to envisage a much broader range of responses, to address the multifaceted challenges raised by slow environmental deterioration. At the local level, Castles (Chapter 16) notes that this could for example include measures to diversify economic activities in order to enable people to better adapt to climate change. More broadly, this would call for incorporating the migration–climate change relationship in existing fields of policy that have so far not only tended to ignore migration, but have also remained quite separate from each other. These notably include development strategies and humanitarian interventions, two well-established fields of effort at all levels (national, regional and international), but that have so far dedicated little energy to climate change, and even less to migration.

In the same vein, environmental migration is also a matter for migration policy at large. If, as argued, environmental factors exacerbate

already existing push factors in less-developed countries, more appropriate migration policies could probably accommodate part of environmental migration through classical schemes such as economic migration
programmes. The IOM thus notes that 'the international community is,
in fact, ignoring labour mobility as a coping strategy for climate stress'
(IOM, 2008, p. 399). Castles also underlines that restricting and crimi-
nalizing migration will do little to improve the policy responses to the
climate change–migration nexus, and that the priority should go to
policies that allow people to move in conditions of safety and dignity.
This echoes the numerous calls for more realistic and flexible approaches
to migration that have been launched in recent years (GCIM, 2005;
UNDP, 2009; see also Pécoud and de Guchteneire, 2007). This also
implies strengthening the legal framework in which international migration takes place, possibly through existing norms such as the UN
Convention on Migrant Workers' Rights (Cholewinski et al., 2009).

Another argument that runs through several chapters is the need
to incorporate potential migrants into policy-making. Hugo
(Chapter 10) argues that resettlement policies target predominantly
poor and powerless people, who are at risk of being voiceless when it
comes to organizing their displacement, and that this is one of the
reasons why such policies may actually make people more vulnerable.
Oliver-Smith (Chapter 7) also states that the displacement of people
has profound socio-cultural and emotional consequences, but is frequently approached as a merely material problem. McAdam adds
(Chapter 5) that in the Pacific islands of Tuvalu and Kiribati some
of the people affected by climate change resist being considered as
refugees, a status which they associate with passivity and helplessness,
and aim at playing an active role in the choice of the policy options
that will shape their future.

This discussion highlights the fact that, even if environmental
migration is regularly presented as a 'new' challenge requiring 'new'
responses, there are actually a number of existing policy fields that
can be relied upon to address the challenges it raises, including
development strategy, humanitarian affairs, post-disaster interventions, or immigration and admission policies. This is not to say
that new normative or policy instruments are irrelevant; rather, it
means that new instruments may not be a prior necessity to address
the needs of populations at risk and that an absence of consensus on
the desirability of such new standards does not imply that nothing
can be done.

Conclusion

The contributions to this volume confirm that climate change has consequences in terms of human migration and mobility, and that its impact can be expected to increase. But they also underline that, given the complexity of the relationship between environmental change and migration, climatic or natural hazards do not automatically lead to displacements. Another core argument is that migration is an adaptation strategy in itself; it is not necessarily the worst scenario and should not be seen as an intrinsically negative outcome to be avoided. Finally, climate change will be experienced very differently around the world and across countries, as the vulnerability to nature is ultimately a product of the socio-economic forces that shape all societies.

The social dimension of vulnerability should be interpreted as an opportunity to increase people's ability to resist climate change. Indeed, if human beings were completely helpless in the face of nature and climate change, very little could be done. But they are not and this opens opportunities for local and international efforts in gathering knowledge, drafting measures and increasing protection. Provided that the necessary financial means are made available, even such an apparently unavoidable threat as rising sea levels could be partially counteracted. It also follows that, if environmental migration is fundamentally a political process, the actual number of people who will move cannot be predicted, but depends on current and future efforts.

This approach also implies going beyond the traditional 'alarmists vs. sceptics' debate and recognizing that, while there are no reasons to exaggerate the threats and inspire ungrounded panic, there are nevertheless good reasons to take the problem seriously. This particularly concerns data gathering. More knowledge is required to address the situation of people affected by environmental change and it is paramount to understand better the kind of patterns that develop from it in order to envisage potentially successful policies. In addition, research on these issues requires increased cooperation between social and natural sciences, for example in the elaboration of complete and comparable databases.

All in all, climate change is a process that exacerbates some of the most pressing issues of our time. It does not take place in a vacuum but is closely associated with underdevelopment, inequalities within and between countries, global justice and the lack of solidarity between states, human rights or human security. Climate change as a policy area may be relatively recent, but most of these issues represent long-standing challenges for states and the international community. It follows that policies that focus on the climate change–migration nexus must be accompanied by renewed efforts to combat the very context that makes people vulnerable in the first place.

References

Afifi, T. and Jäger, J. (eds). 2010. *Environment, Forced Migration and Social Vulnerability.* Bonn: Springer Verlag – United Nations University – Institute for Environment and Human Security.

Afifi, T. and Warner, K. 2008. *The Impact of Environmental Degradation on Migrations Flows across Countries.* Bonn, United Nations University Institute for Environment and Human Security. (UNU-EHS Working Paper 5.)

Anthoff, D., Nicholls, R. J., Tol, R. S. J. and Vafeidis, A. 2006. *Global and Regional Exposure to Large Rises in Sea-Level: a Sensitivity Analysis.* Norwich, UK, Tyndall Centre for Climate Change Research. (Tyndall Centre Working Paper 96).

Arango, J. 2000. Explaining migration: a critical view. *International Social Science Journal*, Vol. 52, No. 165, pp. 283–96.

Arenstam Gibbons, S. J. and Nicholls, R. J. 2006. Island abandonment and sea-level rise: An historical analog from the Chesapeake Bay, USA. *Global Environmental Change*, Vol. 16, No. 1, pp. 40–47.

Barnett, J. and Webber, M. 2009. *Accommodating Migration to Promote Adaptation to Climate Change.* Stockholm, Commission on Climate Change and Development.

Barrios, S., Bertinelli, L. and Strobl, E. 2006. Climatic change and rural–urban migration: The case of sub-Saharan Africa. *Journal of Urban Economics*, Vol. 60, No. 3, pp. 357–71.

Biermann, F. and Boas, I. 2010. Preparing for a warmer world. Towards a global governance system to protect climate refugees. *Global Environmental Politics*, Vol. 10, No. 1.

Bilsborrow, R. E. 2009. Collecting data on the migration–environment nexus. In: Laczko and Aghazarm (eds), op. cit., pp. 115–96.

Black, R. 2001. *Environmental Refugees: Myth or Reality?* Geneva, United Nations High Commissioner for Refugees. (*New Issues in Refugee Research*, No. 34.)

Black, R. and Robinson, V. 1993. *Geography and Refugees.* London, Belhaven.

Boano, C., Zetter, R. and Morris, T. 2008. *Environmentally Displaced People: Understanding the Linkages between Environmental Change, Livelihood and Forced Migration.* University of Oxford, UK, Refugee Studies Centre. (Forced Migration Policy Briefing 1.)

Brettell, C. B. and Hollifield, J. F. (eds). 2007. *Migration Theory – Talking across Disciplines*. London, Routledge.

Brown, O. 2008: *Migration and Climate Change*. Geneva, International Organization for Migration.

Burton, I., Kates, R. W. and White, G. F. 1993. *The Environment as Hazard*. New York, Guilford Press.

Cambrézy, L. 2001. *Réfugiés et exilés – crise des sociétés – crise des territoires*. Paris, Editions des Archives Contemporaines.

CARE/CIESIN/UNHCR/UNU-EHS/World Bank. 2009. *In Search of Shelter – Mapping the Effects of Climate Change on Human Migration and Displacement*. Care International.

Carvajal, L. and Pereira, I. 2008. *Evidence on the Link between Migration, Climate Disasters, and Human Development*. Paper presented at the International Conference on Environment, Forced Migration and Social Vulnerability, Bonn, 9–11 October 2008.

Castles, S. 2002. *Environmental Change and Forced Migration: Making Sense of the Debate*. Geneva, United Nations High Commissioner for Refugees. (*New Issues in Refugee Research*, No. 70.)

Castles, S. and Kosack, G. 1973. *Immigrant Workers and Class Structure in Western Europe*. Oxford, UK, Oxford University Press.

Cholewinski, R., de Guchteneire, P. and Pécoud, A. (eds). 2009. *Migration and Human Rights. The United Nations Convention on Migrant Workers' Rights*. Cambridge, UK/Paris, Cambridge University Press/UNESCO Publishing.

Christian Aid. 2007. *Human Tide: The Real Migration Crisis*. May, London. http://www.christianaid.org.uk/Images/human-tide.pdf

Cohen, R. 1995. *The Cambridge Survey of World Migration*. Cambridge, UK, Cambridge University Press.

Collectif Argos. 2010. *Climate Refugees*. Boston, Mass., MIT Press.

de Haan, A., Brock, K. and Coulibaly, N. 2002. Migration, livelihoods and institutions: contrasting patterns of migration in Mali. *Journal of Development Studies*, Vol. 38, No. 5, pp. 37–58.

Dennell, R. W. 2008. Human migration and occupation of Eurasia episodes. *Journal of International Geoscience*, Vol. 31, No. 2, pp. 207–10.

El-Hinnawi, E. 1985. *Environmental Refugees*. Nairobi, United Nations Environmental Programme.

Findley, S. E. 1994. Does drought increase migration? A study of migration from rural Mali during the 1983–85 drought. *International Migration Review*, Vol. 28, No. 3, pp. 539–53.

GCIM. 2005. *Migration in an Interconnected World. New Directions for Action*. Geneva, Global Commission on International Migration.

Geyer, H. S. 2002: An exploration in migration theory. In: H. S. Geyer (ed.), *International Handbook of Urban Systems: Studies of Urbanization and*

Migration in Advanced and Developing Countries. Cheltenham, UK, E. Elgar, pp. 19–37.

Ghatak, S., Levine, P. and Price, S. W. 1996: Migration theories and evidence: an assessment. *Journal of Economic Surveys,* Vol. 10, No. 2, pp. 159–98.

Gregory, J. W. 1928. *Human Migration and the Future – A Study of the Causes, Effects & Control of Emigration.* London, Seeley, Service & Co.

Hammar, T., Brochmann, G., Tamas, K. and Faist, T. (eds). 1997. *International Migration, Immobility and Development: Multidisciplinary Perspectives.* Oxford, UK, Berg.

Hamza, M. A., Faskaoui, B. E. and Fermin, A. 2008. *Migration and Environmental Change in Morocco: The Case of Rural Oasis Villages in the Middle Drâa Valley.* Bonn, United Nations University Institute for Environment and Human Security. (Case Study Report.)

Harris, J. and Todaro, M. P. 1970. Migration, unemployment and development: a two-sector analysis. *American Economic Review,* Vol. 60, No. 1, pp. 126–42.

Henry, S., Boyle, P. and Lambin, E. F. 2003. Modelling inter-provincial migration in Burkina Faso: the role of socio-demographic and environmental factors. *Applied Geography,* Vol. 23, Nos. 2–3, pp. 115–36.

Henry, S., Schoumaker, B. and Beauchemin, C. 2004. The impact of rainfall on the first out-migration. *A Environment,* Vol. 25, No. 5, pp. 423–60.

Hernandez, J. 2009. The Long Way Home: une catastrophe qui se prolonge à La Nouvelle-Orléans, trois ans après le passage de l'ouragan Katrina. *L'Espace géographique,* Vol. 38, No. 2, pp. 124–38.

Hugo, G. 2008. *Migration, Development and Environment.* Geneva, International Organization for Migration.

Hulme, M. 2008. Geographical work at the boundaries of climate change. *Transactions of the Institute of British Geographers,* Vol. 33, No. 1, pp. 5–11.

Hunter, L. M., White, M. J., Little, J. S. and Sutton, J. 2003. Environmental hazards, migration, and race. *Population & Environment,* Vol. 25, No. 1, pp. 23–29.

ICPD. 1994. *Programme of Action of the International Conference on Population and Development.* Cairo, Egypt, September 1994. (A/CONF.171/13.) http://www.un.org/popin/icpd/conference/offeng/poa.html

IOM. 2008. *World Migration Report 2008. Managing Labour Mobility in the Evolving Global Economy.* Geneva, International Organization for Migration. http://www.iom.int/jahia/Jahia/cache/offonce/pid/1674?entryId=20275

IOM. 2009. *Migration, Climate Change and the Environment.* Geneva, International Organization for Migration. (IOM Policy Brief, May.)

IOM/UNFPA. 2008. *Expert Seminar: Migration and the Environment.* Geneva/New York, International Organization for Migration/United Nations Population Fund. (International Dialogue on Migration 10.) http://www.reliefweb.int/rw/lib.nsf/db900SID/PANA-7FNH38?OpenDocument

IPCC. 2007. Summary for Policymakers. In: *Climate Change 2007: The Physical Science Basis.* Contribution of Working Group I to the Fourth Assessment Report of the Intergovernmental Panel on Climate Change. Cambridge, UK/New York, Cambridge University Press.

IPCC-1. 1990. *Policymakers' Summary of the Potential Impacts of Climate Change.* Geneva, Intergovernmental Panel on Climate Change. (Report from Working Group II to IPCC.)

Isaac, J. 1947. *Economics of Migration.* New York, Oxford University Press.

Jacobson, J. 1988. *Environmental Refugees: A Yardstick of Habitability.* Washington DC, World Watch Institute. (World Watch Paper 86.)

Jäger, J., Frühmann, J. and Grünberger, S. 2009. *Environmental Change and Forced Migration Scenarios: Synthesis of Results.* Synthesis Report for the European Commission, EACH-FOR project. http://www.each-for.eu/

Jakobeit, C. and Methmann, C. 2007. *Klimaflüchtlinge.* Hamburg, Germany, Greenpeace. http://www.foe.org.au/resources/publications/climate-justice/CitizensGuide.pdf/view

Jonsson, G. 2010. *The Environmental Factor in Migration Dynamics – a Review of African Case Studies.* University of Oxford, UK, International Migration Institute. (Working Paper.)

Kabat, P., Fresco, L. O., Stive, M. J. F., Veerman, C. P., van Alphen, J. S. L. J., Parmet, B. W. A. H., Hazeleger, W. and Katsman, C. A. 2009. Dutch coasts in transition. *Nature Geoscience*, Vol. 2, pp. 450–52.

Kälin, W. 2008. *The Climate Change–Displacement Nexus.* Paper presented at ECOSOC Panel on Disaster Risk Reduction and Preparedness: Addressing the Humanitarian Consequences of Natural Disasters. Geneva, United Nations Economic and Social Council.

King, R. (ed.). 2007. *The History of Human Migration.* London, New Holland.

Kliot, N. 2004. Environmentally induced population movements: their complex sources and consequences – a critical review. In: J. D. Unruh, M. S. Krol and N. Kliot, *Environmental Change and its Implications for Population Migration.* Dordrecht, Netherlands, Kluwer Academic Publishers.

Kniveton, D., Schmidt-Verkerk, K., Smith, C. and Black, R. 2008. *Climate Change and Migration: Improving Methodologies to Estimate Flows.* Geneva, International Organization for Migration. (Migration Research Series No. 33.) www.iom.int/jahia/webdav/site/myjahiasite/shared/shared/mainsite/published_docs/serial_publications/MRS-33.pdf

Kniveton, D., Smith, C., Black, R. and Schmidt-Verkerk, K. 2009. Challenges and approaches to measuring the migration–environment nexus. In: Laczko and Aghazarm (eds), op. cit., pp. 41–111.

Laczko, F. and Aghazarm, C. (eds). 2009. *Migration, Environment and Climate Change: Assessing the Evidence.* Geneva, International Organization for Migration.

Leighton, M. 2006. Desertification and migration. In: P. M. Johnson, K. Mayrand and M. Paquin, (eds), *Governing Global Desertification*. London, Ashgate, pp. 43–58.

Leighton Schwartz, M. and Notini, J. 1994. *Desertification and Migration: Mexico and the United States*. US Commission on Immigration Reform Research Paper. San Francisco, Calif., Natural Heritage Institute.

Lonergan, S. 1998. The role of environmental degradation in population displacement. *Environmental Change and Security Project Report*, No. 4, pp. 5–15.

Marx, E. 1990. The social world of refugees: a conceptual framework. *Journal of Refugee Studies*, Vol. 3, No. 3, pp. 189–203.

Massey, D. S., Arango, J., Hugo, G., Kouaouci, A., Pellegrino, A. and Taylor, J. E. 1993. Theories of international migration: a review and appraisal. *Population and Development Review*, Vol. 19, No. 3, pp. 431–66. www.jstor.org/stable/2938462

—— 1998. *Worlds in Motion: Understanding International Migration at the End of the Millennium*. Oxford, UK, Clarendon Press.

McGranahan, G., Balk, D. and Anderson, B. 2007. The rising tide: assessing the risks of climate change and human settlements in low elevation coastal zones. *Environment and Urbanization*, Vol. 19, No. 1, pp. 17–37.

McGregor, J. 1993. Refugees and the environment. In: R. Black and V. Robinson (eds), *Geography and Refugees. Patterns and Processes of Change*. London, Belhaven, pp. 157–70.

McLeman, R., Mayo, D., Strebeck, E. and Smit, B. 2008. Drought adaptation in rural eastern Oklahoma in the 1930s: lessons for climate change adaptation research. *Mitigation and Adaptation Strategies for Global Change*, Vol. 13, No. 4, pp. 379–400.

Meze-Hausken, E. 2004. Migration caused by climate change: How vulnerable are people in dryland areas? *Mitigation and Adaptation Strategies for Global Change*, Vol. 5, No. 4, pp. 379–406.

Morrissey, J. 2009. *Environmental Change and Forced Migration: A State of the Art Review*. Oxford, UK, Department of International Development, Refugee Studies Centre.

Mortreux, C. and Barnett, J. 2009. Climate change, migration and adaptation in Funafuti, Tuvalu. *Global Environmental Change*, Vol. 19, No. 1, pp. 105–12.

Munshi, K. 2003. Networks in the modern economy: Mexican migrants in the U.S. labor market. *Quarterly Journal of Economics*, Vol. 118, No. 2, pp. 549–99.

Myers, N. 1993. Environmental refugees in a globally warmed world. *Bioscience*, Vol. 43, No. 11, pp. 752–61.

Naik, A., Stigter, E. and Laczko, F. 2007. *Migration, Development and Natural Disasters: Insights from the Indian Tsunami*. Geneva, International Organization for Migration. (Migration Research Series No. 30.)

Naudé, W. 2008. *Conflict, Disasters and No Jobs – Reasons for International Migration from Sub-Saharan Africa*. Helsinki, United Nations University, World Institute for Development Economics Research. (UNU-WIDER Research Paper 85.)

Neumayer, E. 2005. Bogus refugees? The determinants of asylum migration to Western Europe. *International Studies Quarterly*, Vol. 49, No. 3, pp. 389–410.

Pais, J. F. and Elliott, J. R. 2008. Places as recovery machines: vulnerability and neighborhood change after major hurricanes. *Social Forces*, Vol. 86, No. 4.

Paul, B. K. 2005. Evidence against disaster-induced migration: the 2004 tornado in north-central Bangladesh. *Disasters*, Vol. 29, No. 4, pp. 370–85.

Pécoud, A. and de Guchteneire, P. 2007. *Migration without Borders. Essays on the Free Movement of People*. Oxford, UK/Paris, Berghahn/UNESCO.

Pedersen, J. 1995. Drought, migration and population growth in the Sahel: the case of the Malian Gourma: 1900–1991. *Population Studies*, Vol. 49, pp. 111–26.

Perch-Nielsen, S., Bättig, M. B. and Imboden, D. 2008. Exploring the link between climate change and migration. *Climatic Change*, Vol. 91, Nos. 3–4, pp. 375–93.

Petersen, W. 1958. A general typology of migration. *American Sociological Review*, Vol. 23, No. 3, pp. 256–66.

Piguet, E. 2008. *Climate Change and Forced Migration*. Geneva, United Nations High Commissioner for Refugees. (*New Issues in Refugee Research*, No. 153.)

2010*a*. *Les apatrides du climat*. Fondation Mémoire Albert Cohen – E-colloque 2010 'L'état de droit'. www.fondationmémoireAlbertCohen.ch

2010*b*. Linking climate change, environmental degradation and migration: a methodological overview. *Wiley Interdisciplinary Reviews: Climate Change*, Vol. 1, July/August, pp. 517–24.

Poncelet, A. 2008. *Bangladesh: 'The Land of Mad Rivers'*. Bonn, United Nations University, Institute for Environment and Human Security. (Case Study Report.)

Portes, A. and Böröcz, J. 1996. Contemporary immigration: theoretical perspectives on its determinants and modes of incorporation. *International Migration Review*, Vol. 23, No. 3, pp. 606–30.

Ravenstein, E. G. 1889. The laws of migration. *Journal of the Royal Statistical Society*, Vol. 52, No. 2, pp. 241–305.

Renaud, F., Bogardi, J., Dun, O. and Warner, K. 2007. *Control, Adapt or Flee: How to Face Environmental Migration*. Bonn, United Nations University, Institute for Environmental and Human Security. (Paper 5/2007, Interdisciplinary Security Connections.) http://www.ehs.unu.edu/file.php?id=259

Reuveny, R. 2008. Ecomigration and violent conflict: case studies and public policy implications. *Human Ecology*, Vol. 36, No. 1, pp. 1–13.

Reuveny, R. and Moore, W. H. 2009. Does environmental degradation influence migration? Emigration to developed countries in the late 1980s and 1990s. *Social Science Quarterly*, Vol. 90, No. 3, pp. 461–79.

Richmond, A. H. 1994. *Global Apartheid. Refugees, Racism, and the New World Order*. Toronto, Oxford University Press.

Rodriguez, J., Vos, F., Below, R. and Guha-Sapir, D. 2009. *Annual Disaster Statistical Review 2008: The Numbers and Trends*. Brussels, Centre for Research on the Epidemiology of Disasters. www.emdat.be

Saldaña-Zorrilla, S. and Sandberg, K. 2009. Impact of climate-related disasters on human migration in Mexico: a spatial model. *Climatic Change*, Vol. 96, No. 1, pp. 97–118.

Salt, J. 1987. Contemporary trends in international migration study. *International Migration*, Vol. 25, No. 3, pp. 241–51.

Scally, R. 1995. The Irish and the 'famine exodus' of 1847. In: R. Cohen (ed.), *The Cambridge Survey of World Migration*. Cambridge, UK, Cambridge University Press, pp. 80–85.

Semple, E. C. 1911. *Influences of Geographic Environment*. New York, Henry Holt and Company.

Sen, A. K. 1981. *Poverty and Famines: an Essay on Entitlement and Deprivation*. Oxford, UK, Clarendon Press.

Shrestha, S. and Bhandari, P. 2007. Environmental security and labor migration in Nepal. *Population & Environment*, Vol. 29, No. 1, pp. 25–38.

Sly, D. F. and Tayman, J. 1977. Ecological approach to migration reexamined. *American Sociological Review*, Vol. 42, No. 5, pp. 783–95.

Smith, K. 2001. *Environmental Hazards, Assessing the Risk and Reducing Disaster*. London, Routledge.

Stark, O. and Bloom, D. E. 1985. The new economics of labor migration. *American Economic Review*, Vol. 75, No. 2, 173–78.

Stern, N. 2007. *The Economics of Climate Change: The Stern Review*. Cambridge, UK, Cambridge University Press.

Stouffer, S. 1940. Intervening opportunities: a theory relating mobility and distance. *American Sociological Review*, Vol. 5, No. 6, pp. 845–67.

——. 1960. Intervening opportunities and competing migrants. *Journal of Regional Science*, Vol. 2, No. 1, pp. 1–26.

Tacoli, C. 2009. Crisis or adaptation? Migration and climate change in a context of high mobility. *Environment and Urbanization*, Vol. 21, No. 2, pp. 513–25.

Taft, D. J. 1936. *Human Migration: A Study of International Movements*. New York, The Ronald Press Company.

UN. 1998. *Recommendations on Statistics of International Migration*. New York, United Nations. http://unstats.un.org/unsd/publication/SeriesM/SeriesM_58rev1E.pdf

UNDP. 2009. *Human Development Report 2009. Overcoming Barriers: Human Mobility and Development.* New York, United Nations Development Programme.

UNHCR. 2009. *Climate Change, Natural Disasters and Human Displacement: a UNHCR Perspective.* Geneva, United Nations High Commissioner for Refugees.

Van der Geest, K., Vrieling, A. and Dietz, T. 2010. Migration and environment in Ghana: a cross-district analysis of human mobility and vegetation dynamics. *Environment and Urbanization*, No. 22, pp. 107–23.

WBGU. 2008. *Climate Change as a Security Risk.* Berlin/London, German Advisory Council on Global Change/Earthscan.

Zelinsky, W. 1971. The hypothesis of the mobility transition. *Geographical Review*, No. 61, pp. 219–49.

Zetter, R. 2009. The role of legal and normative frameworks for the protection of environmentally displaced people. In: Laczko and Aghazarm (eds), op. cit., pp. 385–442.

Zolberg, A. R., Suhrke, A. and Aguayo, S. 1989. *Escape from Violence: Conflict and the Refugee Crisis in the Developing World.* New York, Oxford University Press.

PART 1

Evidence on the migration–climate change
relationship

2

The main climate change forecasts that might cause human displacements

MARTINE REBETEZ

Introduction

A number of basic aspects of climate change must be taken into account when assessing the impact on human population movements. First, realistic forecasts must be made of future changes in climate parameters. Second, often it is not changes in the averages themselves that most affect human populations or ecosystems but changes in the frequency or intensity of extreme events associated with changing averages. Third, these changes have to be placed in the broader context of global change – all the environmental changes generated by the current trends in human society in terms of population growth, pressure on natural resources, air and water pollution and urbanization, among other factors. Other environmental changes may either increase or reduce the effect of climate change on human populations and therefore on potential movements of these populations. Lastly, possible measures to tackle climate change must be considered where knowledge is sufficiently advanced and available to the public and to policy-makers for steps to be taken, not only to reduce greenhouse gas emissions and the associated climate change and temperature rise, but also to avoid or reduce the adverse impact of this change.

Thus, for example, improving the quality of weather forecasting and of information and communication systems will provide communities with adequate warning of some extreme events. While it is clear that the rise in sea level and increase in hurricane intensity – both connected with climate change – constitute, whether individually or combined, a serious risk to many communities, the possibility of predicting the formation and path of a hurricane will enable many of them, for a time at least, to seek temporary shelter from danger while keeping their homes in the same areas. This has been the case in Bangladesh, where hurricanes killed hundreds of thousands of people in the 1970s. Since then, improved weather forecasting has prevented disasters of such magnitude, even though the frequency and intensity of the hurricanes has not diminished – quite the contrary. In the United States in 2005, however, Hurricane Katrina demonstrated that as a result of shortcomings in levee maintenance beforehand, and in protection of the population during and after the event, the disaster reached a scale that could have been avoided with better management.

The expected trends in climate parameters described in this chapter, and their potential impact on population movements, should therefore certainly not be regarded as inevitable. Admittedly, the inexorable rise in sea level will gradually cover land that is today inhabited, entailing the displacement of these populations inland. And it is true that the greater part of the expected rise in the twenty-first century has already been triggered by present temperature trends and the existing increase in greenhouse gas concentrations. In the case of many other parameters, however, such as changes in precipitation patterns, better resource management could help to prevent disasters and population movements.

Temperature change

Temperature is the main parameter reflecting current and future increases in atmospheric concentrations of greenhouse gases. Its rise is linked to these concentrations but is erratic in terms of both time and space. Global warming is uneven, with temperatures rising quickly at some periods, such as the 1980s to the present, and more slowly at others, such as the 1950s to the 1970s. The whole climate system (i.e. not just the atmosphere but also the hydrosphere, biosphere, lithosphere and cryosphere) interacts in an extremely complex manner, and this is reflected in atmospheric temperatures. However, the warming of the climate system since the end of the nineteenth century is *unequivocal* according to the Intergovernmental Panel on Climate Change (IPCC, 2007), as is now evident from observations of increases in global average temperatures not just for the air but also for the oceans, which are in contact with the air and thus absorb some of its energy. Today's and tomorrow's air temperatures would be much higher if the oceans did not absorb both a large part of the atmosphere's additional energy and a large part of anthropogenic carbon dioxide emissions. Together with the rise in air temperatures, we are now recording widespread melting of snow and ice across the Earth's surface, which is also consistent with absorption of a significant part of the atmosphere's excess energy by the cryosphere.

The eight warmest years on record between 1850 (when the instrumental record of global surface temperature began) and 2007 all occurred after 1998, and the fourteen warmest occurred after 1990 (Füssel, 2009). The year 2008 was relatively colder than the immediately preceding years, mainly because the solar magnetic activity cycle (sunspot cycle) was at its lowest level and because of a La Niña event in 2007–2008 (Richardson, 2009). Nevertheless, the long-term trend of rising temperatures is obvious, and changes in global surface temperatures are in the upper part of the range predicted by the IPCC (Füssel, 2009; IPCC, 2007; Richardson, 2009). Warming is taking place at a faster pace: whereas global surface temperatures between 1901 and 2000 rose at a rate of 0.06 °C per decade, the figure for 1906–2005 was 0.074 °C per decade (IPCC, 2007). Average Northern Hemisphere temperatures during the second half of the twentieth century were very likely higher than during any other fifty-year period in the last five hundred years and probably the highest in the past 1,300 years at least (IPCC, 2007). The discovery in 2003 of artefacts dating from the Neolithic, the Bronze Age, Roman times and the Middle Ages – footwear, clothing, tools and coins – at the Schnidejoch glacier in Switzerland, on an Alpine pass used whenever the glacier was small enough to allow people across, shows that the glacier has now receded more than at any time over the past 5,000 years (Grosjean et al., 2007).

The temperature increase is widespread across the globe and is greater at higher northern latitudes, and the land has warmed faster than the oceans (IPCC, 2007). It is important to note that temperature rise in some of the Earth's regions has been observed to be much higher than the global average and this will also be the case in future.

Greenhouse gas emissions over the next few decades are expected to continue at or above current rates, which would cause further warming and induce many changes in the global climate system during the twenty-first century that would very likely be larger than those observed during the twentieth century: a range of emission scenarios suggest global warming of about 0.2 °C per decade for the next two decades (IPCC, 2007). The warming may often be at least twice this figure in many regions. If recent observed trends continue in future decades and for the whole of the twenty-first century, warming is expected to be greatest over land and least over the Southern Ocean and parts of the North Atlantic (IPCC, 2007).

This overall rise in temperature suggests that an increase in the frequency of hot extremes and heatwaves is very likely. It is this rise in extremes, more than the rise in averages, that is likely to affect human populations most directly, as we have seen in the past with the summer 2003 heatwave in Europe, for example, which killed tens of thousands of people and sparked considerable research (Cohen et al., 2005; Heudorf and Meyer, 2005; Rebetez et al., 2006; Robine et al., 2008; Stott et al., 2004). The increase in the intensity and frequency of heatwaves and their impact on the human population is currently the subject of many studies, especially with regard to the urbanized regions of the North (which are often ill-equipped for hot weather) but also areas across the world traditionally affected by this problem (Basu et al., 2008; Bell et al., 2008; Fouillet et al., 2008; Haines et al., 2006; Heudorf and Meyer, 2005; Hutter et al., 2007; Kalkstein et al., 2008; Ostro et al., 2009; Schär et al., 2004; Vescovi et al., 2005). The combined effects of heat and air pollution are also likely to be considerable in urban and suburban areas on account of ozone formation (Lacour et al., 2006; Rosenthal et al., 2007; Semenza et al., 2008).

There will be many indirect impacts of temperature rise on human populations (IPCC, 2007), such as higher crop yields in colder environments and decreased yields in warmer environments. Increased insect outbreaks may also affect both crops and human health directly. Insect outbreaks will gradually occur in more and more environments hitherto protected by the cold, both at altitude and at higher latitudes. Countries currently affected by malaria, for example, will see more and more cases of the disease in mountain areas and high plateaus previously largely unaffected. In sub-Saharan Africa many cities, such as Nairobi and Kigali, have been built at altitude to benefit from the coolness and provide protection against malaria. One degree of temperature rise corresponds to about 150 m altitude, making some areas extremely sensitive to slight changes. If the temperature exceeds a significant threshold, allowing the spread of malaria for example, living conditions may undergo a drastic change.

Among the expected positive impacts of temperature rise are a decline in cold-related mortality and morbidity but such positive impacts will not compensate the negative ones. The increase in heatwaves will have various consequences, such as problems relating to safe drinking water on account of increased demand, evaporation and evapotranspiration and also because of the impact of higher temperatures on water quality. In addition to agriculture, a number of economic activities are likely to be directly affected, including tourism, where a decline in winter business is expected because of lack of snow in many mountain areas as well as a decline in summer business in warm and/or seaside areas such as the Mediterranean, and particularly its southern shore (IPCC, 2007).

Reduction in meltwater

As well as causing a sharp decrease in the global surface subject to snowfall or permanently covered by snow (IPCC, 2007), temperature rises are also expected to affect snowfall altitude and melt. This will have consequences for annual water resources and availability in many areas. The dry season could become unbearable for human communities if the glaciers supplying the valleys disappear. In the meantime, this risk is less apparent than it might be because meltwater is more plentiful and thus the supply seems more secure than ever. The origin of this meltwater ought to be ascertained as soon as possible in order to anticipate its future decline, especially in areas with a dry summer climate. This is the case in the Andes, where a number of glaciers have already disappeared or will do so in coming decades. The city of Lima, for example, depends largely on the melting of snow and ice from the Andes for its water supply.

An assessment of the problems likely to be caused by a reduction in meltwater supply is particularly complex because it has to take into account other parameters that may affect water supply. In the Andes, for example, it is necessary to predict not only the scale of meltwater decrease but also changes in summer precipitation. In the Peruvian Andes, this precipitation is tending to decline, but the volume of this decrease and the time-frame in which it will occur are still unclear. In particular, the substantial interannual variability of precipitation patterns will play an important part, especially in South America, where the El Niño phenomenon has a considerable impact. Precipitation varies substantially between El Niño years and non-El Niño years or, conversely, La Niña years. In this context of extreme variation, a reduction in the volume of meltwater is likely to play a critical role in places where it has hitherto provided sufficient water in years with minimum precipitation. In these areas, as in many other cases, it will therefore be during extreme summers caused by a variety of parameters that water supply problems are likely to arise. And the situation will be all the more critical because water demand is rising as a result of population changes, changing patterns of behaviour and irrigation requirements which are putting more and more pressure on water resources.

Sea level rise

Current increases in sea level rise are consistent with the increase in temperature. Global average sea level has been rising at a rate of 1.8 mm annually since 1961 and 3.1 mm annually since 1993 (IPCC, 2007). This phenomenon is due both to thermal expansion (the fact that warmer water takes up a greater volume) and melting of the ice. The latter factor was to some extent underestimated in earlier projections for the twenty-first century and explains their difference from projections published since 2007. Mountain glaciers and snow cover have now declined in both hemispheres, and melting has already partly contributed to increasing the volume of seawater. The rise observed from the end of the twentieth century is largely due to increased melting in Greenland and Antarctica (Richardson, 2009). The rapid increase in ice loss over the past few years was apparent in 2007, for example, when the loss from the Greenland ice cap was 60% higher than in 1998, which was in fact a particularly warm year because of a strong El Niño event. The twelve summers from 1996 to 2007 all showed a higher loss than the average loss for the years from 1973 to 2007 (Mote, 2007; Rignot et al., 2008b; Van den Broeke et al., 2009).

Furthermore, although it was long thought that the Antarctic peninsula would not sustain any loss for many more years, it was recently shown to have lost considerable ice mass between 1992 and 2006, with mass loss increasing by 75% in ten years (Rignot et al., 2008a). In the Arctic, the loss has been 10% over the past decade compared with 3% over the previous decade (Zhang et al., 2008).

Recent projections (made after the last IPCC report) show that by the end of the twenty-first century, the sea level should rise by at least 1 m (Füssel, 2009; Rahmstorf, 2007; Rahmstorf et al., 2007; Richardson, 2009). This is a large increase, which was not possible to forecast previously and which will have considerable implications for human populations over the course of this century. Furthermore, sea level rise will be a long way from stopping in 2100. The rise in sea temperature, together with ice loss in Greenland and Antarctica, will continue to affect the sea level rise for several centuries at least.

The human consequences of rising sea levels in the twenty-first century must be assessed in conjunction with certain other environmental factors affecting coastal stability, such as the condition of mangroves, wetlands and forests. Furthermore, the issue of sea level rise must be considered not only in terms of disappearing land but also in terms of the salinization of rivers and estuaries. Other factors may have an important impact on river discharge and thus salinization, such as human with-drawals in general, whether for irrigation or power-generation reser-voirs. Many examples concern densely populated deltas, such as the Nile delta and the city of Cairo in Egypt. Like the Aswan Dam, many river dams reduce discharge and increase salinization. In Bangladesh, for example, fishing and farming communities are threatened by the salini-zation of river water due not only to higher sea levels, and therefore to climate change, but also to the impact of upstream dams. Another example includes the decrease in Bangkok's height above sea level, partly because of the rise in sea level but above all, at present, on account of water-table withdrawals. The city, which is built on swamps, thus has certain neighbourhoods that are below sea level and particularly at risk of flooding.

Changes in precipitation patterns

Consistent with rising temperatures, significant changes have been observed in precipitation patterns, and these are expected to increase in future (IPCC, 2007). These changes will, for example, exacerbate desertification in some regions such as the African Sahel. It has been observed that monsoon patterns are currently changing in several regions of the world (IPCC, 2007).

From 1900 to 2005, precipitation increased significantly in eastern parts of North and South America, northern Europe and northern and central Asia but declined in the Sahel, the Mediterranean, southern Africa and parts of southern Asia. The global area affected by drought has probably increased since the 1970s (IPCC, 2007).

In regions where water supply is already problematic, the decline in total annual precipitation will constitute a major problem. In most regions, however, the threat stems mainly from the interannual vari-ability of precipitation. Populations will in future have even more prob-lems with extreme seasons or years, since, quite apart from any reduction in total annual precipitation, the variability of precipitation is generally increasing across the globe, which means that there are to be more and more periods of dryness and of heavy precipitation. These events may occur successively in the same regions. This overall change in precipitation patterns will increase the frequency and intensity of heavy precipitation events, which in turn are likely to generate more flooding, landslides and mud flows. This change will represent a direct and growing threat to human communities as well as an indirect threat to their survival because of soil destruction and damage to crops and infrastructure. Such situations could lead to temporary or permanent displacement of populations at risk.

This is a field with considerable scope for long-term action to avoid disasters. Maintaining or replanting forests to prevent soil erosion, giving back space to rivers and preventing settlements in places at risk are all measures that cannot be improvised at short notice but must be planned far in advance in order to reduce the vulnerability of an area and its inhabitants.

Changes in the frequency and intensity of hurricanes

The rise in air temperature and its corollary, the rise in ocean temperatures, will theoretically increase the intensity of hurricanes, if not their frequency, as there will be more energy in the climate system.

There is observational evidence of a clear increase in intense tropical cyclone activity in the North Atlantic (IPCC, 2007; Richardson, 2009). This trend is less evident elsewhere, but the latest research findings also show an increase in other regions and for other classes of activity (Benestad, 2009; Kossin and Camargo, 2009; Pezza et al., 2009; Usbeck et al., 2010; Yokoi et al., 2009).

Damage, especially in coastal areas, is likely to increase in accordance with an upsurge in storm and hurricane activity over the next few decades. However, actual loss of human infrastructure will depend to a large extent on exposure, the protective measures that may be taken and the state of natural environments with the potential to protect coasts and other vulnerable areas.

Hurricanes and their attendant destruction, together with a rise in the frequency of such damage, could cause temporary or permanent displacement of the human communities affected.

As hurricane damage is partly due to the associated flooding, sea level plays an important part, which will probably increase in future as the level rises. Hurricane-related risks will thus be greater with a higher sea level and if the hurricane coincides with high tides and a high water period. It is thus a combination of various environmental factors and other unforeseen events that will increasingly determine hurricane-related flood risks.

Timescale

The following points concerning future changes in climate parameters likely to have an influence on the displacement of human communities need to be re-emphasized. First, because of the time scales associated with climate processes and feedbacks, the temperature rise and its consequences such as sea level rise should continue well past the end of the twenty-first century into the centuries beyond.

Second, some phenomena will change gradually, such as average temperature rise, whereas others could change abruptly and dramatically, such as some extreme events or the Asian monsoons for example, which we know could diminish very quickly. The impact on human populations and the way they react, including various types of migration, will thus depend on whether changes in climate parameters are gradual or, on the contrary, swift and sudden.

Conclusions

Ultimately, the final impact of climate change will often depend on a combination of human and environmental factors that are to some extent independent of climatic factors. Temperature rise, sea level rise, hurricane intensity, changing precipitation patterns and changes in snow and ice melting patterns may each have their own individual consequences.

But the most problematic consequences will often arise from the conjunction of several factors: increased evapotranspiration (due to higher temperatures), droughts and the disappearance of glaciers. These three factors combined will have a substantial impact on crops, natural ecosystems, water supply and, ultimately, human communities. Moreover, the influence of other human or environmental parameters is often a deciding factor.

To avoid unwanted migration due to climate change, it is therefore necessary to address the problem as a whole. We must not only take action to reduce greenhouse gas emissions – and thus climate change – and the immediate impact of individual climate parameters, but also address the other environmental issues usually associated with them in the event of disasters, including deforestation, soil deterioration, reduced streamflow or mangrove degradation, among many others.

References

Basu, R., Feng, W. Y. and Ostro, B. D. 2008: Characterizing temperature and mortality in nine California counties. *Epidemiology*, Vol. 19, No. 1, pp. 138–45.

Bell, M. L., O'Neill, M. S., Ranjit, N., Borja-Aburto, V. H., Cifuentes, L. A. and Gouveia, N. C. 2008: Vulnerability to heat-related mortality in Latin America: a case-crossover study in São Paulo, Brazil, Santiago, Chile and Mexico City, Mexico. *International Journal of Epidemiology*, Vol. 37, No. 4, pp. 796–804.

Benestad, R. E. 2009: On tropical cyclone frequency and the warm pool area. *Natural Hazards and Earth System Sciences* 9, 635–645.

Cohen, J. C., Veysseire, J. M. and Bessemoulin, P. 2005. Bio-climatological aspects of summer 2003 over France. In: W. Kirch, B. Menne and R. Bertollini (eds), *Extreme Weather Events and Public Health Responses*, Berlin/Heidelberg, Springer Verlag, pp. 33–45.

Fouillet, A., Rey, G., Wagner, V., Laaidi, K., Empereur-Bissonnet, P., Le Tertre, A., Frayssinet, P., Bessemoulin, P., Laurent, F., De Crouy-Chanel, P., Jougla, E. and Hémon, D. 2008. Has the impact of heat waves on mortality changed in France since the European heat wave of summer 2003? A study of the 2006 heat wave. *International Journal of Epidemiology*, Vol. 37, No. 2, pp. 309–17.

Füssel, H. 2009. An updated assessment of the risks from climate change based on research published since the IPCC Fourth Assessment Report. *Climatic Change*, Vol. 97, pp. 469–82.

Grosjean, M., Suter, P. J., Trachsel, M. and Wanner, H. 2007. Ice-borne prehistoric finds in the Swiss Alps reflect Holocene glacier fluctuations. *Journal of Quaternary Science*, Vol. 22, No. 3, pp. 203–07.

Haines, A., Kovats, R. S., Campbell-Lendrum, D. and Corvalan, C. 2006. Climate change and human health: impacts, vulnerability, and mitigation. *Lancet*, Vol. 367(9528), pp. 2101–09.

Heudorf, U. and Meyer, C. 2005. Heat waves and health – analysis of the mortality in Frankfurt, Germany, during the heat wave in August 2003. *Gesundheitswesen*, Vol. 67, No. 5, pp. 369–74.

Hutter, H. P., Moshammer, H., Wallner, P., Leitner, B. and Kundi, M. 2007. Heatwaves in Vienna: effects on mortality. *Wiener Klinische Wochenschrift*, Vol. 19, Nos. 7–8, pp. 223–27.

IPCC. 2007. *Fourth Assessment Report: Climate Change 2007* (AR4). Geneva, Intergovernmental Panel on Climate Change. http://www.ipcc.ch/publications_and_data/publications_and_data_reports.htm

Kalkstein, L. S., Greene, J. S., Mills, D. M., Perrin, A. D., Samenow, J. P. and Cohen, J. C. 2008. Analog European heat waves for U.S. cities to analyze impacts on heat-related mortality. *Bulletin of the American Meteorological Society*, Vol. 89, No. 1, pp. 75–85.

Kossin, J. P. and Camargo, S. J. 2009. Hurricane track variability and secular potential intensity trends. *Climatic Change*, Vol. 97, pp. 329–37.

Lacour, S. A., de Monte, M., Diot, P., Brocca, J., Veron, N., Colin, P. and Leblond, V. 2006. Relationship between ozone and temperature during the 2003 heat wave in France: consequences for health data analysis. *BMC Public Health*, Vol. 6, p. 261.

Mote, T. L. 2007: Greenland surface melt trends 1973–2007: Evidence of a large increase in 2007. *Geophysical Research Letters*, Vol. 34, L22507.

Ostro, B. D., Roth, L. A., Green, R. S. and Basu, R. 2009. Estimating the mortality effect of the July 2006 California heat wave. *Environmental Research*, Vol. 109, No. 5, pp. 614–19.

Pezza, A. B., Simmonds, I. and Pereira, A. J. 2009. Climate perspective on the large-scale circulation associated with the transition of the first South Atlantic hurricane. *International Journal of Climatology*, Vol. 29, No. 8, pp. 1116–30.

Rahmstorf, S. 2007: A semi-empirical approach to projecting future sea-level rise. *Science*, Vol. 315, No. 5810, pp. 368–70.

Rahmstorf, S., Cazenave, A., Church, J. A., Hansen, J. E., Keeling, R. F., Parker, D. E. and Somerville, R. C. J. 2007. Recent climate observations compared to projections. *Science*, Vol. 316, No. 5825, p. 709.

Rebetez, M., Mayer, H., Dupont, O., Schindler, D., Gartner, K., Kropp, J. and Menzel, A. 2006. Heat and drought 2003 in Europe: a climate synthesis. *Annals of Forest Science*, Vol. 63, No. 6, pp. 569–78.

Richardson, K. 2009. *Climate Change, Global Risks, Challenges and Decisions.* Copenhagen, University of Copenhagen.

Rignot, E., Bamber, J. L., Van den Broeke, M. R., Davis, C., Li, Y. H., Van de Berg, W. J. and Van Meijgaard, E. 2008a. Recent Antarctic ice mass loss from radar interferometry and regional climate modelling. *Nature Geoscience*, Vol. 1, No. 2, pp. 106–10.

Rignot, E., Box, J. E., Burgess, E. and Hanna, E. 2008b. Mass balance of the Greenland ice sheet from 1958 to 2007. *Geophysical Research Letters*, Vol. 35, L20502.

Robine, J. M., Cheung, S. L. K., Le Roy, S., Van Oyen, H., Griffiths, C., Michel, J. P. and Herrmann, F. R. 2008. Death toll exceeded 70,000 in Europe during the summer of 2003. *Comptes rendus biologies*, Vol. 331, No. 2, pp. 171–78.

Rosenthal, J. K., Sclar, E. D., Kinney, P. L., Knowlton, K., Crauderueff, R. and Brandt-Rauf, P. W. 2007. Links between the built environment, climate and population health: interdisciplinary environmental change research in New York City. *Annals of the Academy of Medicine Singapore*, Vol. 36, No. 10, pp. 834–46.

Schär, C., Vidale, P. L., Lüthi, D., Frei, C., Häberli, C., Liniger, M. A. and Appenzeller, C. 2004. The role of increasing temperature variability in European summer heatwaves. *Nature*, Vol. 427(6972), pp. 332–36.

Semenza, J. C., Wilson, D. J., Parra, J., Bontempo, B. D., Hart, M., Sailor, D. J. and George, L. A. 2008. Public perception and behavior change in relationship to hot weather and air pollution. *Environmental Research*, Vol. 107, No. 3, pp. 401–11.

Stott, P. A., Stone, D. A. and Allen, M. R. 2004. Human contribution to the European heatwave of 2003. *Nature*, Vol. 432(7017), pp. 610–14.

Usbeck, T., Wohlgemuth, T., Dobbertin, M., Pfister, C., Bürgi, A. and Rebetez, M. 2010. Increasing storm damage to forests in Switzerland from 1858 to 2007. *Agricultural and Forest Meteorology*, Vol. 150, No. 1, pp. 47–55.

Van den Broeke, M., Bamber, J., Ettema, J., Rignot, E., Schrama, E., Van de Berg, W. J., Van Meijgaard, E., Velicogna, I. and Wouters, B. 2009. Partitioning recent Greenland mass loss. *Science*, Vol. 326(5955), pp. 984–86.

Vescovi, L., Rebetez, M. and Rong, F. 2005. Assessing public health risk due to extremely high temperature events: climate and social parameters. *Climate Research*, Vol. 30, No. 1, pp. 71–78.

Yokoi, S., Takayabu, Y. N. and Chan, J. C. L. 2009. Tropical cyclone genesis frequency over the western North Pacific simulated in medium-resolution coupled general circulation models. *Climate Dynamics*, Vol. 33, No. 5, pp. 665–83.

Zhang, X. D., Sorteberg, A., Zhang, J., Gerdes, R. and Comiso, J. C. 2008. Recent radical shifts of atmospheric circulations and rapid changes in Arctic climate system. *Geophysical Research Letters*, Vol. 35, L22701.

3

Climate change, migration and health in Brazil

ALISSON FLÁVIO BARBIERI AND ULISSES
E. C. CONFALONIERI[1]

Introduction

The main purpose of this chapter is to discuss critical linkages between climate change, migration and health, with a particular focus on Brazil. One of the potential impacts of predicted climate change is to induce population displacements, which may in some cases aggravate situations of vulnerability. It is likely, especially in tropical and developing countries, that future population migration induced by climate change may increase population vulnerability given the potential redistribution of endemic infectious diseases.

We describe a case study on Brazil's Northeast (Nordeste) region, which shows potential scenarios of migration and health vulnerability due to the predicted climate changes. The reason for choosing this study area is the fact of its being the second most populated and the poorest region, with an extensive semi-arid area which will be severely impacted by increased temperature and reduced rainfall.

The case study in the Brazilian Northeast also allows us to better understand how population redistribution through migration may impact population health and thus redefine population vulnerability in

[1] The authors would like to thank CPTEC/INPE for the data on climate scenarios and EMBRAPA/UNICAMP for providing the agricultural scenarios. However they take full responsibility for the results and interpretation described here. They would also like to thank the UK Embassy for supporting this research project, which was funded by the Global Opportunities Fund, the UK Ministry of Foreign Affairs, and CEDEPLAR/UFMG and FIOCRUZ for institutional support. Special thanks also to colleagues involved in the research project: professors Bernardo Lanza, Ricardo Ruiz, José Alberto de Carvalho, Edson Domingues and José Irineu Rigotti at CEDEPLAR; and professor Anna Carolina Lustosa Lima at FIOCRUZ.

future scenarios of climate change. For example, droughts in the Northeast have historically induced rural-urban migration and as a consequence caused epidemic episodes of visceral leishmaniasis in the state capitals; and malaria has been 'imported' from the Brazilian Amazon to the Northeast as a consequence of population migration, following a drought year. Great migration processes – particularly to urban areas – also represent pressure over the infrastructure and health systems, especially considering the historically large gap between population demand and supply of infrastructure and health services by the public sector in South America.

The regional persistence of human health problems sensitive to climate variability makes the Northeast region structurally vulnerable to the projected impacts of a changing climate. Although the human population of the region is partially 'adapted' to the droughts in a semi-arid region, climate scenarios project a progressive worsening of the arid condition, which could leave the semi-arid region, which currently has about 21 million inhabitants, inappropriate for human settlements, due to the extreme climate. In view of this, the Northeast became a priority effects for the assessment of the effects of climate change on the economy, society, health and the health-care system.

We focus here on economic factors impacted by climate change which can trigger migration and consequently affect population health and vulnerability. In this regard we consider not only the sanitary implications of population displacements such as the movement of an endemic focus from the origin area of migrants to their destinations, but also the capacity to control and absorb the health-care demand in the destination. We do not address the question of health problems caused by climate change as push factors for migration, as the role of health problems whether or not aggravated by climate change as push factors for migration is, in general, not well established. Economic factors and environmental degradation have historically played a much more important role as triggers of migration in the Northeast, for example, as these are linked to subsistence strategies.

While in our case study we assume that those at higher risk of migration are those facing income deprivation, this may not always be the case (see e.g. Martine and Guzmán, 2002). The precise definition of vulnerable populations must consider the identification of all adaptation mechanisms available to a given population, irrespective of their socio-economic status. The possibility of adopting concurrent adaptation alternatives may be the key mechanism defining migration propensities.

The next section provides a brief literature review on linkages between climate change, migration and health in South America, and is followed by a description of the study area. Then we describe the methodological approach to investigating the linkages between climate change, migration and health in the Brazilian Northeast until 2050. The last two sections present the results, conclusions and policy implications.

Climate change, migration and health

There is scant evidence in the empirical literature on the impacts of climate change on population redistribution through migration and the related effects on population health. Regarding the impacts of climate change on migration, some authors have proposed less comprehensive models to investigate the linkages between climate change and migration, usually focusing on specific impacts of climate change (see examples in Döös, 1997; McLeman and Smit, 2006; Perch-Nielsen et al., 2008).[2] This is the case of studies which explore how droughts induced by climate change affect some specific determinants of population movements in less-developed countries (e.g. Findley, 1994; Meze-Hausken, 2000; Ezra, 2003; Henry et al., 2004; Kniveton et al., 2008). As examples, Henry et al.'s (2004) findings for Burkina Faso indicate that population mobility (rural–rural) from drier regions tends to be higher than mobility from wetter regions, and that short-term rainfall deficits tend to increase this trend, particularly with long-term migration. Based on empirical evidence from case studies, Kniveton et al. (2008) suggest that while droughts can increase the stock of short-term rural–rural migrants, the poorest are not necessarily those more at risk of migration, as they can have alternative adaptation options depending on their socio-economic status.

Perch-Nielsen et al. (2008) propose a conceptual model in which sea level rise and floods are major determinants of migration (see also McGranahan et al., 2007, for a study on the impacts of sea level rise on migration from urban areas). The authors conclude that migration as an adaptation mechanism cannot be considered separately, but as a

---

[2] Adamo (2008) reviews estimates of potential population displacements due to climate change impacts (particularly sea level rise) focusing on less comprehensive methods and finds that several estimates usually reflect populations at risk as a surrogate for population displacements. As a result, these studies suggest a great variation in estimates, depending on methods and data used.

potential response along with several others at multiple spatial scales (households, communities, etc.). In a similar conclusion, McLeman and Smit (2006) propose a conceptual model which investigates population migration as a possible adaptive response to risks associated with climate change. These two studies agree with some theoretical perspectives on migration which assume migration as a concurrent adaptation mechanism in periods of economic depression (Davis, 1963; Bilsborrow, 1987; see also a brief literature review on this subject in Barbieri et al., 2009).

Understanding the determinants of migration and how they may be affected by future climate scenarios is a key requirement for better planning and policies aiming to alleviate the production or reproduction of situations of poverty, particularly that of migrant populations in situations of high socio-economic vulnerability. In this sense, it is important to define in which degree migration may be a mechanism engendering further vulnerability or else a mechanism of adaptation. The *Third Assessment Report* of the Intergovernmental Panel on Climate Change defines vulnerability to climate change as:

> the degree to which a system is susceptible to, or unable to cope with, adverse effects of climate change, including climate variability and extremes. Vulnerability is a function of the character, magnitude, and rate of climate variation and to which a system is exposed, its sensitivity, and its adaptive capacity (IPCC, 2001).

While assuming the diversity of definitions and conceptualizations of the term 'vulnerability' across disciplines, and that there is not necessarily a correct definition (Füssel, 2007), we use this concept to qualify a population degree of exposure and 'resilience' to the adverse effects of climate change on their livelihoods – particularly the impacts on the generation of income and employment. This vulnerability is contingent on a diversity of factors, especially socio-economic, political and institutional, which makes a given population susceptible to external impacts such as increasing temperature and periods of drought. The intensity of vulnerability in a population will depend on the adaptive capacity and the adaptation mechanisms available, as discussed above.

In particular, migrants to urban areas in developing countries are one of the potentially most vulnerable populations in future scenarios of climate change. The International Institute for Environment and Development (IIED, 2007) suggests that in a context of increasing urbanization driven by migration in most of the developing world, the scale of risk to climate change will be affected by infrastructure and housing quality, by the population's ability to cope with changes (proxy of factors such as education, culture, solidarity) and by the quality of institutional responses (e.g. aid and medical care, urban planning).

One of the most dramatic consequences of the relationship between climate change and migration may be population health. The IPCC's *Fourth Assessment Report* (IPCC, 2007) stressed the possibility of global climate change in the coming decades to change the health profile currently observed in different populations, particularly the geographical expansion or intensification of transmission of infectious diseases – especially vector-borne and water-borne infections – and undernourishment in developing countries and regions.

The social and epidemiological implications of human migration, whether at regional or international levels, are well known to public health. 'Health Vulnerability' is determined by an aggregate of factors, besides the epidemiological profile of the populations: socio-economic characteristics such as income, education, habitation, sanitation, institutional capacity and public services, and the increased demand on health care. It is also influenced by demographic aspects such as population density and the age structure of social groups. Other important determinants of vulnerability are geographical in nature, such as settlement in drought-prone areas (drylands) and flood-prone low-lying areas. On the other hand, there are several examples of migrations influencing the health profile of people, both of the migrants and of those living in the destination areas. There are reports for Brazil and for other parts of the world (Barnett and Walker, 2009). Migration, especially if forced and on a large scale, negatively affects the well-being of the migrants and often disrupts patterns of land use, especially in urban areas, facilitating the occupation of risk areas; disrupts local weak economies and overloads services in general, creating social unrest. Therefore it contributes significantly to increased social and health vulnerability.

In this regard, the displacement of human population groups can rearrange spatially the foci of endemic infectious diseases, a phenomenon already observed in several parts of Brazil as well as in other countries; the diseases involved were cholera, malaria, leishmaniasis, schistosomiasis, among others. Another important consequence of human migration is the displacement of a burden of chronic diseases to the areas of destination of the migrants, especially urban areas, resulting in an increased demand for health-care systems, especially the public system.

Among the research publications concerned with climate change and social vulnerability in drylands, the work by Ribot et al. (1996) discusses the fundamental issues and examples of strategies to face the climate change in semi-arid regions; in the Brazilian case it emphasizes the geographical, political, economic and social conditions of the marginalized population. For these authors, the main problem in semi-arid regions is not the harshness of the climate, but the vulnerability of the human population to these processes. Vulnerability is the product of an association of economic, political and social factors and is a function of social and economic status, gender, ethnicity, age and other factors (Ribot et al., 1996). In a specific reference to food security, Downing (1992) described vulnerability as an aggregate measure for a given population or region, related to the underlying factors that influence exposure to the lack of food.

Perhaps one of the regions most vulnerable to climate change in Latin America is the Brazilian Northeast, a mostly semi-arid area (known as the *caatinga* or dryland ecosystem), with poor soils susceptible to salinization, discontinuous and limited land cover and irregular rain rime deposits, with low precipitation (Ribot et al., 1996). Projections by the Centro de Previsão de Tempo e Estudos Climáticos/Instituto Nacional de Pesquisas Espaciais (CPTEC/INPE) suggest changes in temperature and precipitation patterns which will generate an increasing process of 'aridization' in the region, with important impacts on the livelihoods of the poorest (particularly those making their living in the agriculture sector). These scenarios will probably increase situations of socio-economic vulnerability given the persistence of its present status as the poorest region in the country, and the fragility of agricultural systems to climate variations, as well as enhance the desertification process.

The Brazilian Northeast has been historically characterized by the occurrence of periodic droughts associated with annual climate variability (Wang et al., 2004). The agriculture in the semi-arid is mostly based on small subsistence producers, and some studies have shown a loss of up to 80% of agricultural production in periods of long droughts. Historically, these periods of drought have motivated peaks of emigration from the Northeast, particularly to richer areas in the Southeast (Sudeste) region. Franke et al. (2002) show, for example, that the El Niño oscillations in the early 1980s and 1990s induced migration from rural areas to São Luís and Teresina (capitals of the states of Maranhão and Piauí, respectively). Confalonieri (2003) links El Niño oscillations in 1982–83 to migration peaks from the state of Maranhão to the state of

Pará (in the Brazilian Amazon) as a cause of the abrupt increase in imported malaria to Maranhão. It is not clear from these studies, however, if most of the migrants are in fact the most vulnerable among the poorest, or those with some resources (social or financial capital or both) to escape risks and reduce their vulnerability.

Study area

Given the preceding discussion about the vulnerability of the Brazilian Northeast to predicted climate changes, we provide an in-depth case study on potential scenarios for this region between 2025 and 2050. Projected climate changes are analysed in terms of their potential impacts on population migration and increased vulnerability, particularly in terms of population health.

Among the five Brazilian major regions (South, Southeast, Central-West, North and Northeast), the Northeast is the second most populated after the Southeast (where the two Brazilian major metropolitan areas, São Paulo and Rio de Janeiro, are located), with about 49 million individuals in 2000, or 28% of the country's population. The Northeast is within an extensive semi-arid area and a large population share working in the primary sector – mostly agriculture and cattle ranching. Figure 3.1 shows the study area, with its states and metropolitan areas.

The impacts of climate change on urban areas may also have important repercussions in the Brazilian Northeast, considering that it is a highly urbanized area. The urbanization rate in the region jumped from 46% in 1960 to 71% in 2000 and 75% in 2005 according to the Brazilian Institute of Geography and Statistics (IBGE). In addition, in recent years an increasing concentration of the population in the major cities of the region has been observed. This phenomenon has also contributed to increased economic inequality and poverty concentration in major urban areas. The Northeast is characterized by high income inequality compared with other regions (Theil-L of 0.78 compared with 0.60 in the South)[3] and concentration of poor families throughout the region (over 40% families considered to be poor); the UN Human Development Index for the region is 0.57 compared with 0.78 for the South. Furthermore,

---

[3] The Theil-L is one of the most used indexes to measure economic inequality. It is equal to the logarithm of the ratio between the arithmetic and geometric mean of income. In a society with perfect income distribution (everybody has the same income), the index is zero; in the opposite case, the index is one.

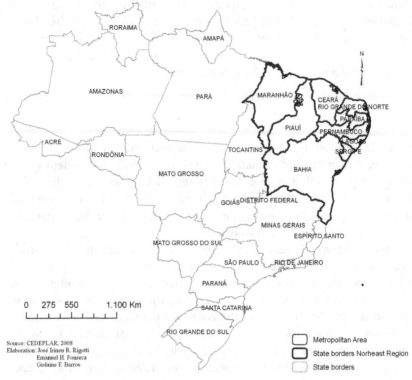

Figure 3.1 State borders and metropolitan areas: Northeast region of Brazil

urban areas in the Northeast are characterized by poor infrastructure: sewerage and treated water cover less than 50% of the population, the region has the lowest ratio of doctors per 100,000 inhabitants and the worst educational performance indicators in the country. All else being equal, or given small improvements in mechanisms that can assure a greater adaptive capacity (e.g. investment in human capital and in infrastructure), these characteristics of urban areas in the Northeast imply an increasing scenario of population vulnerability given the potential increases in temperature over the following decades.

Poor socio-economic indicators associated with periods of drought and demographic pressures have historically motivated peaks of out-migration – with these migrants being known as *retirantes* – from the Northeast region to richer areas in south-east Brazil. During the 1960s and 1970s, a period of increasing industrialization and urbanization in the Southeast, the Northeast's net migration (given by the difference between total number

of immigrants and total number of emigrants in the Northeast) was –
2,166,258 and –3,049,459 individuals (Carvalho and Garcia, 2002). These
figures correspond to net migration rates (NMR – ratio of net migration to
total population in a given year) of approximately –7.6% and –8.7%,
respectively.

The intensity of migration flows from the Northeast has shown a
dramatic decrease since the late 1980s and 1990s, due especially to slower
rates of economic growth in the Southeast. In fact, an analysis of the 1991
and 2000 Brazilian censuses, and the National Household Surveys
(PNADs) between 2001 and 2005 showed a net migration of –138,659
in the period 2000–2005, which corresponds to an NMR of approximately –0.3% (CEDEPLAR, 2007). Furthermore, the largest cities in the
Northeast (particularly the state capitals) have increasingly attracted
migrants from rural or smaller urban areas in the region.

Methodology

In this section we describe the methodology to create migration and
population health scenarios for the Brazilian Northeast between 2025
and 2050.[4] We first describe the regional climate projections for Brazil,
followed by the projected impact of these climate changes on the performance of the agricultural sector and on the economy in the Northeast.
Then we discuss the methodology to estimate migration given climate
change and economic scenarios, and finally we integrate migration and
other factors in an indicator of health and population vulnerability
impacted by climate changes.

Climate change scenarios

We used climate scenarios provided by the Brazilian National Institute of
Space Research (INPE) through the regional model HadRM3P, further
disaggregated by municipality. The model generates the IPCC's A2 and B2
scenarios.[5] The A2 scenario implies high carbon emissions, with temperature increases for the Brazilian Northeast region of up to 4 °C until 2070;

---

[4] This methodology is described in full detail in CEDEPLAR/FIOCRUZ (2008). To the best
of our knowledge no previous studies use this approach of linking climate change,
economic and migration dynamics and health.

[5] The A2 and B2 regional scenarios were the two available from INPE at the time of this
study.

and the B2 scenario implies low carbon emissions, and with temperature increases for the Brazilian Northeast region of up to 1.8 °C until 2070.[6]

INPE scenarios for the Brazilian Northeast (Marengo, 2009; Marengo et al., 2007) suggest an average temperature increase of between 2 °C and 4 °C and 15–20% drier in the A2 scenario until 2070. The figures for the B2 scenarios are between 1 °C and 3 °C and 10–15% drier. The major impacts of these scenarios are (a) loss of biodiversity in the *caatinga* ecosystem; (b) 'aridization'; (c) desertification; (d) great impacts on agriculture (especially subsistence); and (e) impacts on population health.

Economic scenarios: impacts of climate change on agriculture

We then estimated the economic impacts of the A2 and B2 scenarios on the performance of the agricultural sector in the Brazilian Northeast. These impacts are particularly relevant given the importance of agriculture in the regional economy and its strong articulations with the other economic sectors. We used scenarios built by the Brazilian Agriculture Research Agency, EMBRAPA (Pinto and Assad, 2008) on climate impacts on land supply (whether land is suitable for cultivation) for the eight most important agricultural products in the region – rice, beans, corn, cotton, manioc, soybeans, sugar cane and sunflower.[7] Based on these data, the amount of land suitable or not for cultivation for other less important agricultural products and pasture was estimated. Finally, the impacts of the A2 and B2 scenarios on the agricultural sector represent a proxy of land availability for cultivation in the Northeast up to 2050.

Regarding economic scenarios, we used a computable general equilibrium model, IMAGEM-B (Integrated Multi-Regional Applied General Equilibrium Model for Brazil), developed at CEDEPLAR (CEDEPLAR/FIOCRUZ, 2008). Given technological and preference changes, the macroeconomic scenarios and population projections, the model generates economic scenarios (income, employment, gross product, level of consumption of families) for Brazilian states until 2050. The next step was to incorporate the climate impacts in the agriculture on future economic scenarios in the Northeast. In other words, while in the first

[6] The A2 and B2 scenarios discussed here refer to climate scenarios, and not necessarily the socio-economic scenarios implicit in the A2 and B2 scenarios. This distinction is important because economic behaviour in Brazil may not necessarily reflect the trajectory of the global economy (as assumed in the A2 and B2 scenarios). In any case, we assume that the A2 and B2 scenarios provided by the INPE are consistent with national and global trajectories.

[7] For a detailed discussion of agricultural scenarios by EMBRAPA, see Pinto and Assad (2008).

instance we estimated economic scenarios without climate changes in the IMAGEM-B model, later we measured economic scenarios with climate impacts on the agriculture. These impacts are both *direct* (reduction in economic performance of agriculture) and *indirect* (the effects of reduction in economic performance of agriculture on services and industries).

Population and migration scenarios

As discussed above, one of the inputs of the IMAGEM-B model was population projections for Brazilian states until 2050. This baseline population scenario includes the predicted behaviour of fertility, mortality and migration until 2050 without climate change impacts (CEDEPLAR, 2008a).

The migration scenarios impacted by climate change (alternative scenario) are estimated using economic parameters of the IMAGEM-B model. As discussed above, we assume that as a region faces the impacts of climate change on land supply (EMBRAPA scenarios), it suffers variations in income and employment levels (IMAGEM-B model) and consequently population migration. This relationship between employment variation and migration follows our focus on the role of economic factors on population displacements.

Thus, the IMAGEM-B generates a parameter, δ, which measures the effects of variations in employment affected by climate change in relation to the baseline demographic model for each five-year period between 2010 and 2050. This parameter indicates changes in the use of labour as a productive factor impacted by climate changes, and refers only to individuals 15–64 years old (working-age population). Given the interest in estimating migration for the whole population, including those below 15 and over 64 (considering, for example, that migrants under 15 are children following their parents' migration), we developed in another work (CEDEPLAR/FIOCRUZ, 2008) a model that estimates total net migration (for all age groups) from economic and demographic parameters. The model relates the working-age population (15–64) to the dependent population (below 15 and over 64), and how the first is affected by employment variation and, consequently, migration. The sum of the three estimated net migration figures (for age groups below 15, 15–64 and over 64) gives a net migration estimation for the total population.

Besides the two steps above – estimation of employment variation for the population aged 15–64, and estimation of migration for the total population – we included a third refinement in the estimation of migration. We assume that a positive variation in employment levels may be followed by the absorption of local unemployed population (and not necessarily in-migrants), and that non-economic factors such as human capital endowments and household structure may favour the mobility or immobility of the population. For these reasons, we adjusted the net migration to include a tolerance, υ, to the positive or negative variation in the employment level. The tolerance was estimated through a micro-economic model for Brazilian micro-regions (an area which encompasses a set of municipalities), with net migration as a dependent variable, and sex, age, education and fixed-effects as control variables. Thus, an income-elasticity of migration of 0.259 was obtained, meaning that a 1% increase in the wage of a given micro-region may increase by 0.26% the in-migration flow to this region.[8]

The net migration thus obtained is a five-year residual measure of the balance between in-migrants and out-migrants in a given location between t and $t + 5$, corresponding to the period 2010–2050. It represents the net impact of climate change on migration. Therefore, the lack of net impact of climate change on the employment level generates null net migration in the alternative scenario.

Population vulnerability: linking migration and health

Barros (2006) has stressed the importance of indices as metrics of vulnerability. He mentions the work by Confalonieri et al. (2005; 2009) in Brazil that created an aggregate Index of General Vulnerability by combining socio-economic indicators and epidemiological indicators. This was the first work in Brazil to produce policy-relevant indices to support adaptation strategies in the health sector relating to climate change.[9]

---

[8] This result is robust and significant, and consistent with other studies for Brazil using different data and methodologies (see e.g. Lima, 1995).

[9] The general formula for the vulnerability index (IVG) is:

$$IVG\_A2 = \frac{(IVS_p + IVD_p + IVED_{p\_A2} + IVC_{p\_A2})}{4}$$

$$IVG\_B2 = \frac{(IVS_p + IVD_p + IVED_{p\_B2} + IVC_{p\_B2})}{4}$$

where
IVS_p = standardized health vulnerability index
IVD_p = standardized desertification index
$IVED_p$ = standardized economic – demographic index
IVC_p = standardized health-care cost index.
General vulnerability indices were developed for IPCC's A2 and B2 scenarios as the $IVED_p$ and IVC_p were produced for both scenarios.

As already stressed by various authors, there is a need to develop quantitative indicators of vulnerability to guide public policies for human health protection. Here we use empirical information as well as model projections to assess the possible regional social and health impacts of human migration triggered by long-term climate change.

Aiming to summarize in one metric the major social-environmental components of vulnerability to the impacts of climate change, we developed an aggregate index of vulnerability for each state in the Brazilian Northeast. The index includes trends in endemic infectious diseases sensitive to climate factors, trends in desertification and demographic projections, as well as potential impacts of climate-induced migration on health-care costs for major cities.

Results

Migration scenarios

Table 3.1 shows the results of the projected migration for the A2 and B2 scenarios and for the baseline scenario. The net effects of climate change on migration in the Northeast during 2025–2030 are virtually nil: 0.03% in the A2 scenario (representing a volume of 17,752 individuals in-migrating to the Northeast) and –0.01% in the B2 scenario (6,026 individuals out-migrating from the Northeast). The B2 scenario is also associated with only marginal impacts on migration for 2035–2040 and 2045–2050, with migration rates of –0.02% and –0.03%, respectively, showing that this scenario of climate impacts on agriculture is not associated with significant population migration.

The A2 scenario shows stronger impacts on the agricultural sector when compared with the B2 scenario. According to the results for 2025–2030, by affecting more intensely the agricultural sector in the South and Southeast regions (these results are not presented here), the A2 scenario might reduce out-migration from the Northeast. For example, the B2 scenario is less severe in Minas Gerais and Espírito Santo (two Southeast states which border the southernmost states of the Northeast) than in the A2 scenario. The impacts on migration become more significant in 2035–2040 and 2045–2050, and are even higher than that projected by the baseline demographic model. The model suggests a

Table 3.1 *Net migration, migration rate and total population by scenario (Baseline, A2 and B2) – Brazilian Northeast region (2025–2030, 2035–2040, 2045–2050)*

Scenario	Net migration			Net migration rate (%)			Total population/projected		
	2025–2030	2035–2040	2045–2050	2025–2030	2035–2040	2045–2050	2025–2030	2035–2040	2045–2050
Baseline	−192 513	−203 925	−208 781	−0.29	−0.29	−0.29	65 339 961	68 559 267	70 349 764
A2	17 752	−246 777	−236 065	0.03	−0.36	−0.34	65 357 713	68 312 491	70 113 699
B2	−6 026	−13 565	−20 603	−0.01	−0.02	−0.03	65 333 935	68 545 703	70 329 161

Source: CEDEPLAR/FIOCRUZ (2008).

migration rate of –0.36% in the period 2035–2040, which represents the migration of 246,777 individuals, and –0.34% and 236,065 individuals in 2045–2050, respectively, only as a consequence of climate change. Once the expected climate changes are taken into account, the negative net migration would be higher than the baseline scenario between 2035 and 2040, and between 2045 and 2050.

Table 3.2 shows the projected A2 and B2 scenarios of net migration and migration rate for 2025–2030, 2035–2040 and 2045–2050 for metropolitan areas (MAs) and clusters of municipalities according to size. Following the trend shown in Table 3.1, the results are marginal in both scenarios in 2025–2030 except for significant and negative migration rates for the MAs of São Luís, João Pessoa (A2 and B2), Teresina and Salvador (B2). In the following years for the B2 scenario, the migration rates are also marginal except for the MAs of São Luís, João Pessoa, Salvador and Teresina.

The A2 scenario shows consistently negative and significant net migration and migration rates in 2035–2040 and 2045–2050 (except for the MA of Aracajú). The higher net migration occurs in the MAs of Recife and João Pessoa. The MA of São Luís, probably due to its proximity to the Amazon (which may gain population in the future A2 scenario) also shows high negative migration rates, both in the A2 and in the B2 scenarios as discussed above. The MA of Salvador and Teresina will also have significant population loss.

The municipalities over 150,000 inhabitants will probably experience significant net migration and migration rates in the A2 scenario in the three periods of analysis, with higher intensity in 2035–2040 and 2045–2050 (with migration rates above the Northeast average in the period of analysis). On the other hand, municipalities between 70,000 and 150,000 inhabitants in the A2 scenario, and municipalities between 25,000 and 70,000 inhabitants in the A2 and B2 scenarios, will have small positive migration rates in 2025–2030. However, the trend is the same as in the larger municipalities, with negative migration rates in the last two periods of analysis. Finally, the municipalities with less than 25,000 inhabitants also show a trend of negative migration rates in the last two periods, scenario A2.

Overall, the results show that climate impacts mediated by the performance of the agricultural sector may generate loss of income and employment which in their turn may act as a relevant push factor on population migration. These impacts may be reflected through most of the region, from rural areas to smaller or larger urban areas. While these

Table 3.2 *Net migration (NM) and migration rate (MR) for metropolitan areas (MAs) and municipalities according to size in the Brazilian Northeast region – scenarios A2 and B2 (2025–2030, 2035–2040, 2045–2050)*

Metropolitan areas and municipalities	2025–2030				2035–2040				2045–2050			
	A2		B2		A2		B2		A2		B2	
	NM	NMR (%)	NM	NMR (%)	NM	NMR (%)	NM	NMR (%)	NM	NMR (%)	NM	NMR (%)
MA of São Luís	−1 167	−0.06	−5 169	−0.26	−9 529	−0.42	−5 958	−0.27	−5 492	−0.23	−6 849	−0.28
MA of Fortaleza	547	0.01	−131	0.00	−9 462	−0.21	−343	−0.01	−7 576	−0.16	−697	−0.01
MA of Natal	541	0.02	366	0.02	−5 782	−0.22	526	0.02	−7 262	−0.24	715	0.02
MA of João Pessoa	−1 387	−0.08	−1 445	−0.08	−13 728	−0.68	−1 780	−0.09	−16 948	−0.75	−2 223	−0.10
MA of Recife	123	0.00	8	0.00	−47 518	−0.99	61	0.00	−53 005	−1.10	131	0.00
MA of Maceió	436	0.02	74	0.00	−2 236	−0.11	77	0.00	−2 388	−0.11	81	0.00
MA of Aracajú	495	0.04	237	0.02	−406	−0.03	447	0.03	54	0.00	732	0.04
MA of Salvador	−1 286	−0.03	−4 021	−0.08	−12 321	−0.24	−4 877	−0.10	−10 561	−0.21	−5 869	−0.12
Teresina	−422	−0.04	−1 246	−0.12	−5 824	−0.59	−1 236	−0.13	−4 731	−0.58	−1 120	−0.14
More than 250,000 inhabitants*	−101	−0.01	−838	−0.04	−8 355	−0.44	−869	−0.05	−7 448	−0.40	−894	−0.05

150,000 to 250,000**	320	0.01	−883	−0.04	−17 061	−0.67	−826	−0.03	−19 862	−0.77	−788	−0.03
70,000 to 150,000***	3 038	0.07	−647	0.01	−10 987	−0.22	−21	0.00	−7 239	−0.13	−1 435	−0.03
25,000 to 70,000***	7 490	0.05	7 490	0.05	−49 907	−0.34	1 124	0.01	−45 612	−0.32	−2 364	−0.02
Less than 25,000***	9 124	0.05	178	0.00	−53 661	−0.29	110	0.00	−47 995	−0.25	−22	0.00
Total – Northeast region	17 752	0.03	−6 026	−0.01	−246 777	−0.36	−13 565	−0.02	−236 065	−0.34	−20 603	−0.03

*Except the state capitals and municipalities in the MAs. Includes municipalities of Campina Grande, Caruarú, Fiera de Santana and Vitória da Conquista.

**Except the state capitals and municipalities in the MAs. Includes municipalities of Imperatriz, Juazeiro, Sobral and Petrolina.

***Except the state capitals and municipalities in the MAs.

Source: Demographic dynamics, health and population vulnerability.

results reflect a classic migration response, they also represent a new source of population vulnerability assuming that those moving can potentially put pressure on the public health and infrastructure systems in the place of destination and act as potential agents in the redistribution of endemic infectious diseases.

Figures 3.2 to 3.5 show the results for the Standardized Economic–Demographic Index (IVED) and for the General Vulnerability Index (IVG), in the A2 and B2 scenarios. The IVED and IVG indices represent the cumulative percentage difference between the values observed in the A2 and B2 scenarios and the baseline scenario in the period 2005–2030. Regarding the last index, in the worst-case scenario (A2) we observe that the higher values (0.75 or over, in a range from 0.0 to 1.0) correspond to

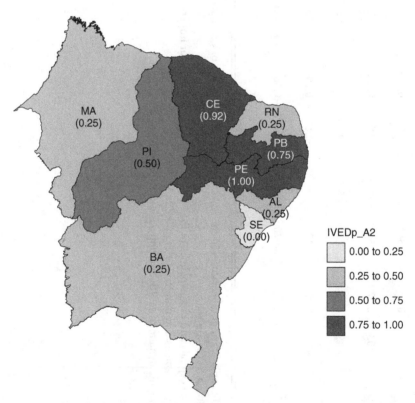

Figure 3.2 Standardized values for the Economic-Demographic Index, for each state in the Brazilian Northeast and for climate scenario A2 (2005–2030)
Source: CEDEPLAR/FIOCRUZ (2008).

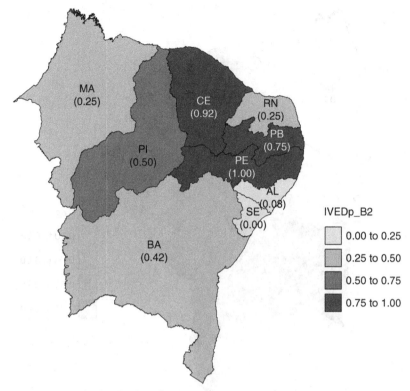

Figure 3.3 Standardized values for the Economic-Demographic Index, for each state in the Brazilian Northeast and for climate scenario B2 (2005–2030)
Source: CEDEPLAR/FIOCRUZ (2008).

the states of Ceará (1.0), Pernambuco (0.89) and Bahia (0.75). In the case of Ceará, all four partial indices – IVS (health), IVD (desertification), IVED (economy-demography) and IVC (health-care costs) – influence the high values of the IVG as they have values equal to or higher than 0.66. Pernambuco (0.89) was basically influenced by the high values of IVED (1.0) and IVD (0.88) while the IVG for Bahia is influenced by the IVS (0.73), IVD (0.88) and IVC (1.0).

High values of the Economic-Demographic Index (IVED) are observed for the states of Pernambuco (1.0), Ceará (0.92) and Paraíba (0.75). The results for the first two states are influenced by extreme values in all three components of the IVED (GDP, employment and population migration), whereas in the case of Paraíba only GDP decrease and employment loss played an important role.

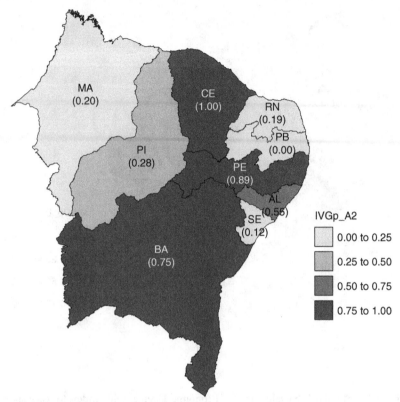

Figure 3.4 Standardized values for the General Vulnerability Index for each state in the Brazilian Northeast and for climate scenario A2 (2005–2030)
Source: CEDEPLAR/FIOCRUZ (2008).

Conclusions and policy implications

This chapter discusses the long-term relationship between climate change, population migration and population health, with a case study on the Brazilian Northeast. If high migration rates in the past were mainly the result of a combination of severe drought periods and better labour opportunities in the Brazilian Southeast, we project a key role of these two factors as important drivers of migration from the Northeast (albeit at a much lower level than observed in the past). These scenarios may also create new foci of endemic diseases due to the mobility of infected people, as well as to increased pressure on urban infrastructure and the public health system.

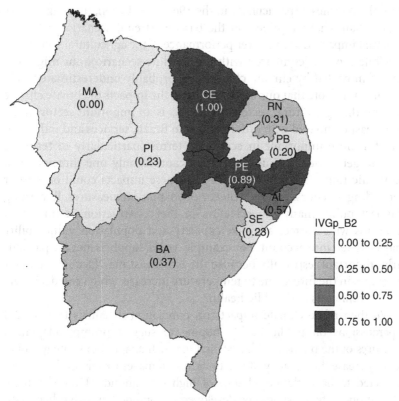

Figure 3.5 Standardized values for the General Vulnerability Index for each state in the Brazilian Northeast and for climate scenario B2 (2005–2030)
Source: CEDEPLAR/FIOCRUZ (2008).

Understanding climate change impacts on migration is important because population adaptation may depend upon their ability to move across space. In the case of the Brazilian Northeast, we assume that this response is associated with those most vulnerable economically. On the other hand, we recognize that the 'migration response' is not only a mechanism of adaptation of the poorest or less favoured in any social or economic dimension, but may also in some circumstances be a mechanism available only for those with sufficient resources or capital (social, financial).

Our case study in the Brazilian Northeast shows that while the B2 scenario does not indicate a significant impact on population migration, the A2 scenario indicates significant population migration from the Northeast after 2030. While still modest compared with historical out-migration from the Northeast (particularly in the 1960s and 1970s), these figures are higher than what we project in the baseline (trend) scenario, and show potential impacts due to poorer performance of the agricultural sector.

While modest (compared with the baseline scenario), our migration results impacted by climate changes are probably underestimated. It is important to note that our model captures the impacts of climate change only on the agricultural sector. While this is an important sector in the Northeast compared with other regions in Brazil, services and industries are still more important in economic terms (particularly in terms of income generation). Thus the results capture only one dimension of economic impacts, and we believe that these impacts could be greater depending, of course, on the efficacy of adaptation measures. Regarding this last point, many other factors in the construction of migration scenarios which affect adaptive capacity and population vulnerability may be taken into account. For example, water supply issues are particularly important, especially because the Northeast may face major water shortages in the future due to temperature increase, which can decisively affect livelihoods and public health.

We then relate climate impacts on population dynamics to potential repercussions on public health, hoping to suggest prompt adaptation measures of the public health system. One such adaptation strategy would be to increase the capacity of health-care systems, especially in those areas projected to have climate change of higher magnitude. However, these adaptation strategies should be developed in conjunction with other adaptation measures relevant for public health, such as the improvement of food security and the management of water resources. The comprehensive Vulnerability Indices provide a reference to prioritize areas for intervention in the context of regional general adaptation policies to climate change.

Even the partial picture given here may provide an understanding of critical linkages between climate change, population mobility and population health. Simulations of scenarios of increased vulnerability of some groups – particularly migrants – can help to promote prompt and strong action in terms of the creation or adaptation of institutional settings on different scales.

References

Adamo, S. 2008. Addressing environmentally induced population displacements: a delicate task. Background paper for the Population-Environment Research Network Cyberseminar on Environmentally Induced Population Displacements. www.populationenvironmentresearch.org

Barbieri, A. F., Carr, D. and Bilsborrow, R. 2009. Migration within the frontier: the second generation colonization in the Ecuadorian Amazon. *Population Research and Policy Review*, Vol. 28, No. 3, pp. 291–320.

Barnett, E. D. and Walker, P. F. 2009. Role of immigrants and migrants in emerging infectious diseases. *Medical Clinics of North America*, Vol. 92, No. 6, pp. 1447–58.

Barros, M. 2006. Clima e endemias tropicais. *Estudos Avançados*, Vol. 20, No. 58, pp. 297–306.

Bilsborrow, R. 1987. Population pressures and agricultural development in development countries: a conceptual framework and recent evidence. *World Development*, Vol. 15, No. 2, pp. 183–203.

Carvalho, J. A. and Garcia, R. A. 2002. *Estimativas decenais e quinquenais de saldos migratorios e taxas líquidas de migração do Brasil*. Belo Horizonte, Centro de Desenvolvimento e Planejamento Regional, Universidade Federal de Minas Gerais (CEDEPLAR/UFMG).

CEDEPLAR. 2007. *Projeção populacional para o Brasil e unidades de federação por idade, 2010–2020, cenário básico. Relatório de pesquisa*. (Research Report.) Belo Horizonte, Centro de Desenvolvimento e Planejamento Regional, Universidade Federal de Minas Gerais.

CEDEPLAR. 2008a. *Cenário tendencial do modelo econômico de projeções territoriais – período 2007/2027*. (Research Report PPA 2008–2011.) Belo Horizonte, Centro de Desenvolvimento e Planejamento Regional, Universidade Federal de Minas Gerais.

CEDEPLAR/FIOCRUZ. 2008. *Mudanças climáticas, migrações e saúde: cenários para o Nordeste brasileiro, 2000–2050. Relatório de pesquisa*. (Research Report.) Belo Horizonte, Centro de Desenvolvimento e Planejamento Regional, Universidade Federal de Minas Gerais.

Confalonieri, U. E. C. 2003. Variabilidade climática, vulnerabilidade social e saúde no Brasil. *Terra Livre*, São Paulo, Vol. 19-I, No. 20, pp. 93–204.

Confalonieri, U. E. C. et al. 2005. *Análise da vulnerabilidade da populacão brasileira aos impactos sanitarios das mudanças climáticas*. MCT-PPA, Programa de Mudanças Climáticas. http//www.mct.gov.br/upd_blob/0014/14534.pdf

Confalonieri, U. E. C., Marinho, D. P. and Rodriguez, R. E. 2009. Public health vulnerability to climate change in Brazil. *Climate Research*, Vol. 40, pp. 175–86.

Davis, K. 1963. The theory of change and response in modern demographic history. *Population Index*, Vol. 29, No. 4, pp. 345–66.

Döös, B. R. 1997. Can large-scale environmental migrations be predicted? *Global Environmental Change*, Vol. 7, No. 1, pp. 41–61.

Downing, T. E. 1992. *Climate Change and Vulnerable Places: Global Food Security and Country Studies in Zimbabwe, Kenya, Senegal and Chile*. Oxford, UK, Environmental Change Unit. (Research Report No. 1.)

Ezra, M. 2003. Environmental vulnerability, rural poverty, and migration in Ethiopia: a contextual analysis. *Genus*, Vol. 59, No. 2, pp. 63–91.

Findley, S. E. 1994. Does drought increase migration? A study of migration from rural Mali during the 1983–1985 drought. *International Migration Review*, Vol. 28, No. 3, pp. 539–53.

Franke, C. R., Ziller, M., Staubach, C. and Latif, M. 2002. Impact of the El Niño: southern oscillation on visceral leishmaniasis, Brazil. *Emerging Infectious Diseases*, Vol. 8, No. 9, pp. 914–17.

Füssel, H.-M. 2007. Vulnerability: a generally applicable conceptual framework for climate change research. *Global Environmental Change*, Vol. 17, No. 2, pp. 155–67.

Henry, S., Schoumaker, B. and Beauchemin, C. 2004. The impact of rainfall on the first out-migration: a multi-level event-history analysis in Burkina Faso. *Population and Environment*, Vol. 25, No. 5, pp. 423–60.

IIED. 2007. *Reducing Risks to Cities from Climate Change; an Environmental or a Development Agenda?* London, International Institute for Environment and Development.

IPCC. 2001. *Third Assessment Report: Climate Change 2001* (TAR). Geneva, Intergovernmental Panel on Climate Change. http://www.ipcc.ch/publications_-and_data/publications_and_data_reports.htm

IPCC. 2007. *Fourth Assessment Report: Climate Change 2007* (AR4). Geneva, Intergovernmental Panel on Climate Change. http://www.ipcc.ch/publications_and_data/publications_and_data_reports.htm

Kniveton, D., Schmidt-Verkerk, K., Smith, C. and Black, R. 2008. *Climate Change and Migration: Improving Methodologies to Estimate Flows*. Geneva, International Organization for Migration. (Migration Research Series No. 33.) www.iom.int/jahia/webdav/site/myjahiasite/shared/shared/mainsite/-published_docs/serial_publications/MRS-33.pdf

Lima, R. 1995. Um exame dos determinantes da migração rural-urbana no Brasil. *Cadernos de Ciência & Tecnologia*, Vol. 12, No. 1/3, pp. 55–67.

Marengo, J. A. 2009. Vulnerability, impacts and adaptation (VIA) to climate change in the semi-arid region of Brazil. In: M. K. Poppe (ed.), *Brazil and Climate Change: Vulnerability, Impacts and Adaptation*. Brasília, Centro de Gestão e Estudos Estratégicos.

Marengo, J. A., Nobre, C. A., Salati, E. and Ambrizzi, T. 2007. Caracterização do clima atual e definição das alterações climáticas para o território brasileiro ao longo do século XXI. Brasília, Min. do Meio Ambiente, SBF/DCBio.

Martine, G. and Guzmán, J. 2002. Population, poverty, and vulnerability: mitigating the effects of natural disasters. *Environment Change & Security Project Report*, No. 8, pp. 45–68.

McGranahan, G., Balk, D. and Anderson, B. 2007. The rising tide: assessing the risks of climate change and human settlements in low elevation coastal zones. *Environment & Urbanization*, Vol. 19, No. 1, pp. 17–37.

McLeman, R. and Smit, B. 2006. Migration as an adaptation to climate change. *Climatic Change*, Vol. 76, Nos. 1–2, pp. 31–53.

Meze-Hausken, E. 2000. Migration caused by climate change: how vulnerable are people in dryland areas? A case study in Northern Ethiopia. *Mitigation and Adaptation Strategies for Global Change*, No. 5, pp. 379–406.

Perch-Nielsen, S. L., Bättig, M. B. and Imboden, D. 2008. Exploring the link between climate change and migration. *Climatic Change*, Vol. 91, Nos. 3–4, pp. 375–93.

Pinto, H. S. and Assad, E. 2008. *Aquecimento global e cenários futuros da agricultura brasileira.* (Research Report.) Campinas, Brazil, Project FCO-GOF: PGL GCC 0214.

Ribot, J. C., Najam, A. and Watson, G. 1996. Climate variation, vulnerability and sustainable development in the semi-arid tropics. In: J. C. Ribot, A. R. Magalhães and S. S. Panagides (eds.), *Climate Variability, Climate Change and Social Vulnerability in the Semi-arid Tropics.* Cambridge, UK, Cambridge University Press, Chap. 1.

Wang, X., Auler, A. S., Edwards, L., Cheng, H., Cristalli, P. S., Smart, P. L., Richards, D. A. and Shen, C. 2004. Wet periods in northeastern Brazil over the past 210 kyr linked to distant climate anomalies. *Nature*, Vol. 432, pp. 740–43.

4

Environmental degradation and out-migration: evidence from Nepal

PRATIKSHYA BOHRA-MISHRA AND DOUGLAS S. MASSEY

Introduction

Although the concept of 'environmental refugees' was introduced in the mid-1980s and remains quite popular in many quarters, there is relatively little empirical work demonstrating the existence and nature of a connection between environmental change and human migration. In this study we contribute to this literature by updating and expanding on recent work done in Nepal using data from the Chitwan Valley Family Study. We use event history data to model local, internal, and international migration as a function of environmental deterioration at baseline while controlling for social, economic, and demographic variables that prior work has shown to affect population mobility. We find a strong and consistent relationship between the likelihood of undertaking a local move and population pressure (measured by neighbourhood density), deforestation (indicated by rising times required to collect fodder and firewood), and declining agricultural productivity. We also find that the environmental effects on local migration are more prevalent for women than for men. We found little evidence that environmental deterioration promoted migration outside the local district, either to other districts in Nepal or to international destinations, though increased time to collect firewood was associated with a higher probability of men leaving Chitwan for other countries or other districts in Nepal. In general, our results suggest that the kind of gradual environmental deterioration studied here is more associated with local than distant population mobility.

The term 'environmental refugees' was introduced by El-Hinnawi (1985) to describe people forced to leave their places of origin, either temporarily or permanently, because of environmental disruptions triggered by human or natural events. Suhrke (1994) has identified desertification, land degradation, deforestation, and rising sea levels as the most important forms of environmental change leading to out-migration, whereas Hugo (2008) lists environmental disasters, environmental degradation, climate change, and disruptions from large-scale human projects as the principal causes of population displacement. Whatever the cause, Jacobson (1988) argues that environmental refugees constitute 'the single largest class of displaced persons in the world'. Despite this bold claim, however, the view that environmental changes induce people to migrate remains a hotly contested topic (Castles, 2002).

On the one hand, Myers and Kent (1995) argue that some 25 million people were environmental refugees in the mid-1990s, and as many as 200 million faced a significant risk of displacement. On the other hand, Black (2001) questions the very concept of environmental refugees as a myth, and argues that such high counts are inflated by including all sorts of migrants under the label. In the only macro-level study to date, Afifi and Warner (2008) found a positive association between the size of migration flows between 172 countries and measures they developed of overfishing, desertification, water scarcity, soil salinization, deforestation, air pollution, soil erosion and soil pollution within sending nations. They also found a positive association between migration and earthquakes, hurricanes, floods and tsunamis.

Most studies to date rely on country-specific data, however. In their study of data from Guatemala and the Sudan, for example, Bilsborrow and DeLargy (1991) found that environmental changes producing either a decline in the productivity of fixed resources (such as land) or lower returns to household resources (such as labour) tended to foster rural out-migration by reducing farm income. Consistent with this view, Kalipeni (1996) showed that internal migrants in Malawi generally moved from densely populated to sparsely populated districts. In her study of population mobility in Soviet Kamchatka, Hitztaler (2004) found that villages experiencing a natural resource crisis sustained greater out-migration than those with a relatively intact resource base.

In their recent study of migration in Nepal's Chitwan Valley, Shrestha and Bhandari (2007) found that decreasing access to firewood increased the likelihood of migration both to domestic and international destinations, controlling for other predictors of migration. In their study of the same region, however, Massey et al. (2007) found that only local moves were predicted by decreasing access to firewood. They also found that local mobility was related to declines in agricultural productivity and decreases in land cover, but that population density and decreasing access to fodder were unrelated to mobility, either over short or long distances. The contradictory results of these two studies might reflect different definitions of migration, however, with Shrestha and Bhandari (2007) considering domestic versus international mobility (i.e. moves within versus outside Nepal) and Massey et al. (2007) focusing on local versus distance mobility (moves within versus outside Chitwan).

The relative paucity of studies analysing the relationship between environmental change and migration partly reflects the focus of prevailing theories on the social and economic roots of population mobility (see Massey et al., 1998) but also stems from a lack of data on the subject. In this chapter we expand on the work of Shrestha and Bhandari (2007) and Massey et al. (2007) by analysing more recent data from the Chitwan Valley Family Study to clarify the effect of environmental degradation on individual migration decisions. Following Suhrke (1994), we hold that if environmental change has any influence on out-migration, it is most likely to be observed in poor agrarian economies where people cannot insure against unexpected natural events, leaving them little choice except to migrate in the face of environmental change. We investigate this hypothesis by specifying and estimating an event history model that links environmental conditions in 1996 to monthly migration decisions made between 1997 and 2006. In an effort to reconcile the contradictory findings of Shrestha and Bhandari (2007) and Massey et al. (2007) we distinguish between three kinds of moves: those within Chitwan, those outside Chitwan but within Nepal, and those outside Nepal.

Study site

Nepal is one of the least-developed countries in the world. A majority of its inhabitants continue to subsist on agriculture and an almost threefold increase in population over the past four decades (from 9.4 million in 1961 to 23.2 million in 2001) has placed severe pressure on land and other natural resources. The traditional Nepalese adage, *hariyo ban Nepal ko dhan* ('green forests are Nepal's wealth') may once have accurately reflected the abundance of forests and other natural resources in the country, but it is fast becoming obsolete because of widespread deforestation, soil erosion, and other forms of environmental degradation associated with rising population pressure.

Nepal's Chitwan Valley offers an ideal setting to study environmental effects on migration because of its rapid transformation through economic and demographic growth. As recently as the early 1950s, the valley was covered by dense forests; but these were subsequently cleared by the national government to make land available for farming and settlement. Given the valley's favourable climate, fertile soil and flat terrain, people from nearby hills and mountains moved in and quickly settled in the valley, placing new pressures on available land and other natural resources. In the late 1970s, Chitwan's largest town was connected by road to major cities throughout the country including Kathmandu, the capital, as well as to India. As a result, the district began to attract investment, government services and new employers.

In spite of this economic growth, the infrastructure in Chitwan is only marginally better than in the rest of Nepal. Except for the national highway, most roads in the district are still unpaved and most jobs are in service-oriented government agencies, with a few more in agricultural industries (Ghimire and Mohai, 2005). Overall, the valley is still home to an agrarian society in which the great majority of households rely on subsistence farming and animal husbandry, supplemented by resources gathered from local forests, for survival. Under these circumstances, declining access to natural resources such as fodder and firewood, deteriorating soil fertility and water quality, and other forms of environmental deterioration are of great concern to the valley residents.

Data and methods

The Chitwan Valley Family Study (CVFS) used a combination of ethnographic and survey methods to gather detailed data on the social, economic and demographic characteristics of individuals, their households and the communities where they reside. The migration data, in particular, come from a monthly panel survey that began in February 1997 and ended in January 2006. Households from 151 neighbourhoods were followed for the entire 108-month study period even if they left sample neighbourhoods. Respondents were lost to follow-up only when the entire household moved out of Nepal and did not return during the study period. Here we focus on those respondents between the ages of 15 and 69 who were residing in the 151 neighbourhoods at the beginning of the panel survey and were followed month-to-month thereafter.

We merge these monthly panel data with data on the characteristics of individuals, households and neighbourhoods. Despite the complexities of merging across three levels, fewer than 2% of all person-months were lost through list-wise deletion, leading us to discount missing data as a significant source of bias in our analysis. We defined migration using a multinomial variable that equalled 0 if the respondent did not move between month t and $t+1$; 1 if the respondent moved to another neighbourhood within Chitwan during this time; 2 if they moved to a district outside Chitwan; and 3 if they moved to another country. All person-months spent outside an individual's survey neighbourhood were excluded from the analysis, meaning that respondents were only considered to be at risk of migration when living in their places of origin. We followed each respondent from 1997 until the final survey date in 2006 or the point at which the person left the valley without returning, yielding a total of 295,635 person-months for analysis.

Our independent variables of interest, along with relevant control variables, are defined in Table 4.1. As can be seen, some variables are fixed effects defined at the baseline survey whereas others are time-varying with values that differ month by month. The environmental variables are all defined at the baseline in 1996. At this time, respondents were asked to answer questions regarding present environmental conditions and those prevailing three years earlier. The responses were then used to derive four potential measures of environmental degradation within Chitwan: change in the time required to collect animal fodder; change in the time required to collect firewood for fuel; change in agricultural productivity; and change in the quality of drinking water. As a final environmental indicator, we included population density in 1996 to assess the effect of population pressure on out-migration. Our selection of these environmental indicators is justified below.

The livelihood of many households in Chitwan depends on access to fodder, as animal husbandry is a common source of livelihood throughout the valley. Although households typically graze livestock on cleared parcels, they also supplement the animals' diet with fodder gathered from nearby forests. The forests, however, have been declining steadily through deforestation since the 1960s (Massey et al., 2007). This depletion of forests increases the time required to gather fodder, and as gathering time increases at some point it becomes easier simply to purchase fodder commercially or to abandon husbandry altogether, making out-migration for wage labour an increasingly attractive alternative. Likewise, the vast majority of households in Chitwan use firewood for heating and cooking and the gathering of firewood is itself a major cause of deforestation throughout the Himalayan region (Ali and Benjaminsen, 2004). Although some households buy firewood, most collect it from local forests. As with fodder, deforestation increases the time required to gather firewood and at some point

Table 4.1 *Definition of variables*

VARIABLE	DEFINITION
OUTCOME VARIABLE	
Migration to three competing destinations	Whether respondent migrates to one of three locations in month t+1: within Chitwan = 1, to other districts = 2, to other countries = 3, and 0 if does not migrate at all
Migration to two competing destinations	Whether respondent migrates to one of two locations in month t+1: within Chitwan = 1, to other districts or other countries = 2, and 0 if does not migrate at all
ENVIRONMENTAL VARIABLES	
Increase in time to collect fodder compared with three years ago	1 if time to collect fodder increased now compared with three years ago, 0 otherwise*
Did not collect fodder in both years	1 if household did not collect fodder now and three years ago, either because it did not own livestock or it chose to buy all the fodder it needed, 0 otherwise*
Increase in time to collect firewood compared with three years ago	1 if time to collect firewood increased now compared with three years ago, 0 otherwise*
Did not collect firewood in both years	1 if household did not collect firewood now and three years ago either because it did not use firewood or bought all the firewood it used, 0 otherwise*
Perception of decrease in crop production compared with three years ago	1 if the household thinks crop production has decreased compared with three years ago, 0 otherwise*
Does not farm	1 if the household does not farm and is not questioned regarding crop production or if the household reports it does not know if the production has changed, 0 otherwise (only 2.3% report this)
Water less clear compared with three years ago	1 if the household thinks that their drinking water has become less clear compared with three years ago (or report they do not know if it has changed – only 1.3% report this), 0 otherwise*

Table 4.1 (*cont.*)

VARIABLE	DEFINITION
Neighbourhood population density	Number of people in the neighbourhood per 100,000 ft^2 (9,290 m^2)*

CONTROLS
Physical Capital

VARIABLE	DEFINITION
Owns farmland	1 if respondent's household owns farmland, 0 otherwise*
Standardized index of household amenities	A composite index of household amenities derived through factor analysis using data on the materials used to build the floor and roof of the respondent's house; and whether household has a toilet, access to own drinking water source, and electricity*
Standardized index of goods owned	A composite index of assets owned derived through factor analysis using data on durables owned by the household such as ownership of a radio, television, bicycle, motorcycle, cart, tractor, pumpset, and bio gas plant*
Standardized index of livestock owned	A composite index of livestock owned derived through factor analysis using data on number of chickens, pigeons, buffaloes, cows, sheep, goats, pigs, etc. owned by household*

Neighbourhood Development

VARIABLE	DEFINITION
Standardized index of neighbourhood development	A composite index of neighbourhood level of development derived through factor analysis using data on average hours on foot to nearest resources such as health care, bus service, school, market, bank, employment and police station*

Human Capital

VARIABLE	DEFINITION
Education	Number of years of schooling completed by the respondent*
Salaried job	1 if respondent holds a salaried job, 0 otherwise*
Age	Respondent's age, monthly event
Age squared	Respondent's age square, monthly event

Table 4.1 (*cont.*)

VARIABLE	DEFINITION
Social Capital	
House member migrated within Chitwan	1 if any member from the respondent's household has migrated within Chitwan in 1996 before the respondents are observed, 0 otherwise
House member migrated to other districts	1 if any member from the respondent's household has migrated to other districts in 1996 before the respondents are observed, 0 otherwise
House member an international migrant	1 if any member from the respondent's household has migrated to other countries in 1996 before the respondents are observed, 0 otherwise
House member migrated to other districts or other countries	1 if any member from the respondent's household has migrated to other districts or other countries in 1996 before the respondents are observed, 0 otherwise
Level of Violence	
Violence from Maoist insurgency	1 for all months after November 2001 to capture the elevated level of violence from Maoist insurgency, 0 for months before that
Demographic Variables	
Female	1 if respondent is female, 0 otherwise
Married	1 if respondent was ever married, 0 otherwise, monthly event
Number of household members	Number of people in the household*
Ethnicity	
Hindu upper caste	1 if Hindu upper caste, 0 otherwise
Hindu lower caste	1 if Hindu lower caste, 0 otherwise
Hill Tibeto-Burmese	1 if Hill Tibeto-Burmese caste, 0 otherwise
Newar and other	1 if Newar or other caste, 0 otherwise
Terai Tibeto-Burmese	1 if Terai Tibeto-Burmese caste, 0 otherwise

*As reported in the baseline survey conducted in 1996.

makes the purchase of firewood using migrant-generated remittances a better alternative.

Many prior studies have used the time required to access natural resources to measure environmental degradation (see Biddlecom et al., 2005; Baland et al., 2006; Shrestha and Bhandari, 2007; Filmer and Pritchett, 1997; Kumar and Hotchkiss, 1988; Massey et al., 2007). The CVFS baseline survey asked respondents to estimate how long it took them to travel to where fodder or firewood was located, collect it, and then bring it home, both at the time of the survey and three years earlier. We took the difference between the two reported times to create a dummy variable indicating whether the collection time increased, leaving no change or a decline as the reference category. We also include in our models separate dummy variables to indicate those households that did not collect fodder or firewood at both or one of the two dates.

The vast majority of households in Chitwan rely on farming for subsistence, and following Ghimire and Mohai (2005) we use data on perceptions of change in agricultural productivity as an additional measure of environmental degradation. The baseline survey asked each respondent: 'Compared to three years ago, do you think crop production has increased, decreased, or stayed the same?' We created a dummy variable to indicate whether productivity was perceived to have declined as well as a dichotomous variable to indicate those respondents who did not farm, with no change or increase in productivity as the reference category. A follow-up question put to respondents asked about reasons for perceived changes, and 56% of those who perceived a decline attributed it to inadequate irrigation, bad weather, pests, disease or poor soil, with another 30% mentioning inadequate or poor manure, thus confirming the perception of productivity decline as a valid measure of environmental degradation.

We also measured environmental deterioration using an item on perceived changes in the quality of drinking water. The specific item asked: 'Compared to three years ago, do you think that the clarity of the water you drink has changed?' Among those who responded positively a follow-up question asked: 'Do you think that the water you drink has become a little more clear, much more clear, a little more unclear, or much more unclear?' As with our other indicators, we created a dummy variable that equalled 1 if the water was less clear than three years ago and 0 if it was clearer or unchanged.

Finally, to indicate demographic pressure we used neighbourhood population density at the time of the baseline survey. Despite mounting pressure on local resources, Chitwan's population has continued to grow, both through in-migration and natural increase, and it is expanding faster than in the rest of the country (Ghimire and Mohai, 2005). Using the household census conducted in 1996, we determined the number of people living in each household and summed across households to derive neighbourhood population, which was then divided by the total area of each neighbourhood and multiplied by 100,000 to convert the ratio into population per 100,000 ft^2 (9,290 m^2) for ease of interpretation.

It is worth noting that with the exception of neighbourhood population density, our environmental measures are based on the individual's own account of environmental conditions, which might raise concerns about possible bias in the assessment of environmental effects. Of course, it is true that if individuals were concerned or aware of environmental degradation in Chitwan, they might be more inclined to give a pessimistic report. However, the questions on which our environmental measures are based do not ask individuals to assess whether environmental conditions became worse or better compared with three years ago. Instead, they are asked to report the amount of time it used to take them to collect firewood or fodder three years ago versus now, quality of water then versus now, and agricultural productivity then versus now, which should generally result in unbiased answers from respondents. Furthermore, only those households who farmed in both years were asked their perception of any change in agriculture productivity and if they perceived any change in production, they were further probed to identify specific reasons for any change. Of course, these facts do not eliminate the possibility of recall bias or biases stemming from changes in average transport times rather than environmental shifts.

In assessing the effect of environmental conditions on migration, we control for a variety of individual, household and neighbourhood characteristics that prior work has shown to influence migration decisions in the Chitwan Valley (see Bohra-Mishra and Massey, 2011; Bohra and Massey, 2009). We measure access to physical capital using four indicators defined from the baseline survey: ownership of farmland, quality of household amenities, possession of consumer goods, and ownership of livestock. We indicate ownership of farmland using a dichotomous variable that equals 1 if farmland was owned in 1996 and 0 otherwise. The remaining three indicators were measured using factor scaling methods. Table 4.2 summarizes each factor model, which relied on principal components analysis to estimate loadings that were applied as weights to z-scores of the constituent variables to create composite scales measuring access to household amenities, consumer goods, and livestock.

Table 4.2 *Index weights for the composite index of physical capital and neighbourhood characteristic variables*

	Index weights
Physical Capital Variables	
Household amenities	
Roof of house is made from slate, tin, or concrete	0.27
Floor of house is made from concrete	0.26
No own drinking water source	−0.21
No toilet	−0.22
Has electricity	0.24
Variance explained by first factor	*0.69*
Goods owned	
Household has a radio	0.19
Household has a television	0.20
Household has a bicycle	0.18
Household has a motorcycle	0.20
Household has a cart	0.16
Household has a tractor	0.18
Household has a pumpset for irrigation	0.14
Household has a bio gas plant	0.21
Variance explained by first factor	*0.46*
Livestock owned	
Number of chickens and ducks	0.08
Number of pigeons	0.24
Number of bullocks	0.30
Number of cows	0.27
Number of male buffaloes	0.21
Number of female buffaloes	0.33
Number of sheep and goats	0.42
Number of pigs	0.14
Variance explained by first factor	*0.22*
Neighbourhood Characteristic Variables	
Neighbourhood development	
Minutes on foot to nearest school	0.19
Minutes on foot to nearest health-care centre	0.23
Minutes on foot to nearest bus service	0.23
Distance by bus to Narayanghat	0.19
Minutes on foot to nearest market	0.18
Minutes on foot to nearest bank	0.25
Minutes on foot to nearest place of employment	0.21
Minutes on foot to nearest police station	0.15
Variance explained by first factor	*0.36*

We also developed a factor scale of neighbourhood development to control for the quality of local infrastructure and access to economic resources. The baseline neighbourhood questionnaire recorded the time it took in 1996 for travel to access various public resources, such as health-care centres, bus stops, schools, markets, banks, police stations and places of employment. In general, the lower the travel time required to these amenities, the greater the local level of development. As before, the various travel times were converted to z-scores and combined using the weights shown in Table 4.2 (again derived from a principal components analysis) to create a composite index of neighbourhood development.

Neoclassical economic theory views migration as a strategy used by individuals to maximize returns to human capital (see Sjaastad, 1962; Todaro and Maruszko, 1987), which we measure using three indicators: years of education in 1996, the holding of a salaried job in 1996 (a measure of occupational status), and age during the person-month under observation (to capture experience), along with a squared term added to capture non-linear curvature in the relationship. According to social capital theory, having a social tie with a current or former migrant reduces the costs and risks of movement to promote migration (Massey et al., 1998). We thus defined three dummy variables to indicate the presence of other household members with migratory experience in 1996: whether anyone in the household had ever migrated within Chitwan, whether anyone had migrated to other districts in Nepal, and whether anyone had migrated outside Nepal. These indicators are expected to yield strong destination-specific effects, with ties to local migrants predicting local moves, ties to internal migrants predicting internal moves, and ties to international migrants predicting international moves.

The time period covered by our analysis is unusual in the sense that Nepal was in the midst of a decade-long civil conflict waged by Maoist guerrillas. The insurgency was launched on 13 February 1996 and finally ended on 21 November 2006 with the signing of a Comprehensive Peace Agreement. Between these dates, a total of 13,347 people were killed by government or rebel forces (Informal Sector Service Center, 2008). Although the insurgency began in 1996, for the first five years the conflict was low intensity and mainly involved guerrillas and the police. After the failure of the peace talks in November 2001, however, the government proclaimed a state of emergency and labelled the Maoist rebels as terrorists, leading to a high intensity conflict that pitted Maoist insurgents against the Royal Nepalese Army (Murshed and Gates, 2005; Bohra-Mishra and Massey, 2011). By the end of 2002, armed fighting was reported in seventy-three of Nepal's seventy-five districts (Kok, 2003).

In their analysis of the effect of violence on out-migration from Chitwan, Bohra-Mishra and Massey (2011) found a threshold effect, such that only after violence escalated in November 2001 was an effect detected. Hence, here we control for the effect of violence using a dummy variable to indicate months after November 2001. Given well-known effects of demographic factors on patterns and processes of migration, we also introduce controls for gender, marital status, household size and ethnicity. Gender is relevant in this context as some environmentally linked tasks (such as the gathering of fodder) are gendered, and also because the Nepalese Government imposed a ban on the migration of female workers to the Gulf in 1998 (see Graner, 2001). Previous studies of migration from Chitwan have also documented significant effects of ethnicity (see Bohra-Mishra and Massey, 2011; Bohra and Massey, 2009) and so we include a series of dummy variables to identify higher caste Hindus, lower caste Hindus, Newar, and the Hill Tibeto-Burmese, leaving Terai Tibeto-Burmese, the indigenous people of the Chitwan Valley, as the reference category.

We measure environmental effects on migration using a multinomial logit model to predict out-migration to one of three possible destinations from the foregoing indicators of environmental degradation while holding constant the effect of the control variables just defined. As already noted, the model contains both fixed and time-varying effects, defined either for 1996 or person-month t, and these are regressed on migratory outcomes defined for month $t+1$. Table 4.3 contains means, standard deviations, and ranges for variables used in the composite indicators of physical capital and neighbourhood development; and Table 4.4 presents means, standard deviations, and ranges for the variables used to predict migration and the frequency of migration to different destinations.

Out of 295,635 person-months contained in the event history file, we observed 1,748 moves within the Chitwan Valley, 1,335 moves to other districts in Nepal, and 357 moves to a foreign country. In terms of environmental conditions in 1996, some 4% of respondents reported an increase in the time required to collect fodder compared with three years earlier, 9% reported an increase in the time required to collect firewood, 50% perceived a decline in crop production, and 23% perceived a decline in water quality. Neighbourhood density varied widely from 0.89 to 1,339 persons per 100,000 ft^2 (9,290 m^2) with a mean of 37 and a standard deviation of 120.

In terms of physical capital, 80% of the households owned farmland. Among variables used in the index of amenities, 49% of all households reported a durable roof of slate, tin or concrete, 25% had a floor of

Table 4.3 *Descriptive statistics for the measures used to create a composite index of physical capital and neighbourhood characteristics*

	Count	Min	Max	SD	Mean
Physical Capital					
Household amenities					
Roof of house is made from slate, tin, or concrete	1391	0	1	0.50	0.49
Floor of house is made from concrete	1391	0	1	0.43	0.25
No own drinking water source	1391	0	1	0.50	0.46
No toilet	1391	0	1	0.48	0.36
Has electricity	1391	0	1	0.47	0.34
Goods owned					
Household has a radio	1391	0	1	0.50	0.52
Household has a television	1391	0	1	0.33	0.12
Household has a bicycle	1391	0	1	0.48	0.62
Household has a motorcycle	1391	0	1	0.18	0.03
Household has a cart	1391	0	1	0.26	0.07
Household has a tractor	1391	0	1	0.09	0.01
Household has a pumpset for irrigation	1391	0	1	0.18	0.03
Household has a bio gas plant	1391	0	1	0.21	0.05
Livestock owned					
Number of chickens and ducks	1391	0	2210	121.64	20.16
Number of pigeons	1391	0	150	7.01	1.15
Number of bullocks	1391	0	7	0.94	0.57
Number of cows	1391	0	8	0.91	0.36
Number of male buffaloes	1391	0	4	0.51	0.16
Number of female buffaloes	1391	0	8	1.35	1.25
Number of sheep and goats	1391	0	24	1.98	1.46
Number of pigs	1391	0	8	0.35	0.06
Neighbourhood development					
Minutes on foot to nearest school	151	0	30	6.55	9.17
Minutes on foot to nearest health-care service	151	0	90	18.06	20.48
Minutes on foot to nearest bus service	151	0	75	14.91	12.31
Distance by bus to Narayanghat	151	0	195	51.74	80.49
Minutes on foot to nearest market	151	0	120	16.41	12.13
Minutes on foot to nearest bank	151	0	150	35.83	58.22
Minutes on foot to nearest place of employment	151	0	180	22.95	20.58
Minutes on foot to nearest police station	151	2	240	38.93	64.32

Table 4.4 *Descriptive statistics for the dependent and independent variables*

	Count	Min	Max	SD	Mean
OUTCOME VARIABLES					
Migration to within Chitwan locations	1,748				
Migration to other districts	1,335				
Migration to other countries	357				
ENVIRONMENTAL VARIABLES					
Increase in time to collect fodder compared with three years ago	1383	0	1	0.20	0.04
Did not collect fodder in both years	1383	0	1	0.46	0.30
Increase in time to collect firewood compared with three years ago	1386	0	1	0.29	0.09
Did not collect firewood in both years	1386	0	1	0.44	0.26
Perception of decrease in crop production compared with three years ago	1391	0	1	0.50	0.50
Does not farm	1391	0	1	0.37	0.17
Water less clear compared with three years ago	1391	0	1	0.42	0.23
Neighbourhood population density	151	0.89	1338.93	120.41	36.97
CONTROLS					
Physical Capital					
Owns farmland	1391	0	1	0.40	0.80
Standardized index of household amenities	1391	−1.18	1.80	1	0
Standardized index of goods owned	1391	−1.22	4.74	1	0
Standardized index of livestock owned	1391	−1.04	10.12	1	0
Neighbourhood Development					
Standardized index of neighbourhood development	151	−2	5	1	0
Human Capital					
Education	295635	0	16	4.21	3.56
Salaried job	295635	0	1	0.24	0.06
Age	295635	15	69	13.37	38.65
Age squared	295635	225	4761	1088.35	1672.80

Table 4.4 (*cont.*)

	Count	Min	Max	SD	Mean
Social Capital					
House member migrated within Chitwan	295635	0	1	0.25	0.07
House member migrated to other districts	295635	0	1	0.30	0.10
House member an international migrant	295635	0	1	0.31	0.11
House member migrated to other districts or other countries	295635	0	1	0.40	0.20
Level of Violence					
Violence from Maoist insurgency	295635	0	1	0.49	0.41
Demographic Variables					
Female	295635	0	1	0.49	0.57
Married	295635	0	1	0.34	0.87
Number of household members	295635	1	26	3.43	6.68
Ethnicity					
Hindu upper caste	295635	0	1	0.50	0.47
Hindu lower caste	295635	0	1	0.30	0.10
Hill Tibeto-Burmese	295635	0	1	0.35	0.14
Newar and other	295635	0	1	0.25	0.07
Terai Tibeto-Burmese	295635	0	1	0.41	0.22

concrete as opposed to dirt, 46% had no source of potable water, 36% had no toilet, and just 34% had electricity. Among variables used in the index of consumer goods, 62% owned a bicycle and 52% owned a radio, but only 12% had a television, 7% a cart, 5% a biogas plant, 3% a motorcycle or irrigation pump and just 1% had a tractor. On average, each household owned 20.2 chickens and ducks, 1.15 pigeons, 0.57 bullocks, 0.36 cows, 0.16 male buffaloes, 1.25 female buffaloes, 1.46 sheep and goats, and 0.06 pigs. Among variables used in the neighbourhood development index, distance by bus to the district capital averaged 80.5 minutes and average foot travel times to the nearest bank and police station were 58.2 and 64.3 minutes respectively. Other resources were more accessible, with average foot travel times of 9.2, 20.5, 12.3, 12.1 and 20.6 minutes to the nearest school, health-care facility, bus stop, market, and nearest place of employment, respectively.

As can be seen, educational levels were generally quite low among respondents. Even though total schooling ranged from 0 to 16 years, the average was just 3.6 years. Likewise, only 6% held a salaried job, and during the typical person-month a respondent was 38.7 years old. With respect to social capital, 7% of respondents lived in a household where someone had migrated within Chitwan by 1996, 10% lived in a household where someone had gone to another district in Nepal, and 11% lived in a household where someone had migrated internationally. Some 41% of the person-months under observation were lived during the period of heightened violence after November 2001, 57% were lived by women, 87% by a married individual, and the average household size was 6.7 persons. Finally, the largest share of respondents were upper caste Hindus as they contributed 47% of the person-months in the event history, followed by Terai Tibeto-Burmese at 22%, Hill Tibeto-Burmese at 14%, and lower caste Hindus at 10%, with the Newar and other castes making up only 7% of the person-months under observation.

One caveat in the analysis arises because the environmental variables were measured in 1996 while the impact of these environmental measures is assumed to influence migration pattern for the subsequent ten years – from 1997 to 2006. Ideally, it would be interesting to predict the effect of monthly or yearly change in the environmental variables on the monthly migration pattern over the ten-year period. However, given data limitations, we have to make the assumption that the environmental factors in the baseline period do a fair job of predicting migration pattern in the subsequent periods.

Effects of environmental degradation on migration

Table 4.5 presents the results of a multinomial logit model estimated to predict the effects of environmental degradation on individual decisions to migrate to one of three possible destinations, along with relevant controls. Of the five environmental indicators, four are significantly related to the likelihood of moving within Chitwan. An increase in the time required to collect fodder raised the odds of undertaking a local move by 25% [exp(0.221) = 1.25]; an increase in the time required to gather firewood increased the odds by 42% [exp(0.348)= 1.42]; a perceived decrease in crop production raised them by 18% [exp(0.168)= 1.18]; and each additional person per 100,000 ft^2 (9,290 m^2) raised the odds of local movement by 0.2% [exp(0.002)= 1.002].

Of all the environmental indicators, only a perceived decline in water quality was unrelated to the likelihood of undertaking a local move

Table 4.5 *Multinomial logistic regression output for predicting the competing risks of taking a trip to three competing locations in month t + 1*

INDEPENDENT VARIABLES IN MONTH t	Within Chitwan		To other districts		To other countries	
	B	SE	B	SE	B	SE
Environmental Variables						
Increase in time to collect fodder compared with three years ago	0.221**	(0.107)	0.294**	(0.124)	-1.359***	(0.456)
Did not collect fodder in both years	0.251***	(0.072)	0.145	(0.089)	-0.186	(0.174)
Increase in time to collect firewood compared with three years ago	0.348***	(0.074)	0.023	(0.096)	0.276	(0.177)
Did not collect firewood in both years	-0.030	(0.070)	0.041	(0.075)	0.201	(0.142)
Perception of decrease in crop production compared with three years ago	0.168***	(0.055)	0.076	(0.063)	-0.153	(0.116)
Does not farm	0.243**	(0.119)	0.267*	(0.146)	0.234	(0.316)
Water less clear compared with three years ago	0.046	(0.058)	0.037	(0.068)	0.087	(0.132)
Neighbourhood population density	0.002***	(0.000)	-0.000	(0.001)	0.001***	(0.001)
CONTROLS						
Physical Capital						
Owns farm land	-0.172*	(0.103)	-0.255**	(0.129)	0.488	(0.298)
Standardized index of household amenities	-0.108***	(0.036)	-0.098**	(0.040)	-0.196**	(0.080)
Standardized index of goods owned	0.008	(0.031)	0.080**	(0.034)	0.126*	(0.070)
Standardized index of livestock owned	-0.033	(0.030)	0.077***	(0.030)	-0.042	(0.072)
Neighbourhood Development						
Distance from essential facilities	0.178***	(0.028)	0.051	(0.037)	-0.000	(0.072)
Human Capital						
Education	0.050***	(0.008)	0.094***	(0.009)	0.049**	(0.018)
Salaried job	0.546***	(0.087)	0.397***	(0.098)	0.086	(0.172)
Age	-0.173***	(0.013)	-0.197***	(0.015)	-0.145***	(0.031)
Age squared	0.002***	(0.000)	0.002***	(0.000)	0.001***	(0.000)

Table 4.5 (*cont.*)

INDEPENDENT VARIABLES IN MONTH t	Within Chitwan		To other districts		To other countries	
	B	SE	B	SE	B	SE
Social Capital						
House member migrated within Chitwan	0.353***	(0.086)	0.016	(0.127)	0.025	(0.291)
House member migrated to other districts	0.061	(0.084)	0.873***	(0.074)	0.349*	(0.196)
House member an international migrant	-0.169**	(0.086)	0.163*	(0.095)	1.502***	(0.137)
Level of Violence						
Violence from Maoist insurgency	-0.886***	(0.065)	-1.439***	(0.090)	-1.853***	(0.204)
Demographic Variables						
Female	0.012	(0.055)	0.007	(0.063)	-1.881***	(0.148)
Married	0.499***	(0.076)	0.541***	(0.084)	0.559***	(0.165)
Number of household members	-0.018*	(0.010)	-0.049***	(0.012)	-0.120***	(0.027)
Ethnicity						
Hindu upper caste	0.165**	(0.079)	0.385***	(0.101)	0.783***	(0.223)
Hindu lower caste	0.116	(0.098)	0.419***	(0.126)	1.125***	(0.238)
Hill Tibeto-Burmese	0.427***	(0.087)	0.602***	(0.112)	0.804***	(0.242)
Newar and other	0.045	(0.128)	0.494***	(0.137)	0.246	(0.331)
Terai Tibeto-Burmese	–		–		–	
Constant	-2.109***	(0.251)	-2.176***	(0.295)	-3.702***	(0.614)
No. of person months			295635			

*** p < 0.01, ** p < 0.05, * p < 0.1

within Chitwan. In addition, those who did not collect fodder and did not farm were all more likely to migrate, suggesting perhaps that people who might earlier have given up these activities owing to environmental deterioration were more likely to become local migrants. In general, the estimates provide strong and consistent evidence that environmental degradation is associated with an increase in local population mobility. In particular, to the extent that deforestation increases the time required to gather fodder and firewood, and to the degree that population pressure increases and farm productivity declines, people can be expected to respond by looking for opportunities elsewhere within the valley.

In contrast, only one environmental factor – increased time to collect fodder – was related to out-migration to other districts in Nepal. According to our estimates, a perceived increase in the time required to collect fodder raised the odds of internal migration within Nepal by 34% [exp(0.294) = 1.34]. None of the other dimensions of environmental change appeared to play a role in fomenting internal migration, although people from households that did not farm were more likely to move to other districts in Nepal, which could be an indirect effect of prior environmental deterioration, but we have little evidence of a direct effect.

Similarly, there is little evidence of a relationship between environmental degradation and international migration. Although an increase in the time required to collect fodder has a strong and significant effect on the likelihood of leaving Nepal, the direction is quite strongly negative. In this case, a perceived increase in the time required to gather fodder yields a 74% reduction in the odds of leaving Chitwan for another country [exp(−1.359)= 0.26]. Although the negative effect of increased time to collect fodder on international migration is somewhat unexpected, one explanation could be that, unlike migration to locations within Chitwan or within Nepal, international migration involves much more preparation and time commitment (e.g. gathering information on available foreign jobs, contacting employment agencies, applying for visas, etc.). An increase in time to gather fodder could take away time individuals could have otherwise devoted to preparing for international migration. The need for additional labour hours to collect fodder might therefore significantly lower the probability of people moving abroad. There is however a significant but very small impact of population density on international migration with each additional person per 100,000 ft^2 (9,290 m^2) raising the odds of international movement by 0.15% [exp(0.0015)= 1.0015].

The control variables mainly function as expected from prior theory and research. Indicators of human capital such as education increase the likelihood of all forms of out-migration, and occupational skill increases the odds of local and internal migration. As expected, social capital has strong destination-specific effects, and the period of violence is associated with reduced migration probabilities across all categories, with the effect growing stronger as distance of the move increases. Marriage raises the odds of out-migration to all destinations, whereas larger household size decreases them, but other things being equal females are much less likely to migrate internationally. Over the observed age range from 15 to 69 the likelihood of migration falls with age at a decelerating rate. Consistent with the earlier studies, different ethnic groups evinced different probabilities of migration. In general, those ethnic groups with prior migratory experience were consistently more likely to migrate than the Terai Tibeto-Burmese, the indigenous people of the valley.

Neighbourhood development increases the likelihood of local mobility but has no effect on migration to destinations outside the valley. Land ownership reduces the odds of migration within Chitwan and to other districts but has no effect on international moves, whereas greater access to household amenities decreases the odds of movement to all destinations. Access to consumer goods raises the odds of movement to other districts and other countries, while the ownership of livestock increases the odds of internal migration within Nepal. These results lend credence to precepts derived from both neoclassical economics and the new economics of labour migration and provide strong support for social capital theory.

In order to assess gender interactions in the process of environment-linked migration, we estimated separate event history models for males and females. Given the very small number of women who migrated internationally, however, we had to collapse the internal and international categories into a single indicator that captured distant migration outside Chitwan. These results are presented in Table 4.6. As can be seen, among women, all the indicators of environmental degradation have strong positive effects on local mobility, again with the sole exception of changes in water quality. Thus, a perceived increase in the time required to collect fodder increases the odds of moving within Chitwan by 28% [exp(0.244)= 1.28], a perceived increase in the time to gather firewood raises the odds by 44% [exp(0.362) = 1.44], and a perceived decline in agricultural productivity raises them by 19% [exp(0.173)= 1.19]. Likewise, rising population pressure as indicated by a one-person increase in the number of residents per 100,000 ft^2 (9,290 m^2)

Table 4.6 *Multinomial logistic regression output for predicting the competing risks of taking a trip to two competing locations for males and females in month t + 1*

	FEMALE				MALE			
	Within Chitwan		To other districts or other countries		Within Chitwan		To other districts or other countries	
INDEPENDENT VARIABLES IN MONTH t	B	SE	B	SE	B	SE	B	SE
Environmental Variables								
Increase in time to collect fodder compared with three years ago	**0.244***	(0.138)	−0.007	(0.175)	0.238	(0.173)	0.150	(0.163)
Did not collect fodder in both years	**0.332***	(0.098)	−0.079	(0.121)	0.131	(0.109)	**0.203***	(0.106)
Increase in time to collect firewood compared with three years ago	**0.362***	(0.095)	−0.128	(0.131)	**0.328***	(0.118)	**0.215***	(0.112)
Did not collect firewood in both years	**−0.201***	(0.096)	−0.074	(0.098)	**0.197***	(0.104)	**0.197***	(0.091)
Perception of decrease in crop production compared with three years ago	**0.173***	(0.074)	−0.101	(0.081)	0.123	(0.083)	0.097	(0.076)
Does not farm	0.205	(0.163)	0.094	(0.185)	**0.298***	(0.177)	0.318	(0.196)
Water less clear compared with three years ago	0.058	(0.077)	0.074	(0.086)	0.042	(0.091)	0.053	(0.085)
Neighbourhood population density	**0.002***	(0.000)	0.001	(0.001)	**0.002***	(0.000)	0.001	(0.000)
CONTROLS								
Physical Capital								
Owns farm land	−0.166	(0.140)	**−0.386***	(0.163)	−0.157	(0.154)	0.071	(0.177)
Standardized index of household amenities	−0.068	(0.048)	0.028	(0.051)	**−0.150***	(0.055)	**−0.279***	(0.050)
Standardized index of goods owned	0.058	(0.042)	**0.087***	(0.044)	−0.055	(0.049)	**0.101***	(0.043)
Standardized index of livestock owned	0.052	(0.039)	**0.078***	(0.041)	**−0.119***	(0.049)	**0.064***	(0.038)
Neighbourhood Development								
Distance from essential facilities	**0.173***	(0.038)	0.005	(0.052)	**0.189***	(0.041)	0.049	(0.043)
Human Capital								
Education	**0.082***	(0.012)	**0.092***	(0.013)	0.008	(0.012)	**0.087***	(0.012)

Table 4.6 (*cont.*)

INDEPENDENT VARIABLES IN MONTH t	FEMALE				MALE			
	Within Chitwan		To other districts or other countries		Within Chitwan		To other districts or other countries	
	B	SE	B	SE	B	SE	B	SE
Salaried job	0.823***	(0.187)	0.488**	(0.221)	0.519***	(0.100)	0.382***	(0.096)
Age	−0.237***	(0.018)	−0.272***	(0.020)	−0.077***	(0.021)	−0.137***	(0.020)
Age squared	0.003***	(0.000)	0.003***	(0.000)	0.001**	(0.000)	0.001**	(0.000)
Social Capital								
House member migrated within Chitwan	0.158	(0.115)	−0.219	(0.166)	0.636***	(0.132)	0.218	(0.163)
House member migrated to other districts or other countries	−0.050	(0.082)	0.559***	(0.083)	−0.156	(0.112)	0.774***	(0.082)
Level of Violence								
Violence from Maoist insurgency	−0.795***	(0.087)	−1.276***	(0.109)	−0.990***	(0.098)	−1.817***	(0.128)
Demographic Variables								
Married	0.899***	(0.101)	1.141***	(0.116)	−0.074	(0.123)	0.197*	(0.108)
Number of household members	−0.028**	(0.013)	−0.058***	(0.016)	−0.009	(0.016)	−0.064***	(0.016)
Ethnicity								
Hindu upper caste	0.183*	(0.109)	0.445***	(0.138)	0.143	(0.116)	0.473***	(0.123)
Hindu lower caste	0.372***	(0.133)	0.552***	(0.167)	−0.227	(0.149)	0.512***	(0.146)
Hill Tibeto-Burmese	0.564***	(0.119)	0.443***	(0.156)	0.169	(0.130)	0.691***	(0.134)
Newar and other	0.168	(0.174)	0.856***	(0.175)	−0.067	(0.190)	0.044	(0.192)
Terai Tibeto-Burmese	−		−		−		−	
Constant	−1.540***	(0.344)	−1.169***	(0.389)	−3.076***	(0.380)	−2.731***	(0.378)
No. of person months	169959				125676			

*** p < 0.01, ** p < 0.05, * p < 0.1

raises the odds of moving within Chitwan by 0.2% [exp(0.002)= 1.002]. None of these environmental indicators has any effect on the odds that a woman would move to more distant locations outside the Chitwan Valley, however. Among women, links between environmental degradation and migration are strong, but only for local moves.

Among men, the principal link between environmental change and migration is through the time required to gather firewood, which is consistent with the fact that in Chitwan the collection of firewood is more of a male than a female task, as opposed to the gathering of fodder, which is stereotypically female (Bhandari, 2004; Kumar and Hotchkiss, 1988). As a result, increasing time to gather fodder has no significant effect on male migration, but an increase in the time required to collect firewood is significant in predicting both local and distant moves by men. A perceived increase in the time to gather firewood is associated with a 39% increase in the odds of male migration within Chitwan [exp(0.328)= 1.39] and a 24% increase in the odds of male migration to other districts within Nepal or to other countries [exp(0.215)= 1.24]. As with females, rising population density increases the odds of local but not distant mobility, increasing the odds of moving within Chitwan by 0.2% with each additional person per 100,000 ft^2 (9,290 m^2).

Summary and conclusion

Although the concept of environmental refugees has been around since the 1980s and remains quite popular with many scholars and activists, empirical demonstrations of environmental effects on population mobility have been rare. In this analysis, we took advantage of newly available data from the Chitwan Valley Family Study to examine how environmental degradation along several dimensions affected the propensity to migrate locally within Nepal's Chitwan Valley, internally to other districts in Nepal, and internationally to foreign destinations. In the baseline survey, respondents were asked to compare conditions in 1996 with conditions three years earlier with respect to four outcomes: the time required to gather fodder, the time required to collect firewood, change in agricultural productivity, and change in water quality. We measured environmental degradation using dummy variables indicating an increase in time required to gather fodder and firewood, a decline in agricultural productivity, and a decrease in water quality. We measured demographic pressure in 1996 by computing neighbourhood density in persons per 100,000 ft^2 (9,290 m^2).

We examined the effect of these indicators on the monthly probability of making a local, internal and international move over the ensuing 108 months, controlling for the effects of human, physical and social capital, as well as demographic background. We found no evidence that changes in water quality had any effect on migration to any destination. However, we did find that increases in the time required to collect fodder and firewood, as well as perceived decline in agricultural productivity and higher population densities, were associated with greater population mobility; but these effects were confined almost entirely to moves within the Chitwan Valley and tended to be more pervasive among women than among men.

Among women, rising collection times for fodder and firewood were both associated with significant increases in the odds of local mobility. As fodder and firewood are gathered from local forests, these results imply that deforestation is a significant cause of increased female mobility within the Chitwan Valley. Female mobility within Chitwan was also predicted by higher population densities and declining agricultural productivity, suggesting that rising pressure on farmland from demographic growth also represents an important cause of local migration by women. Population density also predicted the mobility of men, but declining agricultural productivity and rising collection time for fodder did not, suggesting that farm work and gathering fodder are gendered tasks assigned disproportionately to women. The gathering of firewood, however, is done by males and increase in the time required for this task not only raised the odds of male movement within the Chitwan Valley, but also to other districts within Nepal and to other countries.

Even though we could not estimate a model predicting internal and international trips separately for men and women, the overall model we estimated found very little evidence of a significant effect of environmental deterioration on internal and international migration. Only an increase in time to collect fodder promoted out-migration to other districts in Nepal, while population density somewhat influenced international migration. The only other clear effect of environmental factors on international migration was an increase in the time required to collect fodder, which was strongly negative, sharply reducing the odds of international movement.

In sum, we find strong evidence that deforestation, population pressure and agricultural decline produce elevated rates of local population mobility among women, and to a lesser extent among men, but little evidence that these environmental changes lead to significant increases in internal or international migration, though there is some indication that deforestation may increase internal migration by men by raising the time costs of firewood collection. To the extent that environmental deterioration leads to greater migration, therefore, the effects appear to be highly localized.

Of course, the environmental changes we measure in Chitwan are of a particular type – a slow, gradual depletion of resources through demographic and economic pressure rather than a sudden shift in environmental circumstances as a result of some dramatic human or natural event. Although the term 'environmental refugees' may create images of destitute people clamouring at the gates of the developed world to many in the West, our findings suggest that gradual environmental depredations from processes such as deforestation, desertification, salinization, drought and soil erosion are much more likely to produce local rather than international migrations, simply because the people most affected by these changes – poor agrarian families – lack the resources to finance international trips.

References

Afifi, T. and Warner, K. 2008. *The Impact of Environmental Degradation on Migration Flows across Countries*. Bonn, United Nations University, Institute for Environment and Human Security. (UNU-EHS Working Paper 5.)

Ali, J. and Benjaminsen, T. A. 2004. Fuelwood, timber and deforestation in the Himalayas – the case of Basho Valley, Baltistan Region, Pakistan. *Mountain Research and Development*, Vol. 24, No. 4, pp. 312–18.

Baland, J.-M., Bardhan, P., Das, S., Mookherjee, D. and Sarkar, R. 2006. *Managing the Environmental Consequences of Growth: Forest Degradation in the Indian Mid-Himalayas*. Paper presented at the India Policy Forum 2006. New Delhi, National Council for Applied Economic Research.

Bhandari, P. 2004. Relative deprivation and migration in an agricultural setting of Nepal. *Population and Environment*, Vol. 26, No. 5, pp. 475–99.

Biddlecom, A. E., Axinn, W. G. and Barber, J. S. 2005. Environmental effects on family size preferences and subsequent reproductive behavior in Nepal. *Population and Environment*, Vol. 26, No. 3, pp. 183–206.

Bilsborrow, R. E. and DeLargy, P. F. 1991. Population growth, natural resource use and migration in the third world: the cases of Guatemala and Sudan. *Population and Development Review*, Vol. 16, S125–S147.

Black, R. 2001. *Environmental Refugees: Myth or Reality?* Geneva, United Nations High Commissioner for Refugees. (*New Issues in Refugee Research*, No. 34.)

Bohra, P. and Massey, D. S. 2009. Processes of internal and international migration from Chitwan, Nepal. *International Migration Review*, Vol. 43, No. 3, pp. 621–51.

Bohra-Mishra, P. and Massey, D. S. 2011. Individual decisions to migrate during civil conflict. *Demography*, forthcoming.

Castles, S. 2002. *Environmental Change and Forced Migration: Making Sense of the Debate*. Geneva, United Nations High Commissioner for Refugees. (*New Issues in Refugee Research*, No. 70.)

El-Hinnawi, E. 1985. *Environmental Refugees*. Nairobi, United Nations Environment Programme.

Filmer, D. and Pritchett, L. 1997. *Environment Degradation and Demand for Children: Searching for the Vicious Circle*. Washington DC, World Bank. (World Bank Policy Research Paper 1623.)

Ghimire, D. J., and Mohai, P. 2005. Environmentalism and contraceptive use: how people in less developed settings approach environmental issues. *Population and Environment*, Vol. 27, No. 1, pp. 29–61.

Graner, E. 2001. Labor markets and migration in Nepal. *Mountain Research and Development*, Vol. 21, No. 3, pp. 253–59.

Hitztaler, S. 2004. The relationship between resources and human migration patterns in central Kamchatka during the post-Soviet period. *Population and Environment*, Vol. 25, No. 4, pp. 355–75.

Hugo, G. 2008. *Migration, Development and Environment*. Draft paper for Research Workshop on Migration and the Environment: Developing a Global Research Agenda, Munich, Germany Informal Sector Service Center. http://www.inseconline.org/hrvdata/Total_Killings.pdf

Jacobson, J. 1988. *Environmental Refugees: A Yardstick of Habitability*. Washington DC, World Watch Institute. (World Watch Paper No. 86.)

Kalipeni, E. 1996. Demographic response to environmental pressure in Malawi. *Population and Environment*, Vol. 17, No. 4, pp. 285–308.

Kok, F. 2003. *Nepal: Displaced and Ignored*. Geneva, Norwegian Refugee Council.

Kumar, S. K., and Hotchkiss, D. 1988. *Consequences of Deforestation for Women's Time Allocation, Agricultural Production, and Nutrition in Hill Areas of Nepal*. Washington DC, International Food Policy Institute. (Research Report 69.)

Massey, D. S., Arango, J., Hugo, G., Kouaouci, A., Pellegrino, A. and Taylor, J. E. 1998. *Worlds in Motion: Understanding International Migration at Century's End*. Oxford, UK, Oxford University Press.

Massey, D. S., Axinn, W. G and Ghimire, D. 2007. *Environmental Change and Out-Migration: Evidence from Nepal*. Population Studies Center, University of Michigan, Institute for Social Research.

Murshed, M. S. and Gates, S. 2005. Spatial-horizontal inequality and the Maoist insurgency in Nepal. *Review of Development Economics*, Vol. 9, No. 1, pp. 121–34.

Myers, N. and Kent, J. 1995. *Environmental Exodus: An Emergent Crisis in the Global Arena.* Washington DC, Climate Institute.

Shrestha, S. S. and Bhandari. P. 2007. Environmental security and labor migration in Nepal. *Population and Environment,* Vol. 29, No. 1, pp. 25–38.

Sjaastad, L. A. 1962. The costs and returns of human migration. *The Journal of Political Economy,* Vol. 70, No. 5.2, pp. 80–93.

Suhrke, A. 1994. Environmental degradation and population flows. *Journal of International Affairs,* Vol. 47, No. 2, pp. 473–96.

Todaro, M. P. and Maruszko, L. 1987. Illegal migration and US immigration reform: a conceptual framework. *Population and Development Review,* Vol. 13, No. 1, pp. 101–14.

Refusing 'refuge' in the Pacific: (de)constructing climate-induced displacement in international law

JANE MCADAM[1]

Introduction

Human movement caused by environmental factors is not new. Natural and human-induced environmental disasters and slow-onset degradation have displaced people in the past, and will continue to do so in the future. Such movement is a normal part of adaptation to change. The 'newness' of displacement triggered (at least in part) by climate change is its underlying anthropogenic basis,[2] the large number of people thought to be susceptible to it,[3] and the relative speed with which climate change is to occur, which may hamper people's traditional adaptive patterns that historically were able to develop over time. According to the United Nations High Commissioner for Refugees, it is becoming difficult to categorize displaced people because of the combined impacts of conflict, the environment and economic pressures.[4] While the term 'refugee' describes only a narrow sub-class of the world's forced migrants, it is often misapplied to those who move (or who are anticipated to move) for environmental or climate reasons. As explored below, this is not only erroneous as a matter of law, but is conceptually inaccurate as well.

---

[1] I am grateful to the Australian Research Council for funding this research, including fieldwork in Kiribati and Tuvalu.

[2] That is not to say that 'natural' disasters are without anthropogenic bases: see e.g. Wisner et al. (2004), who argue that few disasters are ever 'natural'; they are a combination of environmental plus socio-economic and political factors.

[3] President of the Global Humanitarian Forum, Kofi Annan, described 'millions of people' being 'uprooted or permanently on the move as a result' of climate change, with '[m]any more millions' to follow (Annan, 2009, p. ii). Debates about numbers remain highly contentious: see e.g. Kniveton et al. (2008); Castles (2002).

[4] See remarks made by High Commissioner Antonio Guterres in an interview with *The Guardian*, in Borger (2008).

In contexts such as the so-called 'sinking islands' of Kiribati and Tuvalu in the South Pacific, movement is less likely to be in the nature of sudden flight, and more likely to be pre-emptive and planned. This does not mean it is not 'forced', but rather that top-down policy responses and normative frameworks that predicate forced migration on a particular notion of exodus may not match up to realities of movement. Furthermore, while 'development-induced displacement' and 'conflict-induced displacement' describe primary motivations for movement in certain contexts, field research in Tuvalu and Kiribati highlights the difficulties of describing human movement from these states as exclusively 'climate-induced'.

A variety of actors has called for a new international treaty on climate change displacement (or a Protocol to the Refugee Convention[5] or the United Nations Framework Convention on Climate Change[6]) to create a new class of refugee-like protected persons. At the state level, for example, Maldives in 2006 proposed amending the 1951 Refugee Convention to extend the definition of a 'refugee' in Art. 1A(2) to include 'climate refugees'.[7] In December 2009, in the run-up to the Copenhagen climate change conference, the Bangladeshi Finance Minister similarly stated: 'The convention on refugees could be revised to protect people. It's been through other revisions, so this should be possible' (see Grant et al., 2009). A Bangladeshi NGO network, Equity and Justice Working Group Bangladesh (EquityBD), has called for a new Protocol to the UNFCCC 'to ensure social, cultural and economic rehabilitation of "climate refugees" through recognizing them as "Universal Natural Persons"'.[8]

Some scholars have also proposed new legal instruments to address climate-related movement.[9] For example, Biermann and Boas suggested

[5] Convention relating to the Status of Refugees (adopted 28 July 1951, entered into force 22 April 1954) 189 UNTS 137, Art. 1A(2), read in conjunction with Protocol relating to the Status of Refugees (adopted 31 January 1967, entered into force 4 October 1967) 606 UNTS 267.

[6] UN Framework Convention on Climate Change (adopted 9 May 1992, entered into force 21 March 1993) 1771 UNTS 107.

[7] Republic of the Maldives Ministry of Environment, Energy and Water, Report on the First Meeting on Protocol on Environmental Refugees: Recognition of Environmental Refugees in the 1951 Convention and 1967 Protocol Relating to the Status of Refugees (Male, Maldives, 14–15 August 2006) cited in Biermann and Boas (2008).

[8] See Shamsuddoha and Chowdhury (2009). EquityBd no longer uses the 'refugee' terminology: author's interview with Md Shamsuddoha and Rezaul Karim Chowdhury (Dhaka, 19 June 2010).

[9] I am adopting the term 'climate-related' movement to denote the multiple factors involved, and that climate change is one of several.

a UNFCCC Protocol on the Recognition, Protection, and Resettlement of Climate Refugees.[10] A group of legal scholars from the University of Limoges published a Draft Convention on the International Status of Environmentally-Displaced Persons.[11] Docherty and Giannini (2009, pp. 349, 350, 373) proposed an 'independent' or 'stand-alone' convention defining 'climate change refugee' and containing 'guarantees of assistance, shared responsibility, and administration'. An Australian-based project also seeks to elaborate 'a draft convention for persons displaced by climate change', which would 'establish an international regime for the status and treatment of such persons'.[12] The Council of Europe Parliamentary Assembly's Committee on Migration, Refugees and Population has suggested 'adding an additional protocol to the European Convention on Human Rights, concerning the right to a healthy and safe environment' as a way of 'enhancing the human rights protection mechanisms vis-à-vis the challenges of climate change and environmental degradation processes'.[13]

All these proposals vary in terms of how they seek to define those displaced, and whether such people would be subject to individual status determination (similar to conventional refugee status determination),[14] or whether protection would be granted on the basis of the objective country of origin conditions from which people flee.[15]

While the underlying basis of each proposal is, presumably, to provide a rights-based framework for people forced to move when the impacts of

[10] Biermann and Boas (2007), see also Biermann and Boas (2008); for criticism of their approach, see Hulme (2008). For another UNFCCC-based proposal, see Williams (2008).

[11] Draft Convention on the International Status of Environmentally-Displaced Persons (CRIDEAU/CRDP, 2008). Art. 2(2) defines 'environmentally-displaced persons' as 'individuals, families and populations confronted with a sudden or gradual environmental disaster that inexorably impacts their living conditions and results in their forced displacement, at the outset or throughout, from their habitual residence and requires their relocation and resettlement'. A 'right to resettlement' is elaborated in Art. 9: States Parties are to establish 'transparent and open legal procedures for the demand and grant or refusal of the status of environmentally-displaced persons based on the rights set forth in the present chapter'.

[12] A Convention for Persons Displaced by Climate Change (http://www.ccdpconvention.com/, accessed 7 December 2009).

[13] The Council of Europe Parliamentary Assembly, Committee on Migration, Refugees and Population, Environmentally Induced Migration and Displacement: A 21st Century Challenge. Doc 11785 (23 December 2008) paras 6.3 and 121 respectively.

[14] For example, Maldives, Bangladesh, Limoges proposals.

[15] For example, Docherty and Giannini proposal, Australian proposal and Biermann and Boas proposal.

climate change render life and livelihoods at home impossible, it is not self-evident that a treaty would presently best serve this end. Accordingly, this chapter provides a partial response to calls to protect 'climate refugees' through an international instrument. It is partial, because it does not engage in a detailed discussion about whether new substantive norms or machinery are needed (on which, see Kälin, 2010; McAdam and Saul, 2010), or respond to the particular detail of each proposal mentioned above. Nor does it examine issues of compensation or responsibility-sharing – matters which might usefully be addressed in a multilateral instrument (and which some of the proposals suggest). Rather, it simply addresses the appropriateness of defining a 'climate displaced person' category within an international protection paradigm. By an 'international protection paradigm', I mean something akin to refugee protection: requiring states, as a matter of international treaty law, not to return people to climate-related harms and to grant them a domestic legal status.

I do this by examining some conceptual and pragmatic difficulties in attempting to construct a refugee-like instrument for people fleeing the effects of climate change, and by critiquing whether there are legal benefits, as opposed to political benefits, to be gained by advocating for such an instrument. This should not be read as an outright rejection of a future treaty regime whereby states accept a duty to assist people displaced in part by climate change, and to agree to responsibility-sharing mechanisms.[16] Indeed, people are already moving in response to environmental changes,[17] and ultimately states will need to develop coordinated responses that acknowledge

---

[16] For example, the Agreement on the Cooperation for the Sustainable Development of the Mekong River Basin (entered into force by signature, 5 April 1995) 2069 UNTS 3 is a regional treaty between four states that establishes a framework for cooperation 'in all fields of sustainable development, utilization, management and conservation of the water and related resources of the Mekong River Basin ... in a manner to optimize the multiple-use and mutual benefits of all riparians and to minimize the harmful effects that might result from natural occurrences and man-made activities' (Art. 1). It also establishes an institutional framework, the Mekong River Commission, 'to provide an adequate, efficient and functional joint organizational structure to implement this Agreement and the projects, programmes and activities taken thereunder in cooperation and coordination with each member and the international community, and to address and resolve issues and problems that may arise from the use and development of the Mekong River Basin water and related resources in an amicable, timely and good neighbourly manner' (Preamble). Thus, at a minimum, it commits states to negotiate on the issues, through 'consultation and evaluation' and 'a dynamic and practical consensus' (Ch. II definitions).

[17] See e.g. *Pacific Island Report* (2009), reporting a story from the *Papua New Guinea Post-Courier* (1 May 2009). More generally, see the special issue of *Forced Migration Review* on climate change and displacement (No. 31, 2008).

the need for cross-border movement in certain circumstances and which regularize the status of those who move, either through humanitarian or migration schemes.

This chapter proceeds with three main lines of argument. First, it critiques some assumptions about the nature of climate change and displacement which underpin advocacy for a protection instrument, and suggests that other measures may achieve (more) desirable outcomes for those affected. Secondly, it queries whether it is desirable, in any event, for any new international instrument to focus on one displacement driver – climate change – rather than poverty, or conflict, or natural disaster. In other words, should displacement be addressed in terms of what drives it, or rather in terms of the needs of those who move? Thirdly these conceptual critiques are linked to a more pragmatic one: that states currently lack the political will to negotiate a new instrument requiring them to provide international protection to additional classes of people, and that even if they did, its ratification, implementation and enforcement could not be easily compelled. While this of itself is not an argument against legal developments, it highlights a significant obstacle to achieving treaty-based solutions (at least in the short to mid-term) and the limitations of a treaty even if negotiated. Furthermore, it relates back to the question of how best to promote the human rights of affected communities: while international human rights law principles should inform any decisions relating to movement, a protection-like response may not necessarily respond to communities' human rights concerns, especially those relating to cultural integrity, self-determination and statehood (McAdam, 2010a). It may also obscure other human rights that need attention.

Together, these concerns suggest that the focus on a multilateral treaty to extend states' international protection obligations may not now be the most appropriate tool to achieve outcomes for populations severely affected by the impacts of climate change. It is important to recognize that migration is a normal adaptation response to environmental change. There is a risk that legally defining a 'climate refugee' category may lead to a hardening of the concept, simultaneously defining groups 'in' or 'out' of need. Given the conceptual difficulties for devising such a definition, and for premising movement as 'flight' in response to certain triggers, it may not best encapsulate the likely nature of movement, especially with respect to slow-onset changes in small island states. Focusing attention on culturally sensitive outcomes for people in particular contexts, which respond to the nature, timing and location of predicted movement within, from and to particular states, and their own views about how they want to be perceived, may ultimately better facilitate a human rights approach to the phenomenon. In advancing this view, particular attention is paid to the case of two small Pacific island states, Kiribati and Tuvalu.

Understanding the phenomenon

The way a phenomenon is understood necessarily determines the way it is regulated. Responses to human trafficking, for example, will differ depending on whether the issue is viewed through a criminal justice or a human rights lens. Similarly, how (and to what extent) international law and institutions respond to climate-related human movement will depend in part on:

- whether such movement is perceived as voluntary or involuntary;
- the nature of the trigger (a disaster versus a slow-onset process);
- whether international borders are crossed;
- the extent to which there are political incentives to characterize something as linked to climate change or not; and
- whether movement is driven or aggravated by human factors, such as discrimination.[18]

At the macro level, there are a number of ways that such movement can be categorized and responded to: for example, as a protection issue, a migration issue, a disaster issue, an environmental issue, or a development issue. Each of these is built around an implicit set of assumptions that motivate different kinds of policy outcomes. Broadly speaking, as a protection issue, the assumption is that movement is forced and should be treated as refugee-like in nature, with binding protection obligations for states (hence calls for a new treaty) with respect to those displaced. As a migration issue, movement is cast as voluntary, and therefore as not compelling the 'international community' to respond. Rather, the assumption is that states can respond as and when they see fit through

---

[18] See e.g. Khan (2009). How an issue is characterized can cut both ways, of course. While some Pacific leaders have highlighted the existential threat that climate change poses to their states (see further below), countries such as the United States, Australia and the Member States of the European Union have highlighted the security threats that climate change poses to themselves, including the threat of climate migration: see e.g. Schwartz and Randall (2003); CNA (2007); Prime Minister Kevin Rudd (2008); Commonwealth of Australia (2009); Council of the European Union (2008).

domestic immigration policy. As a disaster issue, assistance can be provided by *in situ* humanitarian relief and temporary relocation where needed. As an environmental issue, the movement of 'climate refugees' from 'sinking islands' can be used as a potent political image in advocating for the reduction of carbon emissions and the protection of endangered ecosystems. Here, the refugee terminology, while rejected by most forced-migration scholars, contributes to its dramatic effect. As a development issue, foreign aid and investment are seen as the tools that can fund adaptation measures and assist climate-affected countries to 'develop' their way out of poverty, poor governance and so on and thereby enhance their capacity to adapt to climate change. In each of these conceptualizations, the extent to which climate change features as the key issue varies: it is predominant in the protection and environmental discourses; it is one of a number of relevant impacts in the migration, disaster and development characterizations.

From a legal perspective, this is important because the way a phenomenon is characterized determines how law and policy are developed. Questions of conceptualization have clear governance implications, as they inform the appropriate location of environmental migration procedurally (as an international, regional or local, developed and/or developing country concern/responsibility); thematically (for example, within the existing refugee protection framework or under the UNFCCC); and institutionally (such as whether a mandate should rest with the UNHCR, IOM, UNDP, UNEP, or a new organization).

Climate change as driver? A case study of Kiribati and Tuvalu

This section anchors this somewhat abstract discussion in a case study centred on the Pacific island states of Kiribati and Tuvalu, which have become the focus of the 'climate refugee' movement.

Background

Tuvalu and Kiribati are independent small island states in the South Pacific. Despite belonging to different ethnic groups (Polynesian and Micronesian respectively), the British claimed them in 1892 as a single protectorate – the Gilbert and Ellice Islands – which became a Crown Colony in 1916 until independence was achieved some thirty years ago (Tuvalu in 1978, Kiribati in 1979). With an average height of less than 2 m above sea level, they frequently feature in the media and NGO reports as 'sinking islands' that will be uninhabitable by the middle of this century, with their people becoming the world's first 'climate refugees' (see e.g. MacFarquhar, 2009; Bone and Pagnamenta, 2009; Callick, 2009; Lateu, 2008).

Kiribati has a population of around 100,000, while Tuvalu is the world's smallest state (apart from the Vatican), with only 10,000 people. Half of Kiribati's population live in Tarawa, and the population is increasing rapidly.[19] On its southern tip, the population density of the 1.7 km² islet of Betio is greater than that of Hong Kong, but without the high-rise apartments to house it. Sanitation is poor and pollution is high, with beach toileting and washing very common. Only 20% of households have access to a sewerage system; 64% do not use toilets (Sherborne, 2009, citing a Kiribati government report, *The Challenge: Things (Beginning to) Fall Apart*). Septic tanks seep into the groundwater supply, which is often brackish, and the tank infrastructure is too rudimentary to keep up with population growth. The majority of people are unemployed: only a quarter have a regular job, and half of them work in government administration. The average weekly wage in Tarawa is AU$60 (Sherborne, 2009). Of the states threatened by eventual annihilation, Kiribati has the largest population (especially in light of future population growth), and virtually no capacity for long-term internal migration because of the absence of high land.[20] Tuvalu faces similar problems of unemployment, pollution and a general lack of resources, although there each house has a rainwater tank (albeit not always functional). Population pressure is not quite as severe, but there is considerable reliance on employed family members to provide for their relatives.

Climate change is undoubtedly impacting on these low-lying atoll states. Driving along the main road on the central Kiribati atoll of Tarawa, with the lagoon on one side and the ocean on the other, the sense of vulnerability to

[19] A 2005 census put South Tarawa's population at 40,300, an increase of almost 43% over a decade (Sherborne 2009). See generally Stahl and Appleyard (2007).

[20] Ironically, Banaba, which is the only high land, was all but depopulated in the 1950s when the people relocated to Fiji to enable phosphate mining to take place. The President of Kiribati has mentioned the possibility of eventually relocating the government there, to continue a presence on the territory for as long as possible (author interview with President Anote Tong, Kiribati, 12 May 2009).

the environment is palpable – a vulnerability that is magnified when there is a climate crisis such as a cyclone or king tide. But to what extent can climate change be singled out as a driver of forced migration, and is the concept of 'climate-induced displacement' accurate in this context?

The existential threat of climate change in the Pacific

Certainly, some Pacific leaders have highlighted the impacts of climate change as an existential threat.[21] In June 2009, the Pacific island states were among those that sponsored a UN General Assembly resolution on 'Climate Change and its Possible Security Implications'. The delegate from Palau stated that: 'Never before in all history has the disappearance of whole nations been such a real possibility.'[22] The President of the Marshall Islands described the rationale behind the resolution as a 'further pursuit of greater guarantees of our territorial integrity'.[23] Other leaders stressed the impact climate change was having on 'our very existence as inhabitants of very small and vulnerable island nations'.[24]

Some states themselves use this imagery to dramatic effect – a recent underwater Cabinet meeting by the Maldives Government is a good example (see BBC News, 2009). However, while the image of an island disappearing beneath the rising sea provides a potent, frightening basis from which to lobby for global reductions in carbon emissions, it is not necessarily as useful for getting the international community to develop normative frameworks to respond to climate-related movement. Indeed, it may contribute to misunderstandings about the likely patterns, time-scale and nature of such movement. That is not to say this approach is disingenuous, but rather that it is important to be alert to the particular objectives it may promote: raising awareness of climate impacts on small island states, providing pressure for political outcomes in climate negotiations, and making the international community aware that a failure

[21] As early as 1992 the South Pacific Forum 'reaffirmed that global warming and sea level rise are the most serious threats to the Pacific region and the survival of some island states': Forum Communiqué, 23rd South Pacific Forum, Honiara, Solomon Islands (8–9 July 1992) Doc SPFS(92)18, para. 7.

[22] UNGA 63rd session, 9th plenary meeting (25 September 2008) UN doc A/63/PV.9, Mr Chin (Palau).

[23] UNGA 63rd session, 9th plenary meeting (25 September 2008) UN doc A/63/PV.9, Mr Litokwa Tomeing (President of the Marshall Islands).

[24] UNGA 63rd session, 10th plenary meeting (25 September 2008) UN doc A/63/PV.10, Mr Emanuel Mori (Federated States of Micronesia).

to act on global emissions may ultimately lead to serious destruction of human society and structures. Often this sort of advocacy involves simplifying the issues, and partially because of this, Pacific governments cannot agree among themselves on a common approach to the issue of movement.

The Kiribati Government, for example, is keen to secure international agreements in which other states recognize that climate change has contributed to their predicament and acknowledge 'relocation' as part of their obligations to assist (in a compensatory way).[25] By contrast, the governments of Tuvalu and the Federated States of Micronesia have resisted the inclusion of 'relocation' in international agreements because of a fear that if they do, industrialized states may simply think that they can 'solve' problems such as rising sea levels by relocating affected populations, instead of by reducing carbon emissions, something which would not bode well for the world as a whole.[26] In December 2009, the Tuvaluan Prime Minister reiterated that his government rejected resettlement: 'While Tuvalu faces an uncertain future because of climate change, it is our view that Tuvaluans will remain in Tuvalu. We will fight to keep our country, our culture and our way of living. We are not considering any migration scheme. We believe if the right actions are taken to address climate change, Tuvalu will survive.'[27]

In this respect, it is interesting to note that in the last round of the pre-Copenhagen UNFCCC climate change talks in early November 2009, the final draft treaty text included two sections referring to human movement on which agreement had previously been unachievable. They called upon states to implement as part of their adaptation measures '[a]ctivities related to national, regional and international migration and displacement or planned relocation of persons affected by climate

[25] Author interview with President Anote Tong (Kiribati, 12 May 2009). See also the remarks of the Bangladeshi finance minister, Abul Maal Abdul Muhith, who prior to the 2009 Copenhagen climate conference stated: 'We are asking our development partners to honour the natural right of persons to migrate. We can't accommodate all these people': cited in Vidal (2009).

[26] Author interview with Kiribati Solicitor-General David Lambourne (Kiribati, 8 May 2009).

[27] Prime Minister Apisai Ielemia says climate change threatens Tuvalu's survival (European Parliament, 2009), press release quoting the Tuvaluan Prime Minister's comments to the Development Committee on 10 December 2009. See similar comments made by the government of Nauru when it was proposed that its population relocate to Australia in the 1960s, in McAdam (2010).

change, while acknowledging the need to identify modalities of inter-state cooperation to respond to the needs of affected populations who either cross an international frontier as a result of, or find themselves abroad and are unable to return owing to, the effects of climate change'.[28]

These are important statements of principle that identify the need for international cooperation in responding to any movement relating to climate change impacts, but which fall short of articulating the precise measures through which such cooperation would be facilitated.[29] They were ultimately omitted from the final text agreed at Copenhagen (Copenhagen Accord, adopted 18 December 2009; see also Lawton, 2009), but a more watered-down form appeared in the June 2010 negotiating text in Bonn. Finally, in the 2010 Cancún Adaptation Framework, paragraph 14(f) was adopted which 'invites' States to 'enhance action on adaptation ... taking into account their common but differentiated responsibilities and respective capabilities, and specific national and regional development priorities, objectives and circumstances', by undertaking '[m]easures to enhance understanding, coordination and cooperation with regard to climate change induced displacement, migration and planned relocation, where appropriate, at national, regional and international levels'.[30] This is *not* an agreement by states to 'protect' people displaced by climate change. Rather, the provision references human movement within the much broader context of enhancing national action on adaptation; no guidance or mechanism (let alone obligation) is proposed in relation to how to translate enhanced 'understanding, coordination and cooperation' into *international* strategies.

Another disagreement among Pacific governments relates to the extent to which climate change should be pinpointed as a driver of migration. In Tuvalu, the predominant official view is that climate change must remain the focal point in any multilateral or bilateral

---

[28] See Negotiating text, UN doc FCCC/AWGLCA/2009/14 (20 November 2009) para. 12(c), p. 38: http://maindb.unfccc.int/library/view_pdf.pl?url=http://unfccc.int/resource/docs/2009/awglca7/eng/14.pdf (accessed 14 December 2009).

[29] Cooperation may take the means of fiscal as well as practical burden-sharing, and also comprehensive approaches: see Hurwitz (2009, pp. 138–71). However, scholars such as Fitzpatrick have lamented the prevalence of fiscal burden-sharing (as opposed to others) as a 'questionable substitute': see Fitzpatrick (2000, pp. 279, 291, cited in Hurwitz, p. 163).

[30] See: http://unfccc.int/files/meetings/cop_16/application/pdf/cop16_lca.pdf#page=3 (accessed 15 March 2011).

discussions about development, assistance and migration. Officials worry that if they acknowledge the more complex, multifaceted dimensions of pressures facing the population, this will detract from the urgency of the climate change threat. Furthermore, they fear that without a climate change focus, adaptation efforts on the ground will stall. In other words, there is a concern that if climate drivers are overshadowed by other factors such as general poverty,[31] which have traditionally not been seen as giving rise to a protection response by third states, efforts to achieve funding for adaptation and migration options for the future will be stymied.[32]

In contrast, as one government official in Kiribati observed, climate change overlays pre-existing pressures – overcrowding, unemployment, environmental and development concerns – which means that it may provide a 'tipping point' that would not have been reached in its absence.[33] Irrespective of the threat posed by climate change *per se*, the government of Kiribati would be lobbying neighbouring states such as Australia and New Zealand for migration opportunities, given the pressures at home (see 'Migration' section below). However, the spectre of climate change makes those negotiations all the more pressing. Over time, the climate impacts will necessarily affect resource availability, such that it may not be a single extreme weather event that provides the trigger for movement, but rather the longer-term unsustainability of the environment for human habitation as freshwater lenses are contaminated by salt water and it becomes impossible to grow crops (already problematic on the outer islands). Thus, at various points in time, the role of climate change in individual or household decisions to move may be stronger or weaker. But it is a factor that can exacerbate pre-existing vulnerabilities or impede adaptation that, in a more developed country, or in a country with greater natural resources or internal relocation options, might not be as problematic.

---

[31] An interesting suggestion was made at the first session of the UN Human Rights Council's Advisory Committee that the Human Rights Council and the Secretary-General use their good offices to extend the principle of *non-refoulement* to 'hunger refugees': Report of the Advisory Committee on its First Session (Geneva, 4–15 August 2008), UN doc A/HRC/10/2, A/HRC/AC/2008/1/2 (3 November 2008) Recommendation 1/6, 15th meeting (15 August 2008).

[32] This was the impression given in author interview with Enele Sopoaga, Secretary for Foreign Affairs, Tuvalu (25 May 2009).

[33] Author interview with Kiribati Solicitor-General David Lambourne (Kiribati, 8 May 2009).

While people in Kiribati and Tuvalu are aware of climate change, for a variety of reasons they are not necessarily worrying about it now. Religion, lack of education, and a culture of 'living for today and not planning for tomorrow' contribute to a certain degree of complacency about environmental change.[34] In Tuvalu, for example, whereas ten years ago any community meeting related to climate change would draw a large crowd, interest has subsided. The explanation given is that some of the doomsday scenarios predicted a decade ago have not eventuated, and people's immediate fears have subsided. Perhaps also because of this time lag, while people could describe recent changes to the environment, weather patterns and local resources – which they attributed to climate change – they also felt that they could adapt to them over time. In addition, a number of people believe that God's promise to Noah that there would be no more floods could be trusted, and that God would not have put people on low-lying atolls if they were not meant to survive (Loughry and McAdam, 2008, p. 51).

Indeed, the empirical evidence suggests that even in the so-called 'sinking islands', a simple 'climate change' cause and effect is not so straightforward, and motivations for movement even less so. While ambiguity or multicausality may complicate the establishment of parameters for dealing with climate-related movement, this is not unique to displacement situations generally, and is a poor reason to overstate the emphasis of climatic factors (which could backfire).[35] For example, disaster literature questions the extent to which 'natural disasters' are unconnected to social, economic and political factors (see e.g. Wisner et al., 2004).

[34] Author interviews with President Anote Tong (Kiribati, 12 May 2009); Betarim Rimon, Office of the President (Kiribati, 12 May 2009); Tebao Awerika, Deputy Secretary, Ministry of Foreign Affairs and Immigration (Kiribati, 12 May 2009); Church leader in Tuvalu, identity withheld by request (26 May 2009).

[35] See Campbell (2010) on the Carteret Islands; Mortreux and Barnett (2009) on Tuvalu. An article published in *New Scientist* (Zukerman, 2010) suggested that the islands of Tuvalu and Kiribati are growing, not disappearing, referring to research by Paul Kench and Arthur Webb. Some media commentators (e.g. Callick, 2010) used the story to suggest that the small island states now had egg on their faces. This is one of the problems with pinning everything on 'climate change', especially when that is not the only factor impacting on movement: scientific research like this can undermine the related claims. By contrast, acknowledging the multicausal nature of movement means that studies like this do not discredit discussions about projected movements, and do not set back research (and policy development) on the issue.

By way of analogy, in the European Union, Art. 15(c) of the Qualification Directive extends subsidiary protection (a watered-down version of refugee protection) to those who face 'a serious and individual threat to a civilian's life or person by reason of indiscriminate violence in situations of international or internal armed conflict'.[36] There has been considerable scholarly debate and divergence in state practice as to whether 'international or internal armed conflict' must be interpreted in accordance with its international humanitarian law meaning, or whether this imposes a layer of analysis that could, if too rigidly applied, divert the focus from the key inquiry: the risk to the applicant and his or her need for protection (compare Storey, 2008; McAdam, 2010*b*; Goodwin-Gill, 2009). Differing interpretations have resulted in particular conflicts being characterized as within the scope of 'international or internal armed conflict' in some EU Member States, but not in others. For example, in France, Bulgaria and the Czech Republic the situation in Iraq has been treated as an 'internal armed conflict', leading to protection under Art. 15(c), whereas the Swedish and Romanian authorities have not viewed it as such, and in Germany there has been inconsistency across the various state jurisdictions (UNHCR, 2007, p. 76; ECRE/ELENA, 2008, p. 215). As Goodwin-Gill (2009, para. 10) observes: 'Given the object and purpose of Art. 15(c) itself (protection from the risk of indiscriminate violence), the qualifying context ought to be one in which IHL may be illustrative, but cannot be determinative.' In other words, protection needs are better realized by leaving aside the intricacies of international humanitarian law and instead focusing on the risk to fundamental human rights occasioned by indiscriminate violence in situations of conflict (see further McAdam, 2010*b*). A parallel argument can be made in the context of assessing risk from climate-aggravated harms: emphasizing climate change as the principal driver may inadvertently narrow an instrument's protective scope.

Refugee law and the 'refugee' label in the Pacific

This section examines whether the 'climate refugee' notion is embraced or eschewed by Pacific islanders to whom it is ascribed. In Kiribati and

---

[36] Council Directive 2004/83/EC of 29 April 2004 on Minimum Standards for the Qualification and Status of Third Country Nationals or Stateless Persons as Refugees or as Persons Who Otherwise Need International Protection and the Content of the Protection Granted [2004] OJ L304/12.

Tuvalu, it is resoundingly rejected at both the official and the personal levels.[37] This is because it is seen as invoking a sense of helplessness and a lack of dignity which contradicts the very strong sense of Pacific pride. Rather than regarding 'refugees' as people with resilience, who have actively fled·situations of violence or conflict, they are seen as passive victims, waiting helplessly in camps, relying on handouts, with no prospects for the future.[38] Some men explain that being described as a 'refugee' would signal a failure on their behalf to provide for and protect their family. Tuvaluans and i-Kiribati people do not want to be seen in this way. When they speak of their own possible movement to countries such as Australia or New Zealand, they describe the importance of being seen as active, valued members of a community who can positively contribute to it.[39]

In part, their discomfort stems from the fact that refugees flee from their own government, whereas the people of Kiribati and Tuvalu have no desire to escape from their countries. They say it is the actions of other states that will ultimately force their movement, not the actions of their own leaders. Indeed, if anything, the persecutor in such cases might be described as the 'international community', and industrialized states in particular – the very states to which movement might be sought if the land becomes unsustainable – whose failure to cut greenhouse gas emissions has led to the predicament now being faced (IPCC, 1990, p. 8; IPCC, 2007, pp. 5, 6, 12, 13). This de-linking of the actor of persecution from the territory from which flight occurs is the opposite to refugee law: it is a complete reversal of the refugee paradigm. Whereas Convention refugees flee their own government (or actors that the government is unable or unwilling to protect them

---

[37] See remarks by Pelenise Alofa Pilitati (chair, Church Education Association, Kiribati) in *Climate Refugees, Australia Talks* (ABC Radio National, 2009). And yet the language persists: *Climate Refugees in Australia 'Inevitable'* (ABC News, 2009).

[38] The Tuvaluan and i-Kiribati languages do not have a single word for 'refugee', because the concept is foreign to their communal cultures: see Etita Morikao (Tuvalu, 25 May 2009), Isala Isala interview (Tuvalu, 27 May 2009). See also the comment by the Maldives President: 'We do not want to leave the Maldives, but we also do not want to be climate refugees living in tents for decades': http://edition.cnn.com/2008/WORLD/asiapcf/11/11/maldives.president/ index.html (citing the President). See also the comments of Kiribati's Foreign Secretary, Tessie Lambourne: 'We are proud people. We would like to relocate on merit and with dignity': cited in Goering (2009). In the specific context of climate change, the President of Kiribati also invoked the language of responsibility: 'When you talk about refugees – climate refugees – you're putting the stigma on the victims, not the offenders': interview with President Anote Tong (Kiribati, 12 May 2009).

[39] This is not unique: many refugees describe similar feelings.

from),[40] a person fleeing the effects of climate change is not escaping their government, but rather is seeking refuge from – yet within – states that have contributed to climate change. This is another reason why focusing on climate change as a driver of movement is misplaced: in refugee law, the refugee is fleeing the persecutor, and an assessment is required as to whether that particular individual is at risk. Thus, identifying the *cause* of flight is imperative to identifying the protection need. By contrast, the purpose of identifying climate change as a driver is not to attribute responsibility for harm, but rather to (presumably) identify a situation of harm from which a person should be protected. As climate change may be but one of a number of factors leading to that situation of vulnerability, focusing solely on it may result in a skewed line of inquiry.

As I have explained in depth elsewhere (McAdam, 2009), international refugee law is a cumbersome tool for trying to address flight from habitat destruction. It was devised for a very different context and will in most cases be an inappropriate framework for addressing environmental displacement.[41] Despite the 'temptation to start with definitions that would be derivative of existing concepts' (Kälin, 2009, p. 1), it does not adequately address the time dimension of pre-emptive and staggered movement (and the fact that in some cases it will be permanent); the maintenance of culture and statehood; or the fact that the juridical aspect of protection by the home state remains forthcoming. In other (mainly non-Pacific) contexts, it may be inappropriate because movement is only internal, and there the *Guiding Principles on Internal Displacement* will be instructive (UN, 1998). Refugee law (and complementary protection) can only be applied for once a person has arrived on the territory of

---

[40] The language of Art. 1A(2) of the Refugee Convention is that the refugee is 'unable or, owing to such fear, is unwilling to avail himself of the protection of that country'. The drafters stated that '"unable" refers primarily to stateless refugees but includes also refugees possessing a nationality who are refused passports or other protection by their own government': see Report of the Ad Hoc Committee on Statelessness and Related Persons (16 January to 16 February 1950) UN doc E/1618 (17 February 1950) 39. A number of domestic courts have also stated that it extends to situations in which the government is either non-existent, ineffective, or colluding with the persecutors: see e.g. *Zalzali* v *Minister of Employment and Immigration* [1991] FCJ No. 341 (Canada).

[41] For example, Refugee Appeal No. 76374, RSAA (28 October 2009) found the applicant to be a Convention refugee on account of the highly politicized nature of disaster relief work in which she was involved in the aftermath of Cyclone Nargis. While her work responded to an environmental disaster, environmental degradation was not the basis of her claim *per se*.

another state. This may encourage spontaneous arrivals rather than planned, gradual movement, and is likely to be a far more traumatic and uncertain experience than facilitating migration over time (especially if, as is likely, many people from Kiribati and Tuvalu will seek entry to Australia and New Zealand regardless).

A related point is that individualized decision-making processes, the conventional way in which decisions on refugee status are made in countries such as Australia and New Zealand, seem highly inappropriate to the situation of climate-induced displacement.[42] This is well illustrated by decisions of the Refugee Status Appeals Authority (RSAA) in New Zealand and the Refugee Review Tribunal (RRT) in Australia, which have expressly rejected refugee claims by people leaving Tuvalu and Kiribati on the grounds of climate change. In a 2000 New Zealand case, the RSAA stated:

> This is not a case where the appellants can be said to be differentially at risk of harm amounting to persecution due to any one of these five grounds. All Tuvalu citizens face the same environmental problems and economic difficulties living in Tuvalu. Rather, the appellants are unfortunate victims, like all other Tuvaluan citizens, of the forces of nature leading to the erosion of coastland and the family property being partially submerged at high tide. As for the shortage of drinkable water and lack of hygienic sewerage systems, medicines and appropriate access to medical facilities, these are also deficiencies in the social services of Tuvalu that apply indiscriminately to all citizens of Tuvalu and cannot be said to be forms of harm directed at the appellants for reason of their civil or political status.[43]

In a 2009 Australian decision, the RRT rejected the i-Kiribati applicant's argument that an 'element of an attitude or motivation' could be adduced from 'the continued production of carbon emissions from Australia, or

---

[42] While nothing in the Convention mandates individualized decision-making, this is the process used in most industrialized countries because it is thought necessary to analyse the subjective and objective elements of the refugee definition. For an excellent analysis of group-based decision-making, see Albert (2010).

[43] Refugee Appeal No. 72189/2000, RSAA (17 August 2000) para. 13. For other NZ cases, see Refugee Appeal No. 72179/2000, RSAA (31 August 2000); Refugee Appeal No. 72185/2000, RSAA (10 August 2000); Refugee Appeal No. 72186/2000, RSAA (10 August 2000); Refugee Appeal No. 72313/2000, RSAA (19 October 2000); Refugee Appeal No. 72314/2000, RSAA (19 October 2000); Refugee Appeal No. 72315/2000, RSAA (19 October 2000); Refugee Appeal No. 72316/2000, RSAA (19 October 2000); Refugee Appeal No. 72719/2001, RSAA (17 September 2001). For other Australian cases, see 0907346 [2009] RRTA 1168 (10 December 2009); N00/34089 [2000] RRTA 1052 (17 November 2000); N95/09386 [1996] RRTA 3191 (7 November 1996); N96/10806 [1996] RRTA 3195 (7 November 1996); N99/30231 [2000] RRTA 17 (10 January 2000); V94/02840 [1995] RRTA 2383 (23 October 1995).

indeed other high emitting countries'.[44] It also found that while there were 'many potential social groups of which the applicant is a member, the absence of the element of motivation means that persecution cannot be said to be occurring for reasons of membership of any such group'.[45]

If and when states recognize that it is no longer possible for people to continue to live in their traditional homelands, then it would be misplaced, in my view, to require individuals to reach a destination country and show that they meet a particular definition. Rather, as has been the case with schemes such as temporary protection in the EU,[46] group determination in the Netherlands,[47] temporary protected status in the United States,[48] and ad hoc visa regimes in Australia responding to particular crises (Timor-Leste, Kosovo, China),[49] it would seem more appropriate for states to designate particular countries as demonstrating sufficient, objective characteristics that 'justify' movement, thereby obviating the need for people wishing to leave them to show specific reasons why climate change is personally affecting them.[50] *Prima facie* refugee status is similarly predicated on the fact that a person has fled a particular country (generally in conflict), and is deemed on that purely objective evidence to have a protection need.[51]

In the present context, such an approach would enable a holistic assessment of the multiple drivers of movement which, together, render

---

[44] 0907346 [2009] RRTA 1168 (10 December 2009) para. 51.   [45] Ibid., para. 52.

[46] Council Directive 2001/55/EC of 20 July 2001 on Minimum Standards for Giving Temporary Protection in the Event of a Mass Influx of Displaced Persons and on Measures Promoting a Balance of Efforts between Member States in Receiving Such Persons and Bearing the Consequences thereof [2001] OJ L212/12.

[47] The Netherlands had a policy of group protection for specific categories of asylum seekers from countries where there was a situation of indiscriminate and generalized violence. However, on 11 December 2009, the Cabinet adopted a proposal by the State Secretary for Justice to end it, on the basis that it was leading to abuse. The decision was strongly criticized by refugee advocates.

[48] INA § 244, 8 USC § 1254.

[49] See e.g. Migration Regulations 1994 (Cth), Schedule 2, Subclass 448 Kosovar Safe Haven (Temporary).

[50] Others have envisaged a similar mechanism: see e.g. Biermann and Boas (2007; 2008); Australian proposal, op. cit. However, they do so within an international treaty framework, which is not what I am proposing here. Furthermore, they see this mechanism as applicable globally to climate-related displacement, whereas here I refer only to small island states from which movement will, ultimately, need to be permanent.

[51] See Albert (2010). The Migration (Climate Refugees) Amendment Bill 2007, proposed by the Australian Greens (discussed below), suggested a mechanism whereby an individual application for a 'climate change refugee visa' would trigger a requirement for the Minister for Immigration to make a declaration about the 'climate change circumstance' on which the application was based, thus creating a visa pathway for others similarly affected.

a state unsafe for continued habitation. It would avoid the individual examination of claims, thus providing a more efficient process.[52] Although necessarily ad hoc, as it permits states to themselves determine if and when they think such assistance is required, it has the flexibility to respond to particular needs as they arise and avoids the problems of trying to reach international agreement on a treaty (discussed below). To help guide state decision-making about whom to assist, and to avoid haphazardness in the exercise of discretion (see e.g. Bianchi, 2002), consolidating existing law applicable to the movement of people into a single soft law instrument, similar to the *Guiding Principles on Internal Displacement*, might provide states with useful guidance as to the kinds of considerations that might underpin such determinations. Over time, this may facilitate the implementation of such norms into domestic law, or in informing new multilateral instruments.[53]

However, all this still assumes that a protection-like paradigm is appropriate. While it may become so in the absence of other action, in my view it is preferable to first work with affected governments to try to reach other solutions involving a combination of *in situ* adaptation and migration, with the acknowledgement that planned movement is an adaptation strategy. This must be considered within a human rights framework, however. Adaptation cannot occur at all costs – at a bare minimum, it must be adaptation with dignity.[54] In the Pacific island context, the development of labour, education and family migration pathways are better attuned to:

- the desires of people in those countries;
- the likely patterns of climate change on the environment (slow and gradual) and patterns of movement (pre-emptive and gradual, rather than in response to a sudden catastrophic event); and
- the history of movement in the region.

---

[52] For example, in situations where large groups are seriously affected by the outbreak of uncontrolled communal violence, 'it would appear wrong in principle to limit the concept of persecution to measures immediately identifiable as direct and individual': Goodwin-Gill and McAdam (2007, p. 129). See the reference there in fn 364 to *R v Secretary of State for the Home Department, ex parte Jeyakumaran* (No. CO/290/84, QBD, unreported, 28 June 1985). See also Hathaway (1991, pp. 91–92).

[53] See e.g. African Union Convention for the Protection and Assistance of Internally Displaced Persons in Africa (adopted 22 October 2009, not yet in force) Art. 5(4) (Kampala Convention). See also Betts (2010, p. 209).

[54] See e.g. author's interview with Rizwana Hasan, Chief Executive of Bangladesh Environmental Lawyers Association (Dhaka, 16 June 2010).

Migration

The long-term strategy of the government of Kiribati is to secure 'merits-based migration' options to Australia and New Zealand, so that those who want to move have an early opportunity to do so.[55] In this way, the President hopes that 'pockets' of i-Kiribati communities will build up abroad. This would see the gradual, transitional resettlement of i-Kiribati in other countries, so that if and when the whole population has to move, there would be existing communities and extended family networks which those left behind could join. The president hopes that in this way, i-Kiribati culture and traditions will be kept alive, but that his people will also be able to slowly adapt to new cultures and ways of life. By contrast, while Tuvalu is seeking to focus its efforts on adaptation so that people can remain at home, officials also noted the general trend towards securing additional migration pathways for Tuvaluans.[56]

Australia and New Zealand take different approaches to Pacific immigration. New Zealand has long had special concessionary schemes for citizenship or permanent residence, of which the 2002 Pacific Access Category (discussed below) is the most recent. The rationale behind New Zealand's concessionary policies is to promote economic development in Pacific island states, although its original impetus came from a post-war period of industrial expansion. According to Stahl and Appleyard (2007, p. iv), such an approach is unique among developed states. By contrast, Australia maintains a 'non-discriminatory' policy that does not (formally) privilege any national group (ibid., p. v),[57] perhaps as a reaction to its White Australia policy of past and prior exploitation of Pacific labour, such as through 'blackbirding' (forced recruitment). However,

---

[55] Interview with President; see also Goering (2009) referring to remarks by i-Kiribati officials at Copenhagen.

[56] Author interviews with Church leader in Tuvalu, identity withheld by request (26 May 2009); The Rt Hon Sir Kamuta Latasi, Speaker of Parliament (Tuvalu, 27 May 2009); Tito Isala, Secretary Supernumerary, Office of the Prime Minister (Tuvalu, 22 May 2009), although the latter two suggested that this was part of an ongoing historical process of migration from Tuvalu. Other officials described migration as being an option at the back of the government's mind, with adaptation at the forefront: Kelesoma Saloa, Prime Minister's Private Secretary (Tuvalu, 25 May 2009); Enele Sopoaga, Secretary for Foreign Affairs (Tuvalu, 25 May 2009).

[57] Author interview with Sir Kamuta Latasi, Speaker of Parliament (Tuvalu, 27 May 2009), referring to discussions in the mid-1990s with then Australian Prime Minister, Paul Keating.

since 2007 AusAID has funded the Kiribati–Australia Nursing Initiative (KANI), which offers around thirty young i-Kiribati the opportunity to train as nurses at Griffith University in Queensland, and, if successful, remain in Australia. In 2009, Australia implemented a three-year Pacific Seasonal Workers Pilot Scheme, modelled in part on New Zealand's Recognised Seasonal Employer (RSE) scheme.[58] Over three years, up to 2,500 visas will be granted to people from Kiribati, Tonga, Vanuatu and Papua New Guinea, to work in the Australian horticultural industry for between six to seven months in each twelve-month period.[59] To date, no i-Kiribati have been part of the scheme (Ball, 2009). Anecdotal evidence suggests that the programme has been less successful than anticipated, not least because of a lack of job offers from Australian farmers who seem reluctant to provide labour conditions that can be avoided by relying on 'black market' labour.

Strategically, Australia and New Zealand would benefit from a more cooperative approach to migration, especially since many Pacific islanders view movement to New Zealand as the first step towards ultimately reaching Australia: once they obtain New Zealand citizenship,[60] they can freely travel to and work in Australia.[61]

In 2002, New Zealand created a visa called the Pacific Access Category, which was based on an existing scheme for Samoans and replaced previous work schemes and visa waiver schemes for people from Tuvalu, Kiribati and Tonga (Dalziel, 2001). This visa has mistakenly been hailed as an immigration response to people at risk of climate-induced displacement in the Pacific, both in media and academic circles.[62] Although the scheme was extended to citizens of Tuvalu after a plea from that country's government for special immigration assistance

---

[58] See e.g. Ramasamy et al. (2008). The RSE scheme was introduced in April 2007 and currently includes Kiribati, Samoa, Tonga, Tuvalu, Vanuatu and, shortly, Solomon Islands: see http://www.dol.govt.nz/initiatives/strategy/rse/index.asp (accessed 19 December 2009).

[59] See http://www.deewr.gov.au/Employment/Programs/PSWPS/Pages/default.aspx (accessed 12 December 2009).

[60] For people who obtained permanent residence post-21 April 2005, the waiting period is five years. Before that time, it is three years: http://www.teara.govt.nz/en/citizenship/3.

[61] 1973 Trans-Tasman Travel Arrangement; see also the Special Category Visa (SCV) for New Zealand citizens since 1994. See generally http://www.immi.gov.au/media/fact-sheets/17nz.htm.

[62] For example, it is relied upon in Boano et al. (2008) citing Gemenne (2006). See also Corlett's critique (2008). It appears that the misunderstanding was perpetuated by Al Gore's film, An Inconvenient Truth.

to enable some of its citizens to relocate, it is a traditional migration programme rather than one framed with international protection needs in mind.[63]

The scheme permits an annual quota of seventy-five citizens each from Tuvalu and Kiribati and 250 each from Tonga (and previously Fiji), plus their partners and dependent children, to settle in New Zealand.[64] Eligibility is restricted to applicants between the ages of 18 and 45, who have a job offer in New Zealand, meet a minimum income requirement and have a minimum level of English. Selection is by ballot. The programme is well known in Tuvalu and Kiribati: almost every person interviewed referred to and welcomed it, although they noted that some improvements could be made.[65]

Although New Zealand does not formally have any humanitarian visas relating to climate change and displacement, it is developing a general policy on environmental migration. In 2008, in light of the fact that '[t]he perceived problem posed by the potential for environmental migrants – and the perceived need for action – has gained traction within several Pacific island countries',[66] it revised its approach away from simply 'correct[ing] misperceptions about New Zealand's position on the environmental migrants issue [i.e. no agreement to resettle people from Tuvalu], while outlining New Zealand's current commitment to climate change adaptation efforts in the Pacific region'.[67] Instead, its focus is now to:

---

[63] Interestingly, programmes like this may ultimately be the basis on which veiled assistance is afforded to those at risk of climate-induced displacement, as this may be politically more palatable than an explicit scheme to address the issue.

[64] http://www.immigration.govt.nz/migrant/stream/live/pacificaccess/ (accessed 8 February 2007).

[65] People interviewed commented on difficulties in securing a job offer in New Zealand, and the fact that eligibility is only assessed after the ballot has been drawn. Although I did not encounter this view in my own interviews, one community leader reportedly condemned the scheme as a new type of 'slavery immigration', whereby educated Tuvaluans renounce stable, white-collar government employment at home to end up as cleaners or fruit-pickers in New Zealand: quoted in Shen (2007, pp. 18, 19).

[66] New Zealand Ministry of Foreign Affairs and Trade, Climate Change and the Issue of Environmental Migrants: A Proposed Revised Approach (8 August 2008) (document circulated to the Prime Minister, the Minister Responsible for Climate Change Issues, the Minister for the Environment, and the Minister of Immigration, released pursuant to an Official Information Act request), p. 5.

[67] Background: Environmental Migrants/Relocation/Displacement, New Zealand Government Poznan Delegation Brief for UNFCCC COP14, 343 (released pursuant to an Official Information Act request).

- acknowledge the concerns of Pacific island countries in relation to this issue;
- stress that current climate change efforts in the Pacific should continue to focus on adaptation, and should be underpinned by the desire of Pacific peoples to continue to live in their own countries; and
- reaffirm that New Zealand has a proven history of providing assistance where needed in the Pacific, and that its approach to environmentally displaced persons would be consistent with this.[68]

This includes a commitment to 'respond to climatic disasters in the Pacific and manage changes as they arise'.[69]

In contrast, the Australian Government does not have a policy on environmental migration. This is despite a proposal made by the Labor government, when in opposition, to create a Pacific Rim coalition to accept climate change 'refugees', and to lobby the United Nations to 'ensure appropriate recognition of climate change refugees in existing conventions, or through the establishment of a new convention on climate change refugees' (ALP, 2006, p. 10). The gap between rhetoric and action was evident, however, when in June 2007 Greens Senator Kerry Nettle proposed the Migration (Climate Refugees) Amendment Bill 2007 to create a new visa class for people fleeing 'a disaster that results from both incremental and rapid ecological and climatic change and disruption, that includes sea level rise, coastal erosion, desertification, collapsing ecosystems, freshwater contamination, more frequent occurrence of extreme weather events such as cyclones, tornadoes, flooding and drought and that means inhabitants are unable to lead safe or sustainable lives in their immediate environment'. The Labor Party was quick to note that its idea of an international response meant that without a collaborative approach with other countries, adopting such an obligation would be a unilateral act and therefore inconsistent with its idea of international action.[70]

---

[68] Ibid., p. 344.

[69] Ibid., p. 343. The President of Kiribati has noted that so far, the country most receptive to his plea for more migration has been Timor-Leste: see remarks quoted in Morton (2009). This accords with comments made by the President, Dr Jose Ramos-Horta, at the Diplomacy Training Programme 20th Anniversary Public Lecture (Faculty of Law, University of New South Wales, 23 July 2009) http://tv.unsw.edu.au/video/dr-jose-ramos-horta-dtp-20th-anniversary-public-lecture (accessed 14 December 2009).

[70] See debates following Second Reading Speech in Parliament of Australia, Senate: Official Hansard (9 August 2007), p. 95ff.

Since forming government in November 2007, the Labor Party has not acted on its earlier ideas. A recent Senate inquiry revealed that '[w]hen asked about the possibility of forced re-location from Pacific island countries such as Kiribati and Tuvalu, the Department of Foreign Affairs and Trade (DFAT) informed the committee that it was not aware of any government consideration of this matter. Invited to comment again on whether these two islands were under consideration, DFAT replied no.'[71] Drawing on submissions by the present author, the Committee recommended 'that the Australian Government consider whether it may be necessary to review the legal and policy framework required in the event that regional communities may be forced to resettle as a consequence of changes in climate'.[72] It expressed its concern:

> about the lack of government attention to formulating policy around the possibility that some Pacific island communities may have to re-locate because of rising sea levels or related environmental changes. The committee believes that the Australian Government should allow ample time to consider closely and carefully the legal and policy framework that may be required should such an eventuality arise. The committee believes that Australia could also make a valuable and significant contribution in practical ways to prepare those most at risk of having to resettle. It notes that the Government of Kiribati wants their people to be competitive and marketable. Australia could be a vital partner with countries such as Kiribati by helping with research, training, education and labour mobility arrangements to equip people, should they have to move, to take up productive positions in their new location. It believes that should migration be necessary from these Pacific island countries, the basic principle underpinning the formulation of Australia's policy should be their 'migration with merit and dignity'.[73]

Finally, it may be instructive to situate present discussions about movement within the longer history of mobility in the South Pacific. As Silverman has observed in the context of Pacific relocations, even if a particular movement 'seem[s] unique in the history of a single group, we

---

[71] Senate Foreign Affairs, Defence and Trade References Committee, Economic Challenges facing Papua New Guinea and the Island States of the Southwest Pacific (Commonwealth of Australia, Canberra, November 2009), para. 6.60 (making reference to Committee Hansard, 21 November 2008, p. 28). On 21 November 2009, a spokesperson for the Climate Change Minister, Penny Wong, was reported as acknowledging that permanent migration may eventually be the only option for some people, which will need to be dealt with by governments in the region: Morton (2009).

[72] Senate Foreign Affairs, Defence and Trade References Committee, op. cit., para. 6.62 (Recommendation 3).

[73] Ibid., para. 6.61, referring to UN News Centre (2008).

might find them to be recurrent as we enlarge the scale of analysis to a colonial system or a regional mobility system' (Silverman, 1977, p. 7). While the driver of 'climate change' may be new (or at least labelling it such), the types of movement under consideration are not unfamiliar in the Pacific context. Viewing migration as one of a range of adaptation tools, and one that is frequently utilized by Pacific islanders generally, helps to place current policy debates within a specific framework and may also help to diffuse some of the more sensational approaches that are at times invoked.

Given that most climate impacts in the Pacific will be slow-onset, interim migration measures that permit temporary and circular movement, on the understanding that a permanent migration outcome will ultimately be possible once relocation is imperative, may appeal to affected and receiving countries alike.[74] In this way, a small but sustained migration response may enable communities to remain living in their homes for longer, with certain members of the household working temporarily abroad to generate income that is fed back into the home community (and to assist with adaptation), new diaspora communities forming, and receiving states adapting over time. It is important that any such migration is reinforced by local adaptation mechanisms, as the migration of skilled workers may further deplete local human resources (although they may make a significant economic contribution through remittances, thereby increasing family resilience for those who remain) (Pelling and Uitto, 2001, p. 49). Such an approach builds on the historical migration patterns between Pacific countries and New Zealand. Going into the future, migration schemes might be constructively developed as part of broader bilateral partnerships, such as New Zealand's five-year Strengthened Cooperation Programme with Niue from 2004 to 2009, and through enhancing regional cooperation agreements, such as those adopted at the Pacific Islands Forum.

The overarching aim should be to avoid the protection discourse needing to be engaged at all, by developing other methods for movement that give more choice to i-Kiribati and Tuvaluans about if and when they wish to move. Paradoxically, however, the protection system may have to be resorted to if no action is taken in the interim to secure safe and early migration options for those who wish to move. The principle of *non-*

---

[74] It may also be more palatable for governments to absorb some migrants in traditional labour categories, rather than to acknowledge the drivers behind the movement. Part of the challenge is to 'sell' the solution domestically, within both the country of origin and the host country.

refoulement (non-return) in human rights law precludes states from returning people to places where they would face a substantial risk of torture or cruel, inhuman or degrading treatment, or arbitrary deprivation of life (see further McAdam and Saul, 2010; McAdam, 2007). Although the House of Lords in the United Kingdom has acknowledged that, in theory, *any* sufficiently serious human rights violation could give rise to such an obligation[75] – a proposition that remains open to testing in the courts – current practice suggests that the accepted limits of the principle of *non-refoulement* are relatively narrowly circumscribed and the threshold for demonstrating 'inhuman or degrading treatment' is high. While it is far from clear-cut that complementary protection would ever assist a person displaced for reasons of climate change, the jurisprudence is constantly evolving. Indeed, human rights treaties are generally viewed as 'living instrument[s]' that 'must be interpreted in the light of present-day conditions'.[76] With this in mind, the European Court of Human Rights has reclassified acts that in the past were regarded as 'merely' inhuman or degrading treatment as amounting to the higher threshold of 'torture'.[77] Similarly, the Inter-American Court of Human Rights held that the distinctions and gradations of treatment are not rigid, but rather evolve with increased protection of fundamental rights.[78] Ironically, then, a wait-and-see approach with respect to movement from Kiribati and Tuvalu could ultimately stimulate a dynamic interpretation of human rights law so as to provide a remedy for people whose homes have become uninhabitable. This, in turn, may create a precedent for accepting people from other affected states (with much larger populations, such as Bangladesh).

On the other hand, it is sobering to recall that despite existing international treaty obligations, the courts have limited the protection capability

---

[75] *R v Special Adjudicator ex parte Ullah* [2004] UKHL 26, paras 24–5 (Lord Bingham), 49–50 (Lord Steyn), 67 (Lord Carswell).

[76] *Tyrer v United Kingdom* (1979–80) 2 EHRR 1, para. 31; see also *Soering v United Kingdom* (1999) 11 EHRR 439, para. 102. The House of Lords described the Refugee Convention in this way: *Sepet and Bulbul v Secretary of State for the Home Department* [2003] UKHL 15, [2003] 1 WLR 856, para. 6 (Lord Bingham); see also *Applicant A v Minister for Immigration and Ethnic Affairs* (1997) 190 CLR 225, 293 (Kirby J); *Suresh v Canada (Minister of Citizenship and Immigration)* [2002] 1 SCR 3, para. 87.

[77] *Selmouni v France* (1999) 29 EHRR 403. See also *Henaf v France*, App. No. 65436/01 (European Court of Human Rights, 27 November 2003) para. 55: 'it follows that certain acts previously falling outside the scope of Article 3 might in future attain the required level of severity'.

[78] *Cantoral-Benavides v Peru*, Series C No. 69, Judgment of 18 August 2000.

of human rights law when it comes to deprivation of socio-economic rights –
for thinly disguised policy concerns about 'opening the floodgates'. In *N v
United Kingdom*, the European Court of Human Rights held that in removal
cases concerning a person 'afflicted with any serious, naturally occurring
physical or mental illness which may cause suffering, pain and reduced life
expectancy and require specialised medical treatment which may not be so
readily available in the applicant's country of origin or which may be
available only at substantial cost',[79] the *non-refoulement* aspect of Art. 3 of
the European Convention on Human Rights (protection from torture or
inhuman or degrading treatment or punishment) will only be triggered in
highly exceptional circumstances, such as if 'the applicant was critically ill
and appeared to be close to death, could not be guaranteed any nursing or
medical care in his country of origin and had no family there willing or able
to care for him or provide him with even a basic level of food, shelter or
social support'.[80] The court observed that while many ECHR rights 'have
implications of a social or economic nature', the instrument 'is essentially
directed at the protection of civil and political rights'.[81] It continued:

> Furthermore, inherent in the whole of the Convention is a search for a
> fair balance between the demands of the general interest of the commu-
> nity and the requirements of the protection of the individual's funda-
> mental rights (see *Soering* v. *The United Kingdom*, judgment of 7 July
> 1989, Series A no. 161, § 89). Advances in medical science, together with
> social and economic differences between countries, entail that the level of
> treatment available in the Contracting State and the country of origin
> may vary considerably. While it is necessary, given the fundamental
> importance of Article 3 in the Convention system, for the Court to retain
> a degree of flexibility to prevent expulsion in very exceptional cases,
> Article 3 does not place an obligation on the Contracting State to alleviate
> such disparities through the provision of free and unlimited health care
> to all aliens without a right to stay within its jurisdiction. A finding to the
> contrary would place too great a burden on the Contracting States.[82]

In dissent, Judges Tulkens, Bonello and Spielmann highlighted the policy
considerations that influenced the majority's approach, noting that its
rationale:

---

[79] *N v United Kingdom*, App. No. 26565/05 (Grand Chamber, 27 May 2008), para. 45.
[80] Ibid., para. 42. This was the case in *D v United Kingdom* (1997) 24 EHRR 423, the only
'health' case in which an applicant has succeeded before the European Court of Human
Rights.
[81] Ibid., para. 44.   [82] Ibid.

... reflects the real concern that they had in mind: if the applicant were allowed to remain in the United Kingdom to benefit from the care that her survival requires, then the resources of the State would be over-stretched. Such a consideration runs counter to the absolute nature of Article 3 of the Convention and the very nature of the rights guaranteed by the Convention that would be completely negated if their enjoyment were to be restricted on the basis of policy considerations such as budget-ary constraints. So does the implicit acceptance by the majority of the allegation that finding a breach of Article 3 in the present case would open up the floodgates to medical immigration and make Europe vulner-able to becoming the 'sick-bay' of the world. A glance at the Court's Rule 39 statistics concerning the United Kingdom shows that, when one compares the total number of requests received (and those refused and accepted) as against the number of HIV cases, the so-called 'floodgate' argument is totally misconceived.[83]

Conclusion: which way forward?

The concerns expressed above by i-Kiribati and Tuvaluans highlight some of the central failures of the international protection system. Most notably, their fear about languishing in camps is a real one, given that an absence of political will to implement the principle of burden-sharing is currently leaving millions of refugees in protracted situations with no hope of durable solutions.[84] Indeed, this is a key pragmatic argument against the creation of a new protection treaty. Given the legal obligations that states already have towards Convention refugees, and the fact that some 10 million refugees today, not to mention other displaced people numbering some 34 million in total (UNHCR, 2008, Table 23), have no durable solution in sight, why would states be willing to commit to, and realize protection for, people displaced by climate change?[85]

In responding to the first of these questions, it could be argued that states might be prepared to adopt such an instrument precisely to call for shared responsibility. For example, an individual state might perceive a need to respond to potential arrivals of 'climate refugees', but be unwill-ing to unilaterally create legal avenues for their protection. Were it to elicit the support of other states in adopting a treaty, however, then its

---

[83] Ibid., Dissenting judgment, para. 8 (omitting fn citing statistics on such cases).

[84] On burden-sharing, see Hurwitz (2009, pp. 138–71).

[85] UNHCR, among others, argues that there is a risk that if the Refugee Convention is opened up for renegotiation, we could see a reduction in protection overall: see quotes in Grant et al. (2009). However, this could be avoided by creating a Protocol rather than renegotiating the existing treaty text.

humanitarian impulse could be coupled with mutual self-interest, in that it can call on other states to share the responsibility of caring for such people.[86] However, it is in response to the second question, why states would be willing to *realize* such protection, that real difficulties arise. As we see with the present refugee regime, problems of implementation – and durable solutions – stem predominantly from a lack of political will, rather than an absence of law. Despite the 147 States Parties to the Refugee Convention and/or Protocol, the plethora of soft law relating to refugees, and an international agency with a strong field as well as institutional presence, the displacement of millions remains unresolved.

A treaty is sometimes posited as the answer to climate-related displacement, but it is dangerous to see it in this way. Any treaty is necessarily an instrument of compromise, and even once achieved there needs to be political will of individual states to ratify, implement and enforce it. While international law provides important benchmarks and standards to regulate state action, they must be supported by political will and action to be fully effective. As Aleinikoff (2007, p. 476) argues, 'there can be no monolithic approach to migration management. Some areas might well benefit from norms adopted by way of an international convention; guiding principles might work best for areas in which a consensus is further away.'

Perhaps part of the problem is the disciplinary constraints of international law and international relations. At their very core lies the objective to universalize – to create norms that take the 'particular' to a level of general applicability, that make individual rights 'human rights' at one and the same time. The danger, of course, is that if this is done without sufficient empirical understandings or foresight, we arrive at a level of generality that is too vague, and which cannot be translated into practical, rational policies and normative frameworks. It is clear that legal gaps exist,[87] but they should be first addressed by a dispassionate, careful appraisal of the empirical evidence, rather than motivated by an assumption that existing frameworks should be extended. There is a risk of (prematurely) concentrating the diverse impacts of climate change on human movement into calls for treaties and the like. The local and the particular do not always speak well to an international law or governance

---

[86] That said, some of the states that host the largest numbers of refugees are not parties to the Refugee Convention or Protocol.

[87] Although as Kälin (2010) notes, perhaps fewer than some believe, given that a lot of movement will be internal.

agenda, where the 'cascading' effect requires broad, universalizing statements.

On the other hand, international law retains sufficient flexibility to respond to particular scenarios through bilateral and regional agreements. In my view, this is where attention would best be focused initially. At this stage, it seems more probable that the development of regional soft-law declarations, such as the Niue Declaration on Climate Change,[88] will provide a more effective springboard for developing responses, than will a new international instrument aiming to take into account the interests of all states in a wide variety of contexts. At the normative level, we already have clear frameworks to guide such actions – the human rights law regime is the most relevant and important.

For these reasons, this chapter should not be interpreted as rejecting a treaty-based regime altogether, or the underlying basis of such a regime: that states ought to provide assistance to certain people who are unable to remain in their homes. International cooperation on climate-related movement is sorely needed.[89] Rather, my purpose is to caution against squeezing all forms of apparently forced movement into a protection paradigm, as this may not best address the patterns or needs of those who move.[90] Responses might better be achieved by focusing on states' burden-sharing obligations to each other, and their responsibility to the international community as a whole. Of course, this sidesteps the much larger issue of whether the maintenance of a privileged legal status for certain categories of displaced people is ethically and/or legally defensible,[91] a matter that is beyond the scope of this chapter.

---

[88] Annex B to Forum Communiqué, 39th Pacific Islands Forum, Alofi, Niue (19–20 August 2008) Doc PIFS(08)6.

[89] The UN High Commissioner for Refugees suggests that: 'A development-oriented approach is now required in response to displacement, emphasizing the inclusion of the most vulnerable and marginalized sections of society in efforts to ensure that they benefit from the livelihoods, services and security to which they are entitled' (Guterres, 2009).

[90] As the UN High Commissioner for Refugees has noted, given that most displacement is predicted to be internal, primary legal responsibility for ensuring people's rights will lie with the states concerned (Guterres, 2009). For a recent assessment of the nature of movement as internal, see Laczko and Aghazarm (2009).

[91] See e.g. the recent work of Matthew Price, who argues that limiting asylum to people who face persecution is ethically justifiable: Price (2009); Hathaway (2007); cf. Shacknove (1985, p. 274).

Finally, from an advocacy perspective, lobbying for a 'climate refugee' treaty may successfully generate attention and mobilize civil society such that the issue of climate-related movement becomes one that
states cannot ignore. Policy itself may be generated because of the
lobbying process, and having the maximalist option of a treaty on
the table may paradoxically encourage states at least to negotiate
more minimalist responses, as a compromise or fallback position.
Nevertheless, it is imperative that advocacy is well-informed, because
if there is an absence of rigorous analysis and empirical evidence to
support claims being made (see e.g. Christian Aid, 2007), it will not
achieve its ends. Indeed, messy work may lead to a backlash and
attempts to discredit the phenomenon of climate-related movement
altogether.

References

ABC News. 2009. *Climate Refugees in Australia 'Inevitable'*. 11 December. http://
www.abc.net.au/news/stories/2009/12/11/2769403.htm?site=news (Accessed
14 December 2009.)

ABC Radio National. 2009. Pelenise Alofa Pilitati (chair, Church Education
Association, Kiribati) in *Climate Refugees, Australia Talks*, 3 August. http://
www.abc.net.au/rn/australiatalks/stories/2009/2641241.htm (Accessed 14
December 2009.)

Albert, M. 2010. Governance and *prima facie* refugee status determination: clarifying
the boundaries of temporary protection, group determination, and mass influx,
Refugee Survey Quarterly, Vol. 29, no. 1, pp. 61–91.

Aleinikoff, T. A. 2007. International legal norms on migration: substance without
architecture. In: R. Cholewinski, R. Perrechoud and E. MacDonald (eds),
International Migration Law: Developing Paradigms and Key Challenges.
The Hague, Netherlands, TMC Asser Press.

ALP. 2006. *Our Drowning Neighbours: Labor's Policy Discussion Paper on
Climate Change in the Pacific*. Canberra, Australian Labor Party.

Annan, K. 2009. *The Anatomy of a Silent Crisis*. Human Impact Report: Climate
Change. Geneva, Global Humanitarian Forum. http://ghfgeneva.org/
Portals/0/pdfs/human_impact_report.pdf (Accessed 7 December 2009.)

Ball, R. 2009. *The Pacific Seasonal Workers Pilot Scheme (PSWPS) and Implications
for Pacific Development*. http://peb.anu.edu.au/pdf/2009/PNG/ppp/
Rochelle_Ball_Pacific_Update_081209.pdf (Accessed 18 December 2009.)

BBC News. 2009. *Maldives Cabinet Makes a Splash*. 17 October. http://news.bbc.
co.uk/1/hi/8311838.stm (Accessed 10 December 2009.)

Betts, A. 2010. Towards a soft law framework for the protection of vulnerable irregular migrants. *International Journal of Refugee Law*, Vol. 22, No. 2, pp. 209–36.

Bianchi, A. 2002. Ad-hocism and the rule of law. *European Journal of International Law*, Vol. 13, No. 1, pp. 263–72.

Biermann F. and Boas, I. 2007. *Preparing for a Warmer World: Towards a Global Governance System to Protect Climate Refugees.* (Global Governance Working Paper 33.)

2008. Protecting climate refugees: the case for a global protocol. *Environment*, November–December.

Boano, C., Zetter, R. and Morris, T. 2008. *Environmentally Displaced People: Understanding the Linkages between Environmental Change, Livelihoods and Forced Migration.* University of Oxford, UK, Refugee Studies Centre. (Forced Migration Policy Briefing 1.)

Bone, J. and Pagnamenta, R. 2009. We are sinking, say islanders, but there is still time to save the world. *The Times*, 23 September. www.timesonline.co.uk/tol/news/environment/article6845261.ece (Accessed 10 December 2009.)

Borger, J. 2008. Conflicts fuelled by climate change causing new refugee crisis, warns UN. *The Guardian*, 17 June. http://www.guardian.co.uk/environment/2008/jun/17/climatechange.food (Accessed 2 December 2009.)

Callick, R. 2009. Don't desert us, say sinking Pacific islands. *The Australian*, 30 July. www.theaustralian.com.au/news/dont-desert-us-say-sinking-pacific-islands/story-0-1225756097220 (Accessed 10 December 2009.)

Callick, R. 2010. Coral islands left high and dry. *The Australian*, 11 June. http://www.theaustralian.com.au/news/features/coral-islands-left-high-and-dry/story-e6frg6z6-1225878132101

Campbell, J. 2010. Climate-induced community relocation in the Pacific: the meaning and importance of land. In: McAdam (ed.), op. cit., pp. 37–80.

Castles, S. 2002. *Environmental Change and Forced Migration: Making Sense of the Debate.* Geneva, United Nations High Commissioner for Refugees. (*New Issues in Refugee Research*, No. 70.)

Christian Aid. 2007. *Human Tide: The Real Migration Crisis.* May, London. http://www.christianaid.org.uk/Images/human-tide.pdf

CNA. 2007. *National Security and the Threat of Climate Change.* Alexandria, Va., CNA Corporation.

Commonwealth of Australia. 2009. *Defending Australia in the Asia Pacific Century: Force 2030.* (Defence White Paper.)

Copenhagen Accord. 2009. Adopted 18 December. http://unfccc.int/files/meetings/cop_15/application/pdf/cop15_cph_auv.pdf (Accessed 19 January 2010.)

Corlett, D. 2008. Tuvalunacy, or the real thing? *Inside Story*, 27 November. http://inside.org.au/tuvalunacy-or-the-real-thing/print/ (Accessed 27 November 2008.)

Council of the European Union. 2008. Climate Change and International Security. Report from the Commission and the Secretary-General/High Representative to the European Council, 3 March. (Doc No 7249/08.)

CRIDEAU/CRDP. 2008. Draft Convention on the International Status of Environmentally-Displaced Persons. Faculty of Law and Economic Science, University of Limoges. *Revue européenne de droit de l'environnement*, No. 4, p. 375.

Dalziel, L. 2001. Government Announces Pacific Access Scheme. 20 December. http://www.beehive.govt.nz/node/12740 (Accessed 8 December 2008.)

Deng, F. 1998. *Guiding Principles on Internal Displacement*, 11 February. (E/CN.4/1998/53/Add.2.)

Docherty, B. and Giannini, T. 2009. Confronting a rising tide: a proposal for a convention on climate change refugees. *Harvard Environmental Law Review*, Vol. 33, No. 2, pp. 349–403.

ECRE/ELENA. 2008. *The Impact of the EU Qualification Directive on International Protection*. October. Brussels, European Council on Refugees and Exiles/European Legal Network on Asylum.

European Parliament. 2009. Prime Minister Apisai Ielemia says climate change threatens Tuvalu's survival. Press Release, 10 December. http://www.europarl.europa.eu/news/expert/infopress_page/028-66101-341-12-50-903-20091207IPR66100-07-12-2009-2009-false/default_fn.htm (Accessed 13 December 2009.)

Fitzpatrick, J. 2000. Temporary protection of refugees: elements of a formalized regime. *American Journal of International Law*, Vol. 94, No. 2, pp. 279–306.

Gemenne, F. 2006. *Climate Change and Forced Displacements: Towards a Global Environmental Responsibility? The Case of the Small Island Developing States (SIDS) in the South Pacific Ocean*. Université de Liège, Les Cahiers du CEDEM. http://www.cedem.ulg.ac.be/m/cdc/12.pdf (Accessed 8 December 2008.)

Goering, L. 2009. Kiribati officials plan for 'practical and rational' exodus from atolls. *Reuters AlertNet*, 9 December. http://www.alertnet.org/db/an_art/60714/2009/11/9–181804–1.htm (Accessed 13 December 2009.)

Goodwin-Gill, G. S. 2009. *Challenges to the Protection of Refugees and Stateless Persons: Compliance with International Law*. Asylum Law Seminar, Blackstone Chambers, London, 31 March. http://www.blackstonechambers.com/applications/dynamic/papers.rm?barrister_id=461&id=374&x=48&y=25 (Accessed 9 December 2009.)

Goodwin-Gill, G. S. and McAdam, J. 2007. *The Refugee in International Law*, 3rd edn. Oxford, UK, Oxford University Press.

Grant, H., Randerson, J. and Vidal, J. 2009. UK should open borders to climate refugees, says Bangladeshi Minister. *The Guardian*, 4 December. http://www.guardian.co.uk/environment/2009/nov/30/rich-west-climate-change/print (Accessed 8 December 2009.)

Guterres, A. 2009. Bracing for the flood. *New York Times*, Op-Ed, 10 December. http://www.nytimes.com/2009/12/11/opinion/11iht-edguterres.html?_r=1&emc=eta1 (Accessed 14 December 2009.)

Hathaway, J. C. 1991. *The Law of Refugee Status*. Toronto, Butterworths.

2007. Forced migration studies: could we agree just to 'date'? *Journal of Refugee Studies*, Vol. 20, No. 3, pp. 349–69.

Hulme, M. 2008. Commentary: Climate refugees: cause for a new agreement? *Environment*, Vol. 50, No. 6, pp. 50–51.

Hurwitz, A. 2009. *The Collective Responsibility of States to Protect Refugees*. Oxford, UK, Oxford University Press.

IPCC. 1990. *Climate Change: The IPCC Scientific Assessment: Final Report of Working Group I*. New York, Cambridge University Press.

2007. *Climate Change 2007: Synthesis Report – Summary for Policymakers*. Intergovernmental Panel on Climate Change.

Kälin, W. 2009. *Climate Change, Migration Patterns and the Law*. Keynote Address at the International Association of Refugee Law Judges 8th World Conference, Cape Town, 28 January.

2010. Conceptualising climate-induced displacement. In: McAdam (ed.), op. cit., pp. 81–104.

Khan, I. 2009. *The Unheard Truth: Poverty and Human Rights*. New York, W. W. Norton and Co.

Kniveton, D., Schmidt-Verkerk, K., Smith, C. and Black, R. 2008. *Climate Change and Migration: Improving Methodologies to Estimate Flows*. Geneva, International Organization for Migration. (Migration Research Series No. 33.) www.iom.int/jahia/webdav/site/myjahiasite/shared/shared/mainsite/published_docs/serial_publications/MRS-33.pdf

Laczko, F. and Aghazarm, C. (eds). 2009. *Migration, Environment and Climate Change: Assessing the Evidence*. Geneva, International Organization for Migration.

Lateu, J. 2008. That sinking feeling: climate refugees receive funds to leave islands. *New Internationalist*, March. www.newint.org/columns/currents/2008/03/01/climate-change/ (Accessed 10 December 2009.)

Lawton, C. 2009. What about climate refugees? Efforts to help the displaced bog down in Copenhagen. *Spiegel Online*, 17 December. http://www.spiegel.de/international/europe/0,1518,druck-667256,00.html (Accessed 19 January 2010.)

Loughry, M. and McAdam, J. 2008. Kiribati – relocation and adaptation. *Forced Migration Review: Climate Change and Displacement*. No. 31, October, pp. 51–53.

MacFarquhar, N. 2009. Refugees join list of climate-change issues. *New York Times*, 29 May. http://www.nytimes.com/2009/05/29/world/29refugees.html?_r=1&pagewanted=print (Accessed 1 December 2009.)

McAdam, J. 2007. *Complementary Protection in International Refugee Law*. Oxford, UK, Oxford University Press.

2009. Review essay: From economic refugees to climate refugees? *Melbourne Journal of International Law*, Vol. 10, No. 2, pp. 579–95.

(ed.). 2010a. *Climate Change and Displacement: Multidisciplinary Perspectives*. Oxford, UK, Hart Publishing.

2010b. 'Disappearing states', statelessness and the boundaries of international law. In: McAdam (ed.), op. cit., pp. 105–30.

2010c. The impact of the standard of proof on complementary protection claims: comparative approaches in Europe and North America. In: J. C. Simeon (ed.), *Critical Issues in International Refugee Law: Strategies for Interpretative Harmony*. Cambridge, UK, Cambridge University Press.

McAdam, J. and Loughry, M. 2009. We aren't refugees. *Inside Story*, 30 June. http://inside.org.au/we-arent-refugees/

McAdam, J. and Saul, B. 2010. An insecure climate for human security? Climate-induced displacement and international law. In: A. Edwards and C. Ferstman (eds), *Human Security and Non-Citizens: Law, Policy and International Affairs*. Cambridge, UK, Cambridge University Press.

Morton, A. 2009. Land of the rising sea. *Sydney Morning Herald*, 21 November. http://www.smh.com.au/environment/land-of-the-rising-sea-20091120-iqub. html (Accessed 27 November 2009.)

Mortreux, C. and Barnett, J. 2009. Climate change and adaptation in Funafuti, Tuvalu. *Global Environmental Change*, Vol. 19, No. 1, pp. 105–12.

Pacific Island Report. 2009. *Carteret Islanders Become First Climate Refugees: PNG Relocates Families as Island Home Disappears*, 4 May. http://archives. pireport.org/archive/2009/may/05-04-09.htm

Pelling, M. and Uitto, J. I. 2001. Small island developing states: natural disaster vulnerability and global change. *Environmental Hazards*, Vol. 3, No. 2, pp. 49–62.

Price, M. E. 2009. *Rethinking Asylum: History, Purpose, and Limits*. Cambridge, UK, Cambridge University Press.

Ramasamy, S., Krishnan, V., Bedford, R. and Bedford, C. 2008. The recognised seasonal employer policy: seeking the elusive triple wins for development through international migration. *Pacific Economic Bulletin*, Vol. 23, No. 3, pp. 171–86. http://peb.anu.edu.au/pdf/PEB_23_3_Ramasamy%20et%20al_WEB.pdf (Accessed 18 December 2009.)

Rudd, Prime Minister Kevin. 2008. *The First National Security Statement to the Australian Parliament*, 4 December. http://www.theaustralian.news. com.au/files/security.pdf (Accessed 11 December 2008.)

Schwartz, P. and Randall, D. 2003. *An Abrupt Climate Change Scenario and Its Implications for United States National Security*. Report commissioned for the US Department of Defense. San Francisco, Calif., Global Business Network.

Shacknove, A. 1985. Who is a Refugee? *Ethics*, Vol. 95, No. 2, pp. 274–84.

Shamsuddoha, Md and Chowdhury, R. K. 2009. *Climate Change Induced Forced Migrants: In Need of Dignified Recognition under a New Protocol.* http://www.equitybd.org/English/Press%20040409/English%20Position%20paper.pdf (Accessed 10 November 2009.)

Shen, S. 2007. Noah's Ark to save drowning Tuvalu. *Just Change: Critical Thinking on Global Issues*, No. 10, pp. 18–19.

Sherborne, C. 2009. Sinking sandbanks. *The Monthly*, March. http://www.themonthly.com.au/node/1472 (Accessed 12 September 2009.)

Silverman, M. G. 1977. Introduction: Locating Relocation in Oceania. In: M. D. Lieber (ed.), *Exiles and Migrants in Oceania*. Honolulu, University of Hawaii.

Stahl, C. W. and Appleyard, R. T. 2007. *Migration and Development in the Pacific Islands: Lessons from the New Zealand Experience.* AusAID web-based report. http://www. ausaid.gov.au/ publications/pdf/migration.pdf

Storey, H. 2008. EU Refugee Qualification Directive: A Brave New World? *International Journal of Refugee Law*, Vol. 20, No. 1, pp. 1–49.

UN News Centre. 2008. *Small Island Nations' Survival Threatened by Climate Change, UN Hears*, 25 September. http://www.un.org/apps/news/story.asp?NewsID=28265 (Accessed 29 April 2009.)

UNHCR. 2007. *Asylum in the European Union: A Study of the Implementation of the Qualification Directive.* Brussels, United Nations High Commissioner for Refugees.

2008. *Total Population of Concern to UNHCR: Refugees, Asylum-Seekers, IDPs, Returnees, Stateless Persons, and Others of Concern to UNHCR by Country/ Territory of Asylum, End-2008.* United Nations High Commissioner for Refugees. http://www.unhcr.org/pages/4a0174156.html

Vidal, J. 2009. Migration is the only escape from rising tides of climate change in Bangladesh. *The Guardian*, 4 December. http://www.guardian.co.uk/environment/2009/dec/04/bangladesh-climate-refugees/print (Accessed 8 December 2009.)

Williams, A. 2008. Turning the tide: recognizing climate change refugees in international law. *Law & Policy*, Vol. 30, No. 4, pp. 502–29.

Wisner, B., Blaikie, P., Cannon, T. and Davis, I. 2004. *At Risk: Natural Hazards, People's Vulnerability and Disasters*, 2nd edn. London, Routledge.

Zukerman, W. 2010. Shape-shifting islands defy sea-level rise. *New Scientist*, Vol. 206, No. 2763, 2 June, p. 10.

6

Critical views on the relationship between climate change and migration: some insights from the experience of Bangladesh

ALLAN FINDLAY AND ALISTAIR GEDDES

Introduction

Climate change is perceived by many to be a driver of future population mobility. One example of this surfaced at the 2009 Copenhagen Climate Summit when a Bangladeshi Government minister stated: 'Twenty million people could be displaced [in Bangladesh] by the middle of the century. We are asking all our development partners to honour the natural right of persons to migrate. We can't accommodate all these people' (Grant, 2009). Bangladesh's suggestions were upheld by others at the summit, including the chair of the International Panel on Climate Change as well as representatives from the global South, especially those concerned about the effects of sea level rise on vulnerable coastal populations. Governments of wealthier nations were less supportive. Responding to the Bangladeshi view, the UK's International Development Minister instead encouraged a greater focus on supporting local climate change adaptation. The response continued: 'It's absolutely legitimate for Bangladesh and the Maldives to make a lot of noise about the very real risk of climate migration – they hope it will make us come to their rescue. But reopening the 1951 Convention (on refugees) would certainly result in a tightening of its protections' (Grant, 2009). Funding local adaptation to climate change rather than redefining the UN definition of 'refugee' to accommodate 'environmental refugees' was therefore posed as the solution. This brief cameo relating to Bangladesh provides the basis for the terrain that we wish to examine in more detail in this chapter in relation to climate change, population mobility and 'environmental refugees'.

On the one hand, the concept of 'environmental refugees' has been accepted as part of the popular vocabulary associated with discussions of environmental change. On the other hand, terms such as 'environmental refugees', 'climate refugee' and 'environmental migration' used to describe the relationship between climate change and human mobility have discomfited many social scientists. At least four difficulties have been identified with such terms. A first difficulty is that they elevate climate-related environmental change to a position where it is regarded as the primary factor driving human movements in certain parts of the world. Second, reading population mobility as an inability to adapt to climate change is simplistic. Third, the extent to which movement is treated as an 'option' for all who might be affected by climate change is clearly also contentious. Finally, and in some ways most seriously, the lack of sound empirical support or verification for estimates of the numbers of affected persons is especially problematic, given that such estimates appear to have been accepted by many policy-makers. The most notorious example of such acceptance concerns the environmental refugee estimates made by the influential environmentalist Norman Myers (1993a). In spite of having been strongly critiqued elsewhere (Black, 2001), Myers' figures have been quoted both in the Stern report to the UK Treasury (Stern, 2007) and by many non-governmental organizations.

The objections that have been raised on these fronts indicate the difficulties of understanding human mobility in the context of contemporary climate change. Without clarity on who moves and why in response to changing environmental factors, social researchers have also eschewed refining estimates of the numbers of affected people, even where conjoint maps of climate change impacts and population distributions have been produced (Warner et al., 2009b). Instead migration researchers have sought to understand environment–movement linkages in relation to a number of different typologies (Renaud et al., 2007). For example, environmental events have been conceptualized as part of a spectrum of processes ranging from rapid-onset changes ('hazards' and 'disasters') to more gradual processes of resource depletion. Similarly, distinctions have been made between forced movements (displacement) and voluntary movements, between short- and long-distance (international) movements (Hugo, 1996; Bates, 2002; Stojanov, 2004), and between migration and 'circulation' (Kliot, 2004). These typologies begin to capture something of the complexity of the interactions that may occur at the population–environment interface.

Following on from these concerns, this chapter plots the widening usage of the term 'environmental refugee' based on recent bibliometric data. It then considers why the term has become widespread when there remains little empirical evidence to support claims of rising numbers of environmental refugees. This is followed by a short synthesis comparing views of scientists, who tend to promote the term in relation to the primacy they give to physical processes in their understanding of the world, with the arguments of some social scientists who recognize that all forms of mobility are multicausal and that at best climate change is just one force driving population mobility. Reflections on the authors' field research in Bangladesh are provided to underscore the point that an understanding of the drivers of *specific* migration systems is perhaps the best approach to making suggestions about future trends in environmentally induced mobility. The last main section contrasts top-down policy perspectives on 'environmental refugees' with the approach of political ecology (Zimmerer, 2007) that would give greater support to bottom-up responses to environmental conflicts.

What's in a name? A bibliometric exploration of the term 'environmental refugee'

Bibliometric data confirm, in a convenient if admittedly relatively crude way,[1] the growth in use of the term 'environmental refugee' over the last quarter century. The term can be traced to the work of Lester Brown of the Worldwatch Institute as early as the 1970s as well as to El-Hinnawi's (1985) research in Kenya (Jacobson, 1988). It was not until the 1990s, however, that political and academic use of the term mushroomed (Figure 6.1). This appears to have followed publication of estimates of the global scale of potential environmental refugee flows.

The frequencies shown in Figure 6.1 were obtained using Google Scholar to measure use of the term 'environmental refugee' in journal papers (search completed 4–10 February 2010). Books and other non-journal publications were excluded from the search. Fifty-one journal papers referred to environmental refugees in 1990–91, rising to eighty-eight and 167 in 2002–03 and 2008–09 respectively. The citation analysis shows that the term is most widely found in the journals classed in the literature as social science, rather than those in physical sciences where

---

[1] There are of course many pitfalls in using bibliometric data in assessing research (see e.g. Richards et al., 2009).

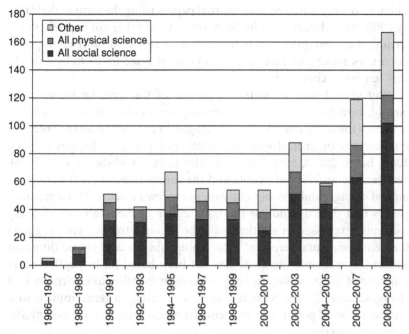

Figure 6.1 Number of academic journal papers citing the term 'environmental refugee(s)'

climate and environmental change scientists might have been expected to report their research. Some might find this pattern surprising since, as noted above, it is social scientists who have been the most critical of the term, suggesting alternatives such as 'climate change induced mobility' and 'environmentally induced population displacements'.[2] It appears from Figure 6.1 that criticism of the term 'environmental refugee' has been ineffective not only in curtailing its general use but also in limiting its use within the social science community.

Figure 6.1 counts the overall number of citations. However more probing bibliometric analysis was also conducted, considering whether the term 'environmental refugees' was used in the title of a journal article or as a keyword. Results from this narrower search carried out using Thomson Reuters ISI (Web of Knowledge) suggest only modest numbers of papers in which the concept formed the central focus of research.

---

[2] The title for a 2008 Population-Environment Research Network cyberseminar: http://www.populationenvironmentresearch.org/seminars082008.jsp.

There were only nineteen such journal papers using the term in their title by 2000–01 and thirty-one by 2008–09. It would therefore appear that a term that is used quite widely by academics and even more so in policy discourses about climate change, has been analysed in depth by remarkably few researchers.

Why should such a contrast in usage of the term be found? One possible interpretation is that while many authors have been comfortable to deploy the term 'environmental refugee' in expanding on the potential implications of critical environmental and political changes (Myers, 2002; Bates, 2002), very few academics have established a causal link between the mobility of people and the forces responsible for environmental change, and consequently there are few evidence-based research papers that directly address the issue either in their titles or as a key concepts. Expressed in another way, the bibliometric evidence suggests that 'environmental refugees' is an ambiguous concept whose diffusion within the academic literature has not been based on in-depth understanding of the processes linking population mobility to environmental changes. Instead its success has been as a catch-all term applied to a diverse range of population movements whose origins have potentially been very diverse.

Given that the evidence appears to be weak supporting use of the term, the question might be asked whether the conceptual underpinnings of environmental refugee estimates are consistent with the understanding of those who specialize in migration research. The substance of most research in the 1980s and 1990s, including the work of the International Geographical Union's Commission on Population and the Environment, is that complex causalities link population and environmental change processes (Dutton et al., 1998). Population mobility is certainly not an inevitable outcome of environmental change. The social constructions of 'environment' and 'nature' have received sustained attention more recently, resulting in fresh significance being attached to investigating 'context' (place, space and time) as central to understanding population–environment and nature–society relations (Whatmore, 2002; Bailey, 2005). Geographers have also argued that the relations between population mobility and environmental change need to be approached in terms of linkages across global to local scales (Carr, 2005; Neumann, 2009). In short, the answer to the question posed above would seem to be a clear 'no'. Large-scale international population displacements are not an inevitable result of global warming or any other environmental change.

What does population mobility in places experiencing rapid environmental change tell us?

A different perspective on current understanding of the links between climate change and migration may be gained by turning to the work of those who have engaged directly in research on human mobility in contexts where significant rapid environmental change has either been occurring, or is anticipated to occur. An extensive review of the emerging literature has been offered by Piguet (2008) and Warner et al. (2009a), the latter drawing on field studies conducted under the European Commission-funded Environmental Change and Forced Migration Scenarios (EACH-FOR) project. Our intention here is not to repeat the work of these researchers, but rather to highlight five key features that seem to emerge from the recent literature.

First, research underscores the wide spectrum of mobility 'events' that exist, ranging from short-term displacements stimulated by natural hazards such as floods and droughts (Basset and Turner, 2007) through to longer-term displacements associated with the breakdown of 'livelihoods dependent on ecosystem services, such as agriculture, herding and fishing' (Warner et al., 2009b, p. 3). Examples of the need for permanent relocation of populations can be found in those locations threatened by sea level rise, as is the case in certain island communities and densely populated river deltas such as the Mekong, Ganges and Nile deltas (Laczko and Aghazarm, 2009). Our Bangladeshi case study provides more details on one country faced with this challenge. Among those studying human mobility in these contexts, there has been a tendency to refer to such movements as 'environmentally induced mobility' (Wood, 2001) or (following the International Organization for Migration position) 'climate change migrants' (Laczko and Aghazarm, 2009). IOM researchers have been particularly eager to consider migration as one of a range of adaptation strategies for populations facing environmental change, rather than portraying mobility as a negative outcome and evidence of a failure to adapt.

A second point found in many investigations is that environmentally induced mobility usually occurs over short distances. Reviewing the then state of empirical research more than a decade ago, Hugo concluded that 'the bulk of migrants displaced by environmental disruption move within national boundaries and international migration has been very limited', although not without speculating 'there are a number of arguments which could be advanced that over the next decade environmental

factors will become more significant in impelling international movements' (Hugo, 1996, p. 118). Ironically, at the end of the first decade of
the new millennium this still has not happened and the situation remains
the same (Hugo, 2008), with empirical research pointing only to short-
distance displacements and to very little international mobility linked to
climate change, but with ongoing concerns that the near future will be
different (Renaud et al., 2007; Pachauri, 2009).

We began this chapter with a recent example of this continued dis-
juncture, in terms of the debate at the 2009 Copenhagen Climate Summit
about strategies to cope with the anticipated future mobility of 20 million
Bangladeshis faced with losing their homes and livelihoods as a result of
sea level rise. The proposal of the Bangladeshi Government that Western
economies should help to accommodate these people seems unlikely to
succeed. It seems very improbable that international flows of this magnitude will occur, for at least two reasons: displaced people remain
unlikely in the first instance to seek settlement outside their own country; and possible host countries are not willing to offer settlement
opportunities to these groups, although they might finance livelihood
adaptation strategies to allow vulnerable populations to relocate within
their country. There is certainly no evidence of any sympathy for the
view that the International Convention on Refugees should be changed
to accommodate environmentally linked migration (Black, 2001;
Conisbee and Simms, 2003).

A third point is recognition that environmental change is just one of
many drivers of mobility. To quote Henry et al. (2004, p. 417), 'unfavourable environmental conditions are unlikely to be the "unique" cause
pushing people to migrate'. Indeed most migration research suggests
that the environment has the greatest impact on mobility where there are
deeper underlying causes associated with political, economic, sociological and demographic processes (Hugo, 2008). This is particularly true of
slow-onset environmental change where, for example, gradual drought-
related land degradation does not impact suddenly, and where livelihood
spaces deteriorate in the long term. In such contexts the mediating role
of 'social system characteristics including social networks' (Warner et al.,
2009b) is essential in understanding the decisions taken in a community
as to whether household members should leave to seek new livelihoods
or remain to participate in strategies geared to climate change
adaptation.

Fourth, migration research suggests that where population mobility is
linked to environmental change, it is more likely to be influenced by

slow-acting processes such as land degradation that contribute to a wider pressure on communities to consider alternative livelihood options, than by sudden-onset events that are perceived to be periodic and that lead only to short-term displacement (Henry et al., 2004). Henry et al., working on mobility trends in Burkino Faso, also make the important distinction between a first migration move and subsequent moves. They point out that attachment to a place of first residence is stronger than to other places, and that once a population has become mobile (perhaps because of underlying conditions associated with poverty), then they are less likely to tolerate unfavourable environmental conditions in a subsequent place of residence. Mobile populations searching for a new location are therefore more likely to select a place where economic, social and environmental circumstances are favourable to longer-term settlement.

The fifth and final point that we wish to draw from the migration literature is that environmentalists have given insufficient attention to pre-existing mobility patterns. Building on the above point, Kniveton et al. (2008) have explored the environmental change–migration nexus in situations where high levels of mobility are already prevalent. It appears that in areas where propensities for out-migration are low, even the impact of severe environmental events may not result in any significant long-term population mobility. In contrast, where a mobility culture predates environmental disruption, awareness through social networks of other livelihood options may prompt other outcomes. Migration research findings therefore tend to bear out the importance attached to social networks in either favouring a community's desire to remain *in situ* and to seek to adapt to climate change or to consider other options (Warner et al., 2009b). This supports the view that more research attention should be given to the contexts (local and regional, social and cultural) in which people affected by environmental change make decisions about their future mobility (Kniveton et al., 2008).

Some observations from fieldwork in Bangladesh

We now turn to a case study based on personal insights from the first author's research experience in Bangladesh – part of the globe generally perceived to be acutely affected by contemporary climate change processes (Nicholls and Mimura, 1998; Mimura et al., 1998[3]). The peoples of

---

[3] Repeated in the report of Working Group II: *Impacts, Adaptation and Vulnerability of the IPCC Third Assessment Report: Climate Change 2001.*

Bangladesh have a long history of living with flooding. Coastal areas have been affected by cyclones and tidal surges while riverine flooding of agricultural lands along the Padma, Meghna and Jamuna rivers is an expected part of the annual monsoon cycle (Nicholls et al., 2007).[4]

Of course 'living with flooding' is an over-optimistic way to describe the scale of population and other losses experienced both at the time of, and after, many flood events. Among the most devastating human impacts in recent history were recorded in 1970 when some 300,000–500,000 Bangladeshis died in the south of the country as a result of coastal flooding. In 1991 another 140,000 died after a cyclone hit the Bengal coast, and in November 2007 cyclone SIDR resulted in a further 3,700 deaths. In addition, sea level is estimated to have risen by 20 cm over the twentieth century, while there are predictions that a further sea level rise of 45 cm may result in the loss of 10.9% of Bangladesh's land area if adequate protective measures are not created (Tickell, 1991; Myers, 1993b). The scale of impacts indicated by such figures has led Piguet (2008, p. 6) to conclude that Bangladesh may indeed be a rare example of a settled nation-state in which environmental change may be reaching a point where it drives significant human movement. While this claim should be treated carefully, given all that is noted in the preceding section, Bangladesh nevertheless presents a compelling if not urgent case in which environment-population mobility linkages may be studied, and addressed.

Research with Bangladeshi communities impacted by riverine flooding has included field visits, undertaken in 1996 and again in 2005. The focus of these visits was three communities in different parts of the country: a village in north-western Bangladesh, outside the town of Rangpur; another village in central Bangladesh, near Ramshankarpur, and a third village in eastern Bangladesh, near Moulivibazaar. Fieldwork in all villages consisted of focus groups and meetings with community leaders, exploring the experiences and consequences of flooding in general, and the impact of the major monsoon-related floods of 2004 in particular. Altogether, about eighty people engaged in these meetings, as part of which community leaders were asked about how members of their community had been displaced, about people who had chosen to leave (migrants), and also about the duration (short or longer term) of such movements. Fieldwork also included interviews with non-governmental organizations having responsibilities for

---

[4] The Padma and Jamuna may be better known by the names given to them as they flow through neighbouring India – the Ganges and Brahmaputra respectively.

emergency relief work, as well as with members of the Bangladeshi national Disaster Emergency Response committee. The 2004 monsoonal flooding inundated thirty-nine of Bangladesh's sixty-four districts. At its peak the flood covered 35,000 km^2, home to 36 million people (about 25% of all Bangladeshis), and made 10 million homeless. Although the area affected was one of the largest ever recorded, the death toll of 800 persons was considered low by comparison with the human loss in previous and later floods (1988, 2007).

In 2004 the monsoonal rains were particularly intense in late June. Flooding started in the Brahmaputra basin in north-east India as well as smaller rivers flowing from the highlands of Meghalaya into Bangladesh. Continued rainfall at the time when these flood waters were entering Bangladesh accentuated the flood problems, initially causing inundations of farmland and settled areas along the Jamuna and Meghna rivers as well as in north-east Bangladesh in Sylhet and neighbouring districts. A second wave of flooding peaked in late July with flooding spreading into central Bangladesh as rivers burst their embankments and flooded huge areas of the country including parts of Dhaka and later some southwestern districts. After two months of flooding came a third downpour in mid-September when Bangladesh experienced the heaviest period of sustained rainfall for fifty years. Waterlogging behind protective embankments in the later stages of the flood became severe, with the problems being now most acute in the more central and southerly areas. Neighbouring countries, in trying to resolve their flood problems, released further waters from reservoirs in West Bengal, adding to Bangladesh's plight. Furthermore, in the south-west the build-up of flood waters coincided with high tides, resulting in flooding in Jessore district.

The floods destroyed infrastructure (field boundaries, roads, bridges, river embankments). Long-term effects on the development effort included damage to schools and medical facilities, with 1,200 primary schools washed away altogether. Official estimates suggest that the floods caused the third greatest level of economic damage at over US$2.2 trillion (Care Bangladesh, 2005).

Whereas the three village areas where research was conducted are in different locations, all were flooded in 2004, with field research providing clear indications of the severity of impacts. Flood water was of course not the only problem experienced by the villagers. The flood waters also carried huge quantities of sand and debris, burying farmland and laying to waste large tracts for the foreseeable future. In addition, farmers' seeds and fish from ponds were lost, threatening future harvests, and the long duration of the floods meant that the disruption to the agricultural year was much greater than usual.

The majority of those impacted by the floods were landless labourers whose normal livelihoods had been completely lost over the sustained period during which flood waters inundated the lands on which they normally sought work and for several months thereafter. This inability to earn a livelihood to meet basic needs for themselves and their families was compounded by flood-borne pollution, contributing to increased incidence of illness and contamination of tube-wells and other water and food supplies. It was at this stage that food stocks reached their all-time low. Interviewees noted that it was still several months before the next harvest could bring them either work or cheaper food. In northern Bangladesh this season of hunger is known as *monga* and regularly equates with a period of widespread malnutrition and hunger. The impact of the flood on the local population proved to be greater in areas of high food insecurity rather than necessarily being found in the locations that were inundated to the greatest depth or for the longest period. Thus it was longer-term impacts of the floods on people's ability to feed themselves that proved to be the most important consideration.

The site visits provided evidence of how people in different parts of Bangladesh thought about mobility in distinct ways in relation to their awareness (or not) of mobility opportunities and relative to the specificity of the livelihood structures in which they were enmeshed. In northern Bangladesh many knew of people who had left to seek opportunities to live in India (especially north-east India where some had tribal links), but many of these migrants had returned disappointed or having been evicted by the Indian authorities. In the east of Bangladesh, people were more likely to know of migrants from Sylhet province that had previously emigrated to the UK, not as a result of environmental change but in relation to a multitude of other socio-economic drivers. An international move such as this was, however, perceived to be very expensive and was not considered a short-term option at a time of flooding. Most rural dwellers simply could not afford to contemplate such a move, even where they were part of a local network with strong links to Sylhetis living abroad. Indeed most villagers seemed determined to stick with their rural livelihood, seeing flooding as a long-established part of their livelihood system and culture, and perceiving temporary local displacement as a well-established alternative coping mechanism.

Unlike the north-west of Bangladesh, the village sites studied in the east included many villagers who had been engaged in small-scale development projects prior to the flooding. Some of these projects had been destroyed, with for example some micro-credit schemes threatened through the loss of uninsured assets. In this area the researcher found that interviewees were more acutely aware of their losses, often having had to sell what little they had left (e.g. surviving poultry or livestock) simply to be able to feed their families. These interviewees were especially conscious of the effects of the flooding in undermining their attempts to move towards greater self-sufficiency. Equally the existence of a local community network, that was set on re-establishing small-scale livelihood projects and adapting them to the probability of future extreme flooding, had resulted in most respondents seeing their village economy as being their best chance of long-term future security. Of course many people had fled the countryside during the worst of the floods, seeking shelter with relatives, or moving to official camps in the towns and cities. Some had explored the possibility of shifting permanently to a subsistence lifestyle in the insecure urban sector of the Dhaka economy, but it was reported that most of those who had been displaced had preferred to return to their homes and communities as soon as possible. Parallel research by Paul (2005) in villages affected by coastal cyclones provides evidence of villagers from another part of Bangladesh returning to their homes soon after the hazard had passed, in preference to settling in other locations.

The field research findings reported above are broadly in line with a parallel small-scale research project conducted by Poncelet (2009). Commenting on the culture of mobility in relation to the monsoon, she observes on the one hand that 'few Bangladeshis can say they have never moved', as most rural populations have at some time been displaced by flood events, yet on the other hand, she notes that 'environmental degradation is not currently considered to be a major cause for migration in Bangladesh' (Poncelet, 2009, p. 18). From her research, based on interviews with rural migrants living in Dhaka after the floods in 2004 (and more recently in 2007), Poncelet found that some had moved not to escape the floods directly, but rather to avoid moneylenders who had been eager to reclaim loans that they had taken out often at exorbitant rates during the period of extended flooding. Yet for the poorest migrants to Dhaka, these extra loans had often been the only means to buy food, repair their homes and to buy seeds for the next planting season. In other words Poncelet's study suggests that broader socio-economic circumstances, in this case debt, were the key drivers of out-migration from rural areas. According to an NGO representative interviewed by one of the current authors 'the greatest effect [of the floods] was the increase in debt levels of the poorest households' (Care Bangladesh, 2005, p. 17), adding some support for Poncelet's work.

In summary, the Bangladeshi case study illustrates some aspects of how environmental events, that seem to have increased in frequency and severity, impact on rural livelihoods. Significant levels of temporary population displacement in rural areas have been part of the local cultural coping mechanisms developed over centuries in relation to monsoonal cultivation. However, this does not appear to be driving large-scale longer-term and longer-distance mobility, although it may be part of the wider multicausal structure underpinning rural to urban transfers within Bangladesh (especially towards Dhaka and other large cities, see Poncelet, 2009). Field research tends to confirm that it is more important to investigate the *additional* burden that flooding may bring to the large numbers who are already impoverished and lack basic food security. In other words, climate-related environmental change is an extra pressure (along with access to social networks, economic pressures and many other forces) for those households considering the migration of a family member as a possible additional livelihood strategy that could help in their struggle to reduce rural poverty. Moreover, the case study affirms the importance of appreciating locally varying cultures of mobility, with different areas having different knowledge of migration opportunities, and with differing assessments being made regarding the desirability of moving, and of the resources necessary to support a permanent migration move by one or all household members.

Policy implications

This chapter has highlighted the idea that migration is a contingent process rather than being 'environmentally determined'. We have argued that understanding 'context' is essential, and this argument carries over into this section, where we turn to policy issues. If environmental change produced population mobility in the simplistic way proposed by Myers (1993a; 2002), then international policies to accommodate 'environmental refugees' would be a logical follow-up, and it would lend support to Bangladesh's recent request to relocate 20 million of its people to other countries. We have suggested that this is not likely to become a policy option on two grounds. First, we have attempted to illustrate the conceptual weaknesses of Myer's arguments about the links between environmental change and migration (i.e. migration is a contingent process). Second, we would argue that, even among those developed countries that seem willing to provide some funding to the global South to help people adapt to climate change, there is little interest in proposals to accept what are perceived to be low-skill international migrants from these regions. If large-scale international relocation of people as proposed by Myers (1993a) and others is not a viable policy option (at least in the current international political environment), then what policy alternatives exist and how do they relate to the 'contextual' perspectives on migration and environmental change that we have privileged here?

Below we compare and contrast the merits of two different policy approaches – one stemming from use of global climate change modelling and the other based on a cultural ecology perspective. Having questioned above the suggestion that environmental change directly causes long-term population displacement, is there any value in seeking to establish top-down policies to increase local adaptation to climate change in place of out-migration? We argue that there is merit in applying external scientific knowledge of probable future environmental change in a variety of ways, but that such an approach should not be taken in isolation from bottom-up perspectives.

The top-down approach typically starts by assessing the scenarios for climate change produced by the Intergovernmental Panel on Climate Change (IPCC) to assess the risk to populations in different locations (usually at state level) of experiencing significant environmental change. A range of social scientists at Columbia University for example have taken up this approach, assessing population sensitivity to climate change, by recognizing that those directly engaged in primary production for their livelihood will be much more affected than say urban populations who gain a living from secondary and tertiary activities (de Sherbinin et al., 2008). Translated into a Bangladeshi context, the rural livelihoods of populations living on the coasts of the Bay of Bengal and those dependent on agriculture in the most flood-prone parts of the major river systems might be seen as being more much more sensitive to the consequences of climate change than those engaged in factory work or the service sector around Dhaka. Put another way, top-down analysis offers the possibility of using external expert knowledge to pinpoint those parts of the Earth that are at greatest risk of becoming less habitable as a result of environmental change. Identifying these locations is a necessary first step in developing policies to assist those living there and targeting external resources to help them.

The next layer of analysis in a top-down perspective is usually the assessment of the adaptive capacity of populations that are at most risk in these environments (Renaud et al., 2007). This is much more complex than drawing up hazard maps of physical events, as it often requires some measurement of a community's capacity to implement externally defined actions to avoid or reduce future risks (such as hazard and disaster preparedness strategies, livelihood diversification, etc.). Valuable as these external recommendations for action may be, the approach risks ignoring local knowledge of environmental events as well as sidelining local coping strategies. Above all the approach normally treats those populations that are socially constructed as 'being vulnerable' (Findlay, 2005) as no more than passive actors who need to be distanced with external support from the risks posed by environmental change (November, 2004). If out-migration of those at risk is ruled out as a policy option, then this perspective is likely to favour 'distancing' the vulnerable population from the external risk through an 'engineered' solution that requires social changes to behaviour in relation to the 'spatiality of the risk' (November, 2008). Usually the strategy will involve this change being imposed on the 'at risk' group (Nicholls et al., 2007). 'Engineering' may be achieved through introducing externally proposed knowledge of environmental risk events (e.g. in flood preparedness planning) or through social engineering aimed at facilitating the adoption of externally proposed livelihood strategies that 'move' the population away from dependence on those livelihood systems perceived to be particularly susceptible to the uncertainties associated with physical environmental change.

The top-down approach to policy intervention, while valuable, is also problematic as we have begun to indicate. Most case studies have shown, for example, that predicting the effects of environmental change on population is very difficult, and this is supported by migration-specific research (Martin, 2009). Environmental change in Bangladesh, as elsewhere, is best understood as one extra factor that adds to many others that impact on rural poverty, and it is this wider socio-economic context that shapes household and community attitudes to mobility as a possible response to multiple livelihood vulnerabilities. Any top-down thinking about the link between environmental change and mobility needs therefore to be situated relative to an understanding of the local social context. This implies recognizing that those affected by environmental change are not simply passive populations onto which externally defined practices need to be imposed to protect them from 'risk'. They are active participants in the spaces of vulnerability in which they live and their primary concerns relate to their future livelihoods. And this recognition may lead to questioning whether environmental change should be taken as the starting point for policy interventions. Perhaps an examination of livelihood–mobility linkages might be a more promising point of departure for policy-makers concerned with assisting communities to adapt to environmental change.

A final problem of the top-down approach to policy formulation in relation to population/mobility issues is that it can lead to a negative view of out-migration. As Kniveton et al. (2008) have noted, top-down environmentally led strategies have tended to favour policies that discourage or avoid out-migration, seeing population mobility as evidence of failure of local peoples to adapt to climate change in other ways. Fortunately there has been increasing recognition of the fallacy of this position. For example the World Bank recently noted:

> The negative portrayal of migration can foster policies that seek to reduce and control its [migration's] incidence and do little to address the needs of those affected by climate hazards. Indeed policies designed to restrict migration rarely succeed, and increase the costs to migrants and communities of origin and destination
>
> (World Bank, 2010, p. 25).

Thus rather than top-down policies seeking to discourage out-migration in terms of 'holding the line', an alternative perspective could be to see migration of some household members as a constructive strategy offering the opportunity to diversify household incomes and to reduce local vulnerabilities to climate change (a point also recognized by Warner et al., 2009a).

A second policy approach to the environmental change–migration nexus is to start by seeing local populations that are impacted by environmental change as purposive actors, with a key role to play in shaping new futures for themselves and for the 'at risk' areas in which they live. Following Zimmerer (2007) the approach tends to privilege study of the local cultural context as the key to effective policy interventions. As such it seeks to avoid a technocratic approach to problem-solving, instead placing emphasis on understanding the political role of different actors in influencing local social and environmental outcomes (Bryant, 1998; 2001). This perspective therefore recognizes decisions on mobility as being strongly influenced by the collective social context in which they are taken and offers an opportunity to integrate environmental issues with other forces that bear on mobility processes. One example of this approach is Carr's (2005) study of the complex interactions between migration, environment, economy and power in Ghana.

With its focus on the importance of contextual analysis and on the differential power of a range of actors within the community, the political ecology approach has the strength of being able to point to the importance of recognizing different scales of intervention – individual, household and community. There is clearly potential for policies to be nested in a way that seeks to increase resilience to environmental change at each level as well as to recognize that mobility decisions usually relate, at least in the global South, to views of the meaning of migration to the household and community in relation to wider concerns about diversifying livelihoods.

If the political ecology approach appears to be helpful in deepening understanding of how environmental factors relate to other drivers of mobility, it also has its limitations. The approach is better at identifying the problems of externally imposed strategies than it is at identifying how policy-makers should act in relation to the challenges to local communities presented by environmental change. This, plus its local orientation, means that intervention strategies based on political ecology seem less likely to be resourced by national and international policy-makers seeking to implement 'one size fits all' strategies for disaster preparedness and to boost resilience in the face of climate change. This should not however be a reason for dismissing the approach, as effective local policy interventions can often become models of best practice for others to replicate.

Integrating top-down and bottom-up approaches to policy is easy to recommend, but hard to achieve. Practice shows that often it is impossible even to get those who claim to represent the interests of local peoples (such as the various local NGOs operating in an area) to agree on a concerted line of action, never mind to sign up as part of a nationally resourced strategy for action. This was made evident to the authors in their Bangladeshi research with each of the NGOs that they surveyed having a different prescription for how external resources should be targeted on the local communities that they served.

For policies to address the environmental change–migration nexus effectively, they have to recognize the failure of the two extreme strategies proposed in the past. On the one hand, proposing an externally engineered mass international out-migration of populations living in areas at heightened risk of environmental change fails to understand the way in which environmental risks and mobility are linked, as well as failing to acknowledge the scale of opposition of Western governments to such an action. On the other hand, environmental change adaptation strategies that either rule out or fail to consider the possibilities of population mobility (Martin, 2009), ignore the critical role of local populations as purposive actors engaged in a process of livelihood adaptation that might include, in some contexts, the possibility of one or more family members working away from the main family home as a risk reduction measure. Between these unhelpful extremes, there is scope for many fruitful policy interventions that draw on external resources and knowledge of environmental change, but which recognize that risk is not only a statistical probability (such as a hazard map) linked to physical environmental events (November, 2008), but it is also a socially constructed and socially mediated phenomenon. It involves purposive local actors from communities with long-standing experiences of how mobility and immobility relate to environmental change, as well as distinctive ambitions, aspirations and capacities to work towards transforming themselves and the environments in which they live.

Conclusions

Having traced interest in so-called environmental refugees, this chapter questioned the value of work that separates environmentally induced migration from other categories of mobility. Mobility is a multicausal phenomenon and the case study material on Bangladesh illustrated that while a culture of temporary population displacement was considered a normal short-run response to extreme monsoonal flooding, longer-term migration was something that often related more to differential patterns of poverty (particularly in relation to *monga*), access to social networks that could facilitate mobility, and household and communal strategies relating to wider thinking about future livelihood options.

The contribution of new primary research on migration and environmental change, as exemplified by the case study reports of the EACH-FOR project, have provided a helpful step forward in understanding environmental change – migration relations. The research that now exists (de Haas, 2008) seems to point conclusively to the view that environmental change is not causing large-scale mass migration at the levels once predicted by Myers

(1993a) and Tickell (1991). Where environmental displacement has occurred it has usually been short distance and often temporary in nature (Warner et al., 2009a). And where large-scale migration flows are occurring from the global South to the rest of the world they seem to be driven by the huge disparities in wealth that have emerged rather than due to forced environmental flight. However, the continued focus on whether specific environmental changes are producing distinctive mobility patterns runs the risk of missing the very useful contribution from Kniveton et al. (2008) that points to the need to understand whether environmental forces are adding significantly to mobility in those local contexts where there is already a strong mobility culture.

More research is needed to establish whether increasing local awareness of likely future trends in climate change results in an increase or a reduction in the probability of mobility among those affected. Equally there is a need to know whether the capacity of external agencies to change livelihoods through development programmes in areas experiencing rapid environmental change will increase the propensity to stay or will simply augment the likelihood of out-movement, as an indirect effect of increasing the value of people's human capital. Another important question that remains unanswered is whether situations can be identified where out-migration and emigration from an environmentally vulnerable context can be seen to benefit all concerned, rather than being of benefit only to the migrant's household.

Recognizing that environmental impacts on vulnerability and mobility can only be analysed in terms of knowledge that is socially constructed (Bailey, 2005) should not lead to a retreat from a search for policy action as long as the relational knowledge that emerges can be shown to be useful. 'Usefulness' of course must be measured not only in relation to those in power, but also to those who have been 'made' vulnerable (Philo, 2005) by social, economic and cultural processes. A greater awareness of these arguments should lead to policy interventions that are more sensitive to local social contexts and thus to the need to allocate resources to those who are often invisible to conventional metrics of vulnerability (Findlay, 2005). Research that gives more attention to the significance of local contexts cannot guarantee this outcome, but it can point towards solutions that might empower the local community. There is certainly a need for environmentally informed mobility policies that take as their starting point an understanding of local perceptions of appropriate coping strategies, including strategies to move, and to unleash the resources of those that move to the benefit of those that do not.

References

Bailey, A. 2005. *Making Population Geography*. London, Hodder Arnold.

Bassett, T. and Turner, M. 2007. Sudden shift or migratory drift? *Human Ecology*, Vol. 35, No. 1, pp. 33–49.

Bates, D. 2002. Environmental refugees? *Environment and Population*, Vol. 23, No. 5, pp. 465–77.

Black, R. 2001. *Environmental Refugees: Myth or Reality?* Geneva, United Nations High Commissioner for Refugees. (*New Issues in Refugee Research*, No. 34.)

Bryant, R. 1998. Power, knowledge and political ecology in the Third World. *Progress in Physical Geography*, Vol. 22, No. 1, pp. 79–94.

——— 2001. Political ecology: a critical agenda for change? In: N. Castree and D. Braun (eds), *Social Nature*. Oxford, UK, Blackwell, pp. 151–69.

Care Bangladesh. 2005. *Flood 2004: Response and Learning*. Dhaka, Care.

Carr, E. R. 2005. Placing the environment in migration: environment, economy, and power in Ghana's Central Region. *Environment and Planning A*, Vol. 37, No. 5, pp. 925–46.

Clarke, J. and Noin, D. (eds). 1998. *Population and Environment in Arid Regions*. Paris, UNESCO/Parthenon Press, pp. 281–303.

Conisbee, R. and Simms, A. 2003. *Environmental Refugees: The Case for Recognition*. London, New Economics Foundation.

de Haas, H. 2008. The myth of invasion: the inconvenient realities of African migration to Europe. *Third World Quarterly*, Vol. 29, No. 7, pp. 1305–22.

de Sherbinin, A., VanWey, L., McSweeney, K., Aggarwal, R., Barbieri, A., Henry, S., Hunter, L., Twine, W. and Walker, R. 2008. Household demographics, livelihoods and the environment. *Global Environmental Change*, Vol. 18, No. 1, pp. 38–53.

Dutton R. W., Clarke, J. I. and Battikhi, A. (eds). 1998. *Arid Land Resources and their Management*. London, Kegan Paul.

El-Hinnawi, E. 1985. *Environmental Refugees*. Nairobi, UN Environmental Programme.

Findlay, A. 2005. Vulnerable spatialities. *Population, Space and Place*, Vol. 11, No. 6, pp. 429–40.

Grant, H. 2009. UK should open its doors to climate refugees says Bangladeshi minister. *The Guardian*, 4 December. http://www.guardian.co.uk/environment/2009/nov/30/rich-west-climate-change

Henry, S., Piché, V., Ouédraogo, D. and Lambin, E. 2004. Descriptive analysis of the individual migratory pathways according to environmental typologies. *Population and Environment*, Vol. 25, No. 5, pp. 397–422.

Hugo, G. 1996. Environmental concerns and international migration. *International Migration Review*, Vol. 30, No. 1, pp. 105–31.

2008. *Migration, Development and Environment.* Paper submitted to PERN Cyberseminar. http://www.populationenvironmentresearch.org

Jacobson, J. 1988. *Environmental Refugees: A Yardstick of Habitability.* Washington DC, World Watch Institute. (World Watch Paper 86.)

Kliot, N. 2004. Environmentally induced population movements: their complex sources and consequences. In: J. D. Unruh, M. S. Krol and K. Nurit (eds), *Environmental Change and Its Implications for Population Migration.* Dordrecht, Netherlands, Kluwer Academic Publishers.

Kniveton, D., Schmidt-Verkerk, K., Smith, C. and Black, R. 2008. *Climate Change and Migration: Improving Methodologies to Estimate Flows.* Geneva, International Organization for Migration. (Migration Research Series No. 33.) www.iom. int/jahia/webdav/site/myjahiasite/shared/shared/mainsite/published_docs/ serial_publications/MRS-33.pdf

Laczko, F. and Aghazarm, C. (eds). 2009. *Migration, Environment and Climate Change: Assessing the Evidence.* Geneva, International Organization for Migration.

Martin, S. 2009. Managing environmentally induced migration. In: Laczko and Aghazarm (eds), op. cit., pp. 354–84.

Mimura, N., Tsutsui, J., Ichinose, T., Kato, H. and Sakaki, K. 1998. Impacts on infrastructure and socio-economic systems. In: S. Nishioka and H. Harasawa (eds), *Global Warming: The Potential Impact on Japan.* Tokyo, Springer-Verlag, pp. 165–201.

Myers, N. 1993*a.* Environmental refugees in a globally warmed world. *Bioscience,* Vol. 43, No. 11, pp. 752–61.

1993*b. Ultimate Security: the Environmental Basis of Political Stability.* New York/London, W. W. Norton.

2002. Environmental refugees: a growing phenomenon of the 21st century. *Philosophical Transactions of the Royal Society of London B,* Vol. 357, No. 1420, pp. 609–13.

Neumann, R. 2009. Political ecology: theorizing scale. *Progress in Human Geography,* Vol. 33, No. 3, pp. 398–406.

Nicholls, R. J., and Mimura, N. 1998. Regional issues raised by sea level rise and their policy implications. *Climate Research,* Vol. 11, No. 1, pp. 5–18.

Nicholls, R. J., Wong, P. P., Burkett, V. R., Codignotto, J. O., Hay, J. E., McLean, R. F., Ragoonaden, S. and Woodroffe, C. D. 2007. Coastal systems and low-lying areas. In: M. L. Parry, O. F. Canziani, J. P. Palutikof, P. J. van der Linden and C. E. Hanson (eds), *Climate Change 2007: Impacts, Adaptation and Vulnerability. Contribution of Working Group II to the Fourth Assessment Report of the Intergovernmental Panel on Climate Change.* Cambridge, UK, Cambridge University Press, pp. 315–56.

November, V. 2004. Being close to risk. *International Journal of Sustainable Development,* Vol. 7, No. 3, pp. 273–86.

November, V. 2008. Spatiality of risk. *Environment and Planning A*, Vol. 40, No. 7, pp. 1523–27.

Pachauri, R. 2009. Copenhagen Climate Summit. *The Guardian*, 4 December. http://www.guardian.co.uk/environment/2009/nov/30/rich-west-climate-change

Paul, B. K. 2005. Evidence against disaster-induced migration: the 2004 tornado in north-central Bangladesh. *Disasters*, Vol. 29, No. 4, pp. 370–85.

Philo, C. 2005. The geographies that wound. *Population, Space and Place*, Vol. 11, No. 3, pp. 441–54.

Piguet, E. 2008. *Climate Change and Forced Migration*. Geneva, United Nations High Commissioner for Refugees. (*New Issues in Refugee Research*, No. 153.) http://www.unhcr.org/47a316182.html

Poncelet, A. 2009. *Bangladesh: A Case Study Report*. EACH-FOR reference 2.3.2.1, European Commission. www.each-for.eu

Renaud, F., Bogardi, J., Dun, O. and Warner, K. 2007. *Control, Adapt or Flee: How to Face Environmental Migration*. Bonn, United Nations University, Institute for Environmental and Human Security. (Paper 5/2007, Interdisciplinary Security Connections.) http://www.ehs.unu.edu/file.php?id=259

Richards, K., Batty, M., Edwards, K., Findlay, A., Foody, G., Frostick, L., Jones, K., Lee, R., Livingstone, D., Marsden, T., Petts, J., Philo, C., Simon, D., Smith, S. and Thomas, D. 2009. The nature of publishing and assessment in Geography and Environmental Studies: evidence from the Research Assessment Exercise 2008. *Area*, Vol. 41, No. 3, pp. 231–43.

Stern, N. 2007. *The Economics of Climate Change: The Stern Review*. Cambridge, UK, Cambridge University Press.

Stojanov, R. 2004. Environmental refugees. An introduction. *Geographica*, Vol. 38, No. 1, pp. 77–84.

Tickell, C. 1991. Global warming and migration. *People*, Vol. 18, No. 4, p. 5.

Warner, K., Afifi, T., Dun, O. and Stal, M. 2009a. Climate change and migration: reflections on policy needs. *MEA Bulletin*, Guest Article No. 64. http://www.iisd.ca/mea-l/guestarticle64.html

Warner, K., Ehrhart, C., de Sherbinin, A., Adamo, S. and Chai-Onn, T. 2009b. *In Search of Shelter: Mapping the Effects of Climate Change on Human Migration and Displacement*. http://www.careclimatechange.org

Whatmore, S. 2002. *Hybrid Geographies, Natures, Cultures, Spaces*. London, Sage Publications.

Wood, W. B. 2001. Ecomigration: linkages between environmental change and migration. In: A. R. Zolberg and P. M. Benda (eds), *Global Migrants, Global Refugees*. New York/Oxford, Berghahn, pp. 42–61.

World Bank. 2010. *World Development Report 2010*. Washington DC, World Bank.

Zimmerer, K. 2007. Cultural ecology (and political ecology) in the 'environmental borderlands': exploring the expanded connectivities within geography. *Progress in Human Geography*, Vol. 31, No. 2, pp. 227–44.

7

Sea level rise, local vulnerability and involuntary migration

ANTHONY OLIVER-SMITH

Introduction

Although the occurrence of climate change is generally accepted, there is still considerable debate in scientific and political forums about the nature and extent of its impacts. The Intergovernmental Panel on Climate Change (IPCC, 2007) asserts that human-induced factors are generating significant increases in temperatures around the world, producing increases in the rate of sea level rise, increases in glacial, permafrost, Arctic and Antarctic ice melt, more rainfall in specific regions of the world and worldwide, more severe droughts in tropical and subtropical zones, increases in heatwaves, changing ranges and incidences of diseases and more intense hurricane and cyclone activity. In addition, these alterations compound each other to accelerate the rates at which they are proceeding and are predicted to impact natural systems globally, producing changes in terrestrial, aquatic and biological subsystems. The actual and potential impacts of changes such as sea level rise are also predicted to generate environmental and social processes that will displace large numbers of people, obliging them to migrate as individuals and families or permanently displacing them and/or relocating them as communities

However, how climate change will actually manifest is still considered by many to fall into the realm of conjecture and therefore discussions about the necessity of adjustments and adaptations in natural and human systems are quite conflicted (Dessai et al., 2007, p. 1). There is considerable debate about the effects of climate change in general, and sea level rise specifically, both at the level of physical impacts and at the level of responses and adaptations in human communities. Indeed, there is as much political as scientific discussion about the projected effects of climate change on human communities (Oliver-Smith, 2009b). Sea level rise in particular has raised important questions about the possibility of major forms of involuntary migration.

The complex of forces constituted in global warming, sea level rise and forced migration cannot be addressed independently of a basic understanding of the interacting roles that local, national and international institutions play in the way human beings adapt to the environments in which they live. The concept of adaptation is the conjunctive concept between human and natural systems and therefore is central to understanding the linkages between sea level rise and migration. Clearly, one of the fundamental tasks that societies must address is some kind of adjustment to the hazardous features of the environment to which they are exposed and vulnerable. The concepts of vulnerability and resilience address the degree to which at a given point in time a society is adapted to the hazards of its environment. Vulnerability describes the degree to which a socio-economic system or physical assets are either susceptible or resilient to the impact of natural hazards. The specific dimensions of social, economic and political vulnerability are related to inequalities, to gender relations, economic patterns and ethnic or racial divisions. Vulnerability thus explicitly links environmental issues, such as hazards, with the structure and organization of society, and the rights associated with membership (Oliver-Smith, 2004).

From a social scientific perspective, adaptation refers mainly to changes in belief or behaviour in response to altered circumstances to improve the conditions of life. Human adaptations to environmental change are largely cultural, social organizational and technological. Mitigation will be an important form of adaptation for people threatened by rising sea levels. Mitigation focuses on strategies that minimize impact. In that sense, mitigation is proactive, in that it increases the resilience of a society prior to onset or impact while adaptation is reactive, adjusting to changes that have already occurred (Wisner et al., 2004).

A further distinction to bear in mind is the difference between adaptation and coping. An adaptation that has developed over years is part of cultural knowledge and practice. A coping strategy is an immediate response to a challenge for which there may be no culturally constructed adaptations. There are human rights implications to this distinction. There may be a significant difference between coping to stay alive and a long-term adaptive process. For example, coping with sea level rise may be a long way from adapting to it. Is migration an adaptation or a coping strategy of last resort? If migration and resettlement merely allow people to survive in an impoverished camp or an urban slum, then it is a far cry from adaptation. Migration, in that case, more closely approximates an immediate coping strategy that enables survival, but it is hardly adaptive. Sea level rise also carries serious human rights implications. It is clear that those who will be most affected, who may be uprooted and forced to migrate, will be among the least responsible for producing the causal features of global climate change.

As sea level rise is a complex physical phenomenon, it will produce an equally complex range of effects. Therefore, there is considerable uncertainty regarding the adaptations that will be necessary to allow people in coastal communities to remain in place. A number of questions arise concerning these impacts. For example, will displacements and migrations be driven solely by environmental factors or by government-organized mitigation efforts? Will people migrate because of economic or political factors simply exacerbated by sea level rise? Will people migrate internally within their nations or will they seek new homes across international borders? How and where will involuntary migration occur because of sudden-onset coastal storms exacerbated by sea level rise and how and where will gradual increases in sea level slowly make environments uninhabitable? Will migrations be the outcome of decisions made by individuals and families, or will local or national governments undertake to displace and resettle entire communities as a form of disaster mitigation?

Today, we do not have satisfactory answers to such questions. However, it is clear that governments and international agencies would be very derelict in their responsibilities if they did not attempt to develop adequate responses to these questions. To adequately address the potential impacts of sea level rise requires blending global projections with local and regional manifestations in the context of local patterns of vulnerability that are socially and economically constructed by local, regional and global processes. Both ecological and social projections must be elucidated in ways that reflect their mutual constitution if we are to frame a coherent research agenda that will inform policy for durable solutions for specific coastal populations facing loss of land, ecosystem services, intensified storms and the possibility of forced migration and resettlement (Oliver-Smith, 2009a; 2009b).

Global climate change and sea level rise

Global climate change has already begun to have serious impacts on socio-ecological systems. Increased average temperatures are producing rises in sea levels around the world and serious impacts in a number of specific localities will soon be felt. Moreover, levels of greenhouse gases already in the atmosphere have moved the process of climate change beyond prevention. Essentially, we must now engage in mitigation and adaptation. Climate change is now happening and will cause sea levels to rise for centuries, even if greenhouse gas emissions are stabilized (IPCC, 2007, p. 17). At this point, stabilization of greenhouse gases may reduce potential impacts, but there will still be the need to adapt to sea level rise (Nicholls and Tol, 2006, p. 1089). The major issue now at hand is how much sea level rise will take place, particularly if efforts at stabilizing or diminishing greenhouse gas emissions are either non-existent or non-effective.

To assess the probable impacts on human populations from sea level rise, including involuntary displacement and migration, we should consider the following issues:

- projected increases in physical exposure to sea level rise;
- socially constructed vulnerability and the impacts of sea level rise on specific exposed populations;
- potential impacts of adaptation and mitigation policies as well as potential migration/displacement and resettlement processes on people affected by sea level rise.

There no longer remains any scientific doubt that, following the laws of physics, increases in carbon dioxide (CO_2) and other greenhouse gases in the atmosphere will cause a global increase in temperature (WBGU, 2009, p. 9). With regard to eustatic or global changes in sea level, the increase in temperature will drive basically two processes: ocean thermal expansion, which increases existing water volume, and Antarctic and Greenland glacial melt, adding water to the oceans (Hemming et al., 2007). Other researchers add the additional factor of changes in terrestrial water storage (Dasgupta et al., 2007). Over the last decade, estimates of sea level rise have varied considerably. Variations in projections are due primarily to differences among the various Global Climate Models employed and the assumptions made about future greenhouse gas emissions in the IPCC's *Special Report on Emissions Scenarios*, SRES (Hemming et al., 2007), making it difficult to synthesize to the larger scales that are necessary for policy discussion (Nicholls and Hoozemans, 2005).

However, recent research on rates of glaciological change in Greenland and Antarctica indicates a need to rethink the upper limits of estimates of sea level rise. Data based on expansion of glacial flow from lower to higher latitudes when combined with surface loss estimates for the Greenland ice sheet indicate a doubling of the rate and twice the loss of the previous decade as estimated in the IPCC *Third Assessment Report* in 2001 (Ringot and Kanagaratnam, 2006; Hanna et al., 2005; Dasgupta et al., 2007). Even greater losses were recorded in Antarctica, in particular for the West Antarctic Ice Sheet (WAIS). Moreover, the stability of the WAIS, resting on bedrock below sea level, is a cause for concern. If global warming resulted in the collapse of the WAIS into the sea, the displacement alone would raise average sea levels by approximately 5 m–6 m (Tol et al., 2006; Dasgupta et al., 2007).

Most recently, Pfeffer et al., combining likely projection methods for Greenland, Antarctic and marine terminating glaciers and ice caps, offer three scenarios that roughly bracket the spread of near-future sea level rise possibilities between 785 mm, 833 mm and 2,008 mm by 2100 (Pfeffer et al., 2008, p. 342). These scenarios suggest that sea level rise may occur at greater volume and in significantly reduced timeframes than those projected by the IPCC (Pfeffer et al., 2008; Dasgupta et al., 2007; Vaughan and Spouge, 2002). According to the German Advisory Council on Global Change (Wissenschaftlicher Beirat der Bundesregierung Globale Umweltveränderungen, WBGU), despite some short-term variation global sea level is rising at rates significantly faster than expected (WBGU, 2009, p. 1). Rates are expected to increase as global warming increases the temperatures of the air, increasing land ice melting and adding to thermal expansion of seawater. In addition, the melting of the Greenland ice sheet is considered to be one of the tipping elements that may lead to catastrophic ecological and human consequences. If the Greenland ice sheet melts completely, it could raise sea levels globally by 7 m. Global warming of more than 1.9 °C for an extended period might be enough to melt the entire ice sheet (IPCC, 2007; WBGU, 2009, p. 2). Our latest research indicates that 'there is only a realistic change of restricting global warming to 2 °C if a limit is set on the total amount of CO_2 emitted globally between now and 2050' (WBGU, 2009, p. 1).

Sea level rise will impact human environments in two general forms: gradual rises in sea level and their effects over time, and rapid-onset events, whose effects will be exacerbated by sea level rise and related phenomena. Gradual impacts will include geomorphic changes (primarily coastal erosion and loss), altered hydrology, habitat and species change, water temperature and chemistry, air temperature and chemistry, impacts on human economy and health, infrastructure, land use changes, variable risk, and ultimately inundation of land and communities (Beever, 2009). Hydrological changes will include salinization of freshwater aquifers, including agricultural land and water supplies. Sea level rise will also make sudden-onset phenomena such as storms, hurricanes, typhoons and monsoon rains more extensive, flooding and inundating occupied terrain more rapidly and further inland. Extreme weather events such as floods, high levels of rainfall and tropical storms are seen to be increasing in many regions and additional warming will probably add to those effects (WBGU, 2009, p. 11). Natural system effects from sea level rise will be inundation, flood and sea surge and backwater effect, wetland loss, erosion, saltwater intrusion into surface waters and groundwater and rising water tables with impeded drainage (Nicholls and Tol, 2006, p. 1075).

Physical exposure to sea level rise

Projections of physical exposure to sea level rise tend to differ somewhat according to scale or unit of analysis, which range from global elevation zones to hypothetical scenarios, varying according to an array of evolving factors such as increases or decreases in greenhouse gas emissions, lack of, or construction of, defensive structures, and demographic growth or movement. These projections, however, cannot represent the uncertainties in coastal data pertaining to population, local vulnerabilities, defensive structures, or storm characteristics and their evolution through time (Hemming et al., 2007).

In one set of physical exposure estimates, McGranahan and colleagues elected to review global population and urban settlement patterns in the Low Elevation Coastal Zone (LECZ) (0 m–10 m above sea level. Currently, 10% of the global population (600 million people) reside in the LECZ (2% of the world's land mass), but there has been and continues to be a steady demographic increase through migration to coastal regions. In addition, 13% of the world's urban population live in the LECZ, in 65% of cities with populations larger than 5 million people. According to their data, Asia contains fully one–third of the land in the LECZ, but the small island states show the highest percentage of land in the LECZ (McGranahan et al., 2007, p. 24).

Researchers at the UK's Hadley Centre for Climate Prediction and Research predict a global average sea level rise of between 0.10 m and 0.15 m by 2030, 0.18 m and 0.30 m by 2050 and 0.41 m and 0.88 m by 2100. The largest relative sea level rises will occur by 2030 in areas such as the eastern seaboard of the United States, the Gulf of Mexico, the southern tip of South America and the Falkland Islands, and the Netherlands. In addition, many of the deltas of large rivers, such as the Mississippi, Rio Grande, Rhône, Nile, Bramaputra, Euphrates/Tigris and Niger are projected to see significant sea level rise also. The Hadley Centre researchers estimate that rates of global land loss will increase from 2,500 km^2 per year between 1990 and 2040 to 17,500 km^2 per year between 2040 and 2100. In addition, they assert that 50,000 km^2 of land could become submerged globally by 2030, increasing to 180,000 km^2 in 2050 and 1,130,000 km^2 by 2100. The Arctic ocean coasts of Canada, Alaska, Siberia and Greenland will experience the greatest land loss by 2030 and 2050. The coasts of Pakistan, Sri Lanka, south-eastern Indonesia and East Africa from Kenya south to Mozambique will experience land losses of between 2,500 km^2 and 5,000 km^2 a year in the Southern Hemisphere. If both population and emissions continue to grow at high rates, the Hadley Centre predicts that the number of people flooded per year will reach 21 million by 2030, 55 million by 2050 and 370 million by 2100. The populations with greatest exposure to coastal flooding from sea level rise are in South and South-East Asia, especially Bangladesh, India, Pakistan, eastern China and southern Indonesia (Hemming et al., 2007).

For a World Bank Policy Research Working Paper on the impact of sea level rise on developing countries, Dasgupta et al. reviewed data on eighty-four coastal developing countries in five regions that correspond to the five regional departments of the bank. The research projected that approximately 0.3% (194,000 km^2) of the territory of the eighty-four developing countries would be affected by a 1 m sea level rise, increasing to 1.2% if the sea rose 5 m. At 1 m, population exposure is predicted to affect approximately 56 million people, increasing to 89 million at a 2 m rise and 245 million for a 5 m rise for the eighty-four developing countries (Dasgupta et al., 2007, pp. 9–10). For land area exposure, at the 1 m increase in sea level, the ten most exposed developing nations are the Bahamas, Vietnam, Qatar, Belize, Puerto Rico, Cuba, Taiwan, China, the Gambia, Jamaica and Bangladesh. In terms of population percentage, the most exposed nations to 1 m of sea level rise are Vietnam, Egypt, Mauritania, Suriname, Guyana, French Guiana, Tunisia, United Arab Emirates, the Bahamas and Benin.

The *Fourth Assessment Report* of the IPCC estimates a rise in sea level of up to 0.6 m or more by 2100. Focusing on geomorphological and biological features in assessing areas of physical exposure to sea level rise, this research reveals that impacts on the systemic dimensions of these natural features have serious implications for both proximate and distant human communities (Nicholls et al., 2007). The research pays special attention to deltas and megadeltas as hot spots for exposure. In terms of population and economies, if there are no upgrades to defensive structures, a 40 cm sea level rise is projected to flood more than 100 million people per year (Nicholls et al., 2007, p. 334). In a study of forty river deltas worldwide, Ericson et al. (2006, p. 63) assert that, at contemporary rates of sea level rise, by 2050 8.7 million people and 28,000 km² could experience inundation and increased soil erosion. Their research also found that direct anthropogenic effects greatly affected estimated sea level rise in the majority of the deltas examined, with relatively little impact from eustatic increases in sea level (Ericson et al., 2006, p. 63). The implications of these findings indicate that human–environment interactions play a major role in the vulnerability of human communities in the delta regions of the world.

In assessing impacts of sea level rise, however, the local change or rate of change in relative sea level is more important than the global or regional average. Relative sea level, or observed sea level, is the level of the sea in relation to the land. Relative sea level is clearly influenced by global or absolute sea level change, but it is also affected by either uplift or subsidence in local terrains, which will vary regionally and locally. Natural processes produce most vertical movements, but human actions such as groundwater extraction or removal of sediments in deltas can increase subsidence (Klein and Nicholls, 1999, pp. 182–83).

However, many of these projections have been criticized recently for being too conservative. At the 2009 Copenhagen Climate Summit research was presented indicating that current high rates of greenhouse gas emissions are driving worst-case IPCC scenarios to greater extremes at faster rates than anticipated. Key parameters of the climate system, including global mean surface temperature, sea level rise, ocean and ice sheet dynamics, ocean acidification and extreme climatic events, have surpassed patterns of natural variability.[1]

Particularly significant for sea level rise, ice sheet and glacial loss, omitted from the 2007 IPCC report, is accelerating, with glacial melt

---

[1] http://climatecongress.ku.dk/newsroom/congress_key messages

constituting two-thirds of the loss (Kintisch, 2009, p. 1546). And compounding these alarming findings, improved methodologies revealed two further trends that will increase temperatures. There are now estimated to be 1.7 trillion tonnes of carbon in permafrost, more than twice the IPCC estimate, and warming temperatures could release even more greenhouse gases, especially methane, thereby adding additional carbon loads to the atmosphere and resulting in even higher temperatures, speeding glacial melt, seawater expansion, and sea level rise.

These recent projections will make assessments of local vulnerability easier or surer only at the extremes. Adequate projections of local impacts will still depend on the careful analysis of local vulnerabilities and this continues to represent a significant challenge. Comprehensive vulnerability assessment at any scale remains difficult because of incomplete knowledge of the processes involved in sea level rise and their interactions. Data on existing conditions are inadequate. In addition, there are significant difficulties in developing scenarios for climate change at local and regional levels. And finally, we lack appropriate analytical methods for some kinds of impact (Nicholls and Hoozemans, 2005, p. 486).

Sea level rise and socially constructed vulnerability

Vulnerability science has established clearly that exposure to hazards alone does not determine where the serious effects of any hazard, including sea level rise, will most probably be felt. It is also clear that we need to develop methods to determine what effects future rises in sea level will have on specific regions and communities This task will continue to be difficult because of the numerous variables and the nonlinearity of their interactions, but we can look at current situations of sea level rise and local effects and adaptations to construct scenarios to help in assessing probabilities of forced displacement and migration.

In addition to determining absolute exposed land and absolute exposed population, it is necessary to specify lands and populations in different socially configured conditions of resilience or vulnerability. General and specific conditions of vulnerability are accelerating rapidly due to increasing human induced pressures on coastal systems. Coastal population growth globally has increased enormously in the twentieth century and is expected to continue to grow in the twenty-first century from 1.2 billion (in 1990) to 1.8 billion to 5.2 billion by 2080 (Nicholls et al., 2007, p. 17). Furthermore, vulnerability to coastal hazards and climate change is in part a function of development levels and per capita income (Nicholls et al., 2007, p. 331). According to McGranahan et al., less-developed nations have a significantly higher proportion of their total populations and their urban populations in the low elevation coastal zone (2007, p. 26), strongly indicating that climate change impacts, including heavier socio-economic costs from climate-related hazards and disasters, will probably impact more severely on coastal regions of developing nations that have fewer resources for mitigation and adaptation (Nicholls et al., 2007, p. 331).

However, assessing exposure of both land and population to sea level rise is substantially more complex. Not only must projected increases in sea level be taken into account, but future projections about various societal and environmental trajectories including greenhouse gas emissions, demographic change, migration trends, infrastructural development, mitigation strategies, adaptive capacities, vulnerabilities and patterns of economic change must also figure in our calculations in all the possible ways they will play out within the political, economic and socio-cultural frameworks of national governments, international organizations and general populations.

To aid in these deliberations, the IPCC's *Special Report on Emissions Scenarios* (SRES), has developed a set of greenhouse gas emission scenarios to present possible socio-economic pathways that would affect global outcomes of climate change, including sea level rise (Nakicenovic and Swart, 2000). The IPCC scenarios are basically alternative possible socio-economic pathways that the world might follow in terms of political, economic, social and technological development. Each scenario is a short narrative of alternatives of future global development, depicting a set of social, economic, political and technical conditions, quantified at a global and regional scale meant to assist in climate change analysis and the assessment of impacts, adaptation and mitigation. The scenarios are based on the assumption that the main driving forces of future greenhouse gas trajectories will continue to be demographic change, social and economic development, and the rate and direction of technological change.

The analysis of these scenarios in the fifteen most vulnerable countries, including eight small island nations and three with large river deltas, assessed in the SRES, projects significant impacts to coastal areas by sea level rise. The analysis also shows that these impacts can be mitigated through climate stabilization and adapted to by protective measures for coastal settlements. Indeed, the authors contend that human attitudes towards the environment may prove to be more significant than sea level rise, suggesting that if coastal populations act judiciously, there may be a wider array of choices than has been assessed. The key factor in this analysis from an economic (cost-benefit) perspective would appear to be the level of development of the affected region, enabling investment in widespread protection rather than abandoning occupied coastal properties (Nicholls and Tol, 2006, p. 1089). Indeed, it is generally agreed that lower growth and levels of development will probably lead to more destructive episodes. This project has evoked considerable debate. The scenarios are all based on calculations at the national scale and uncertain data sets (Nicholls and Tol, 2006, p. 1089). In addition, some of the assumptions on which the scenarios were based are considered dubious. For example, all the scenarios describe futures that are generally more affluent, with reduced income differences among world regions, a trajectory that many hold as unrealistically optimistic.

Social effects of sea level rise

While projections of sea level rise have now acquired considerable confidence, outside areas of total inundation, it remains challenging for current global climate models to specify the range of changes that will occur in particular countries, much less in specific localities (Dow et al., 2006, p. 84). Furthermore, assessing the vulnerability of specific localities is also difficult, particularly in their interactions with larger regional, national or international forces. The identification of specific vulnerable groups spatially and temporally with sufficiently high accuracy for comparative purposes of climate impacts is not currently possible.

Nonetheless, knowledge exists to identify broader regions that are vulnerable to climate change generally and sea level rise specifically. In other words, the areas that will be prone to certain types of climate change and extreme events have been identified. Combining that general knowledge with specific local case studies allows researchers to track current changes, monitor effects and use the information to generate more focused scenarios to develop appropriate policy responses.

Revisiting the initial research questions, how do local environmental conditions, some caused by human action (subsidence, erosion, degradation), combine with social conditions (social vulnerability) to interact with global sea level rise to induce local measures to adapt, mitigate or migrate? It is only possible in the case of complete and permanent inundation to state that people will be unequivocally displaced and forced to migrate. If the land occupied by a community is completely and permanently submerged, migration will be necessary. However, there are many aspects of sea level rise that will affect the sustainability of coastal peoples and communities, but may not pressure people to move. Coastal storm surges, subsidence and erosion, salinization of ground water and rising water tables, and impeded drainage all may seriously impact both residence and economic production in vulnerable communities. Wetlands, estuaries and mangroves, constituting both the ecological and economic base of many coastal communities, may be seriously damaged by sea level rise, requiring adaptation, mitigation or migration.

By the same token, mitigation strategies and adaptive measures may allow communities to avoid displacement and involuntary migration from sea level rise changes falling short of total submergence that will render residential sites uninhabitable. Even in extreme cases of inundation, removal to other locations may remain quite local, constituting more a shift in residence than an actual migration. Mitigation and adaptation that allow for continued residence will be characterized by changes in technology and social (and economic) organization. Restoration of mangroves and wetlands or protective structures (dykes, levees, etc) may afford sufficient defences from storm surges. Or adaptive strategies, such as adoption of new forms of economic activities, may let people adapt to changing conditions and thus avoid the necessity to migrate. There may also be scenarios in which migration may complement or enhance adaptive strategies. Migration, whether permanent or temporary, has always been a traditional response or survival strategy of people who must deal with seasonal or extreme variation in climate conditions (Hugo, 1996). Migration of certain people in a community may, for example, provide either space or resources that will allow other members of the community to remain in their homes. The strategies developed by the Cartaret Islanders, outlined below, offer an interesting case of such a strategy.

Sea level rise: adaptation or migration?

How can we construct scenarios that will aid in the elaboration of strategies that people can use to adapt to rising sea levels? Regional, national and international agencies hoping to assist people coping with sea level rise will need scenarios to craft appropriate policies and practices that let people adapt to the changing environment.

Finan et al. (2002) propose a template for an approach to building local adaptations. According to this template, a long-term process of adaptation to sea level rise can be enhanced by attention to two interrelated features: technology and social organization. Changes in technology frequently require the reorganization of rules and regulations regarding the distribution and use of resources across a population of users. The focus on the interaction of technology and social organization has three dimensions: distributive, institutional and empowerment. The distributive dimension focuses on resource access and resource management strategies, which in effect constitute a livelihood assessment. This livelihood assessment constitutes in turn the 'response space' a capacity to respond, or, in other words, their resilience.

The institutional dimension pertains to the power relationships, forms of dependency, local and global markets (and their distortions) and the effects of outside private and public interventions that focus on the sets of resources and practices that exceed the capacity of the community to mobilize. These aspects include major infrastructure, access to new technology, sources of information and networks. The institutional dimension includes government agencies, local, national and international NGOs, local universities and research centres, and the national disaster response and management systems.

The third dimension addresses the empowerment of local management. To address the vulnerability associated with sea level rise, some combination of technological and social organizational adjustment must enable system-wide responses to its varied impacts. The adaptation process must be localized so that communities can reorganize to address increased environmental uncertainty. In the final analysis, the approach must be organized around local resources and adaptive capacity, outside resources and constraints, and the empowerment of local social organization for decision-making (Finan et al., 2002).

Most of the current research on sea level rise adaptation and migration points to river deltas and small islands, not only as sea level rise hot spots but also as examples from which possible scenarios may be drawn for developing policies and guidelines to assist affected communities. Creating such scenarios is important, not because they are necessarily accurate or true, but because they require a consideration of a broader range of eventualities and responses. The following six cases, three from river deltas and three from small islands, provide an indication of the variety of problems faced by local communities confronted with sea level rise.

Sea level rise in river deltas

Bangladesh

Sea level rise is expected to inflict some of its most grievous impacts on the nation of Bangladesh, one of the poorest and most densely populated countries in the world. As a delta country, most of Bangladesh is well within the LECZ of 10 m or less (Poncelet, 2009, p. 3; McGranahan et al., 2007). As such, roughly half the nation would be flooded if sea level rise were to reach 1 m, well within the most recent estimates (Kintisch, 2009). In addition, the region is prone to cyclones, storm surges and backwater effects (saltwater intrusion). Dasgupta et al. (2007, pp. 37–39) rank Bangladesh first in South Asia for percentages of population, GDP, urban extent, and agricultural extent impacted; and third for wetlands impacted from 1 m to 5 m of sea level rise. In Bangladesh two-thirds of the total population of 150 million people, more than half of whom subsist on less than a dollar a day, are likely to be affected. In addition, 10% of fertile land will be ruined, and the unique biodiversity of the Sunderbans mangrove forests imperilled (Finan, 2009, p. 178). Moreover, fishing and farming livelihoods of the vast majority of the people in vulnerable zones will be seriously compromised.

The coastal area of Bangladesh occupies over 36,000 km^2 and is very densely populated. In this context, Finan's research on beels, small open ponds located in lowland depressions that are used for aquaculture of shrimp, offers a fruitful context to develop scenarios for economic adaptation to sea level rise (Finan, 2009). Shrimp cultivation in beels, formed by earthen walls (*ghers*) are built to enclose areas within the beel. Subsidiary industries in fry collection and feed provision have also emerged. Vegetables also may be grown along the dyke walls, which women largely care for and market. Over the last two decades, the export of shrimp aquaculture has grown into a 350 million dollar industry, and has dramatically improved local livelihoods, increasing income-earning possibilities for poor farm families, in particular for women.

Sea level rise also represents a major source of uncertainty for this form of livelihood, largely in the form of cyclone-driven flooding and backwater saline intrusion as seawater is forced further up into the delta region. Cyclone-driven floods may compromise the *gher* walls, thus destroying an industry that helps to sustain half a million people, among them aquaculturalists, fry collectors and food providers, in a region of dire economic need. To address the vulnerability associated with sea level rise, Finan suggests that some kind of social organizational adjustment is indicated, very likely in the direction of common property. The beel, as a hydrological system, might come under collective management, while each farmer would maintain control over individual production, much as lobster stocks are managed by fishermen in the north-eastern United States. Beel management can then be organized through a framework of participatory decision-making that will enable system-wide technological responses to saltwater intrusion and storm surges. The adaptation process must be organized around local resources and adaptive capacity, albeit perhaps enhanced by outside resources and hindered by outside constraints, as well as the empowerment of local social organization for decision-making.

Vietnam

The risks of displacement and resettlement have been experienced by villagers in the Mekong Delta of Vietnam. Vietnam is one of the nations slated to be most impacted by sea level rise. Dasgupta et al. rank Vietnam among the top five most impacted nations in the world and it ranks first in East Asia for impact on area, exposed population, GDP, urban area, agriculture and wetlands (2007, pp. 29–33). Vietnam's exposure is considerable because of its long coastline, vulnerability to storms and high seas and extensive low-lying areas in the southern Mekong Delta, the elevation of which is only 0.5 m to 4.0 m above sea level. Compounding the physical exposure, the Mekong Delta, as home to 18 million people (22% of the total population), is one of the most densely populated regions on Earth. The IPCC has calculated that 1 m of sea level rise will affect more than 1 million people in the delta (2007, p. 327).

Furthermore, as the Mekong Delta is also the principal rice-growing area, producing half of that staple food for the nation, any significant alteration in the status of the wetlands will have serious implications for the economy and the health and nutrition of the people of Vietnam. The fertility of the delta depends on continued soil replenishment through the distribution of upstream sediments by the river, especially during regular slow-onset seasonal flooding (July–November). It is also an important breeding ground for fish and shrimp, and the source of 60% of annual national production.

Although not developed to deal with sea level rise effects, current government programmes for flood management and environmental sanitation that have resettled people point to challenges that will be faced when resettlement policies are designed for sea level rise. One programme resettled landless people who were dependent on day-to-day employment as wage labourers. The resettlement project broke up the social networks these households depended on for daily employment. Furthermore, to move to the residential clusters, people were required to buy a plot of land and a basic house frame structure in the resettlement area with a government-provided five-year interest-free loan. However, households will also need further loans to complete construction around the basic frame provided. Thus, resettlement puts people further in debt, facing the risk of unemployment, enduring the lack of access to infrastructure such as waste water treatment facilities, health care and schooling, and suffering the loss of support of their social networks. The residential projects are also designed as semi-urban with side-by-side plots for people who formerly lived in comparatively dis-persed households along the riverbanks (Dun, 2008, pp. 7–8). The resettlement process has thus resulted in further impoverishment of already poor people, perhaps removing them from exposure, but doing little to lessen their vulnerability.

US Gulf Coast

The increased frequency and intensity of coastal storms will potentially account for future displacement and resettlement. Although the devas-tation caused by Hurricane Katrina is not directly attributable to climate change or sea level rise, that disaster can be used as a proxy to illustrate the impact of storm intensification and higher sea levels. The Gulf Coast region, particularly surrounding New Orleans, is close to or below sea level, leaving it vulnerable to storms and hurricanes. Climate change-driven sea level rise may well be just as devastating on the Gulf Coast because of the flat terrain, land subsidence, environmental degradation and increased coastal development (Button and Peterson, 2009).

Grand Bayou, a Native American fishing community about 160 km south of New Orleans in Plaquemines parish, was almost entirely destroyed by Hurricane Katrina. All twenty-five families, 125 individuals in total, were displaced by the storm. Although traditional environmental knowledge enabled the community to be resilient to the impact of previous hurricanes, the two storms of 2005 obliterated all but one house in the community. However, that traditional local knowledge enabled them to save all their boats and preserve the principal tools of their economic life. And no one perished in the storm that killed more than 3,000 people elsewhere in the region. Today they continue to live in the parish, many still in mobile homes provided by the Federal Emergency Management Agency (FEMA), and they are determined to rebuild their community and sustain their culture.

Grand Bayou's survival prompted the organization of a collaborative Participatory Action Research (PAR) project between the community and the University of New Orleans on the value of local knowledge for understanding and analysis of environmental problems. In contemporary society, science reigns supreme while local knowledge, based on experience, narrative and tradition, has often been summarily dismissed or relegated to the category of 'alternative' interpretations, thus silencing local perspectives. The PAR project has demonstrated clearly to the scientific community that local knowledge can lessen vulnerability. In Grand Bayou, scientists came to agree that it was far more efficient and effective to tap local knowledge about hydrologic and environmental processes than to try to generate projections for research projects through models. At the same time, local residents, now facing radically altered environmental conditions, including future sea level rise, have come to value scientific insights that may assist them in facing these novel problems. For example, traditional safe harbour locations for boats have been destroyed, and local fishermen are working with their scientific partners in developing new risk assessments and future secure sites for boats during storms (Button and Peterson, 2009).

Sea level rise on small islands

Tuvalu

While the adaptations available to communities threatened by sea level rise will be varied, drawing on combinations of different local social and technological capacities as well as external assistance where available, other populations will face a restricted set of adaptive options. From the point of view of exposure, there would seem to be few places on Earth more exposed to sea level rise than the Small Island Developing States (SIDS) of the Pacific, particularly Tuvalu and Kiribati, and their exposure is followed closely by that of islands in the Indian Ocean, particularly Maldives, although the *Third Assessment Report* of the IPCC asserted that all island states would be negatively impacted by increases in sea levels (IPCC, 2001; Pelling and Uitto, 2001, p. 56).

The environmental threats from climate change to Tuvalu are considerable. Agriculture is extremely difficult because of the salinity of the soil, which is now considerably increasing, making the cultivation of taro, the main crop, ever more difficult. In addition, the increasing intensity and frequency of extreme weather events are issues that Tuvaluans must contend with, especially in the context of sea level rise, which would exacerbate the already serious impacts of these storms. Increased intensity, but decreasing frequency, of rainfall regimes also represent a threat to life on these water-scarce islands, where both salt-water flooding and droughts have reduced supplies for the growing population (Gemenne and Shen, 2009, p. 9). Other effects of climate change include changes in surface and subsurface ocean temperatures, ocean acidification and coral bleaching, pest infestations, reef fisheries deterioration, increase in communicable diseases and infrastructural damage (Lazrus, 2009b, p. 242).

As Nicholls and Tol (2006) point out, local attitudes towards the environment will play a central role in adjustments to climate change. Local knowledge will be instrumental in the patterns of adaptation and mitigation of climate-induced sea level rise, including migration. Lazrus, for example, documents a cultural tradition of narratives, stories and legends on the threatened island of Tuvalu that recount how the challenges of island life are endured and survived, providing mythical metaphors for understanding and adapting to environmental changes (2009a, p. 243). As temporary labour migration is part of their overall livelihood strategies, Tuvaluans recognize the possibility of permanent migration due to sea level rise in the future, but their current strategy is to develop locally based responses and adaptations rather than wholesale migration.

For island nations under threat from sea level rise, it is clear that effective adaptations, whether moderate or extreme, must be based on, or at the minimum consistent with, local interpretations and values. And it is equally clear that any scenario constructed for use in guiding both policy and strategy in dealing with sea level rise, must allow traditional knowledge and local interpretation to frame the challenges faced if adaptations that are culturally acceptable as well as effective are to be developed. Local knowledge and control are essential if outside agencies expect to work effectively with local communities in adapting to sea level rise and climate change in general. As Bankoff points out (2004, p. 35), local knowledge must be respected, not merely tolerated as an alternative discourse, if serious missteps with potentially calamitous results are to be avoided in coping with sea level rise (see also Lazrus, 2009a, p. 247).

Alaska

The assertions of the people of Shishmaref on the Seward Peninsula in Alaska, facing similar threats from sea level rise and coastal erosion, confirm this position (Marino and Schweitzer, 2009). Climate change is happening more rapidly in the Arctic than in any other region and is having serious impacts on both environment and people. Known as Polar Amplification, the rapid changes are driven by an increase in temperature from 2 °C to 3.5 °C (IPCC, 2007). The increase in temperature has accelerated the melting of sea ice, reducing both extent and thickness, which creates the potential for increased wave generation on exposed coasts. In addition, sea level rise for low-relief shorelines is driving rapid erosion, exacerbated by melting permafrost that traditionally has bound coastal soils, warmer ground temperatures, increased thaw and more subsidence linked to ground ice melting (IPCC, 2007, p. 320).

Indeed, sea level rise and coastal erosion threaten 86% of the coastal villages in Alaska. One of them, Shishmaref, is as emblematic in the popular press of the impacts of global climate change and sea level rise as Tuvalu. Four other nearby villages, Kivalina, Newtok, Koyukuk and Shaktoolik, are also facing eventual displacement and resettlement (Bronen, 2009; Marino, 2009, p. 3). For most of these villages there is no higher ground and few distant sites where they could move (Bronen, 2009, p. 4). Fully cognizant of the dangers they face from erosion and flooding, the villagers of Shishmaref have now officially voted to relocate and have chosen a resettlement site. However, the villagers' plans have been frustrated by the lack of clear responsible agencies and a systematic strategy for resettlement on the part of state and federal authorities. There is, in fact, no lead agency responsible for relocation planning and the coordination of all the various agencies working on housing, transportation, community infrastructure, education, health and other related needs (Bronen, 2009, p. 7). This confusion and lack of expertise and coordination have produced resettlement budgets that range between

US$100 million and US$200 million for a village of roughly 600 people – between US$165,000 and US$330,000 per person.

Local people are very sceptical of outside-led resettlement strategies and relocation authorities report that there is contention in the village over who is in charge of the planning process (Marino, 2009, p. 7). Shishmaref and several other villages have taken their case into the public forum, lobbying extensively at state and federal levels and working the media in a sophisticated fashion (Marino, 2009, p. 8). The fundamental issues for the people of Shishmaref are continuity of culture as a discrete village on their own land and local control over resettlement decision-making. Disempowerment and potential impoverishment have proven to be strong incentives for resistance to resettlement in a variety of contexts (Oliver-Smith, 2010). Furthermore, bureaucratic inconsistencies, agency contradiction, and planning and procedural rigidities, so typical of development-forced displacement and resettlement, seem to be part of future scenarios for people facing displacement by sea level rise, unless significant progress in the field is achieved.

Papua New Guinea

Like the people of Shishmaref, the 3,500 residents of the Cartaret Islands and 2,500 islanders from three other nearby atolls (Mortlock, Tasman and Nuguria islands) are facing the need to resettle on Bougainville. Their island habitats are increasingly unable to sustain them due to coastal erosion and land loss, saltwater inundation and growing food insecurity. In the Cartarets, in particular, the situation is deteriorating quickly. In 2006 only three families saw the need to resettle. Two years later, in 2008, thirty-eight families wished to resettle in Bougainville. It is estimated that eventually 300 families or roughly 1,750 people will eventually require resettlement if land loss continues at present rates. It is further expected that while a small population may be maintained on the diminished Cartarets, most of the population will eventually resettle on Bougainville (Displacement Solutions 2009).

A local organization, Tulele Peisa, meaning 'Sailing the Waves on our Own', has been formed to promote and organize the resettlement of these families. Preliminary estimates by Tulele Peisa calculate that roughly 14 million kina (US$5.3 million) will be needed between 2009 and 2019 for the resettlement project. The Cartaret Islanders do not have sufficient economic resources to purchase land and will depend on the Papua New Guinea Government and international aid. The main

challenge facing the resettlers was identifying and obtaining the approximately 1,500 ha of land required for the 300 Cartaret families and an
additional 1,500 ha for the resettlers from the three other affected atolls.
For the community to be sustainable, Tulele Peisa proposed that each
family receive rights to 5 ha of land with 2 ha devoted to livelihood
purposes, 1 ha for homes and gardens, and 1 ha set aside for reforestation
purposes. As of December 2008 only 81 ha donated by the Roman
Catholic Church had been identified for use by the resettlers, leaving
more than 1,400 ha still to be obtained (Displacement Solutions 2009).

As with many resettlement projects, obtaining clear title to land for
uprooted people is difficult. In Bougainville, access to land is frequently
disputed, with competing claims from customary landowners. In many
contexts, there are at least four levels of land rights holders: traditional
owners, the government, the formal title holder and the user. In addition
to land-use rights granted them by the Church, the Carteret communities have initiated conversations with traditional landowners to establish good relations and a smooth integration process after resettlement.
However, they still lack clear title to the land in question. Difficulties
notwithstanding, the Cartaret islanders, through their organization,
Tulele Peisa, have directly undertaken their own resettlement process,
working purposefully and carefully to ensure the basic elements of
successful resettlement: land, housing, infrastructure, environment and
livelihood in ways that reflect and affirm their communities and cultural
values (Displacement Solutions, 2009). Moreover, the relocation strategies that they have devised will allow some portion of their community
to remain on their beloved islands. In all likelihood, ties between those
who remain and those who migrate to Bougainville will remain strong,
thus limiting the sense of loss and dislocation experienced by the
migrants (Displacement Solutions 2009).

All the cases cited point to variable responses and adaptations, some
allowing for continued residence despite sea level rise, others clearly
requiring migration and resettlement of all or parts of the affected
communities. In some sense, we are dealing with what may be only the
first examples of impacts on communities. For example, there are currently at least five communities in Alaska, one of them, Shishmaref, that
will have to move, probably in the next fifteen years, and an undefined
number that will need to move in the next hundred years. There are 189
communities that experience flooding problems (Marino, 2009).
However, these projections are based on best current estimates of the
rate of sea level rise, which could change. As sea level rise is expected to

increase, land, soils and water sources will be correspondingly affected, prompting partial, staged or complete migrations from affected communities. Some of these migrations may allow a portion of the population to remain, as the example from the Cartaret Islands indicates, but as much smaller communities.

However, the question remains as to whether migration is a form of adaptation or a form of coping. Migration is variously framed as a form of adaptation, an adaptive failure and an impact of climate change. Before migration can be unequivocally defined as an adaptation, a set of issues concerning the concept of adaptation needs to be clarified. Is adaptive capacity an index of the viability of a community facing profound change? Or is it a coping effort of last resort? Can environmentally induced migration and its relationship to adaptation be observed and measured? Does migration meet the standards of other adaptive strategies? Under what conditions does migration become adaptation and for whom? Our current research is grappling with these questions. However, notwithstanding the unresolved conceptual issues and the incipient condition of research on sea level rise and migration, it is clear that sea level rise constitutes a potential threat of considerable proportions. It is equally clear that displacement and migration will constitute a multidimensional crisis for affected communities.

Displacement, forced migration and recovery

If sea level rise engulfs coastal communities, either gradually through erosion and coastal land inundation or suddenly through storm surges exacerbated by higher sea levels, and adaptation and mitigation measures are inadequate, national governments and international organizations must be prepared to assist the displaced. Involuntary displacement involves far more than just physical movement. It is important to understand what is lost in displacement and what needs to be recovered if effective policy is going to be developed to avoid creating permanent refugee camps and dependent populations. Here human rights concerns focus not only on the injustice of differential vulnerability, but also on the injustice of inadequate resettlement that is generally considered a secondary disaster, leaving people destitute and disempowered. Poorly implemented resettlement both compounds and makes permanent many of the losses incurred in displacement (Scudder and Colson, 1982).

Environmentally displaced people will face a complex series of events most often involving dislocation, homelessness, unemployment, the dismantling of families and communities, adaptive stresses, loss of privacy, political marginalization, a decrease in mental and physical health status and the daunting challenge of reconstituting livelihood, family and community (Cernea, 1996; 1990; Scudder, 1981). The dispersal of family members that often occurs in displacement fragments not only a household, but erodes the social cohesion of a community as well, shredding those networks of relationships that form the basis of personal and social identity, setting people adrift, without those ties that anchor the self in the social world. Cultural identity is also often placed at risk in uprooted communities. Uprooted people generally face the daunting task of rebuilding not only personal lives, but also those relationships, networks and structures that support people as individuals and which we understand as communities. The disarticulation of spatially and culturally based patterns of self-organization, social interaction and reciprocity constitutes a loss of essential social ties that affects access to resources, compounding the loss of natural and man-made capital (McDowell, 2002).

In general, however, the process of reconstruction has been approached largely as a material problem. Material aid is often donor-designed as largely a transfer process and frequently delivered in content and form in ways that compound the social and psychological effects of destruction and displacement by undermining self-esteem, compromising community integrity and identity and creating patterns of dependency. Reconstructing and reconstituting community is an idea that needs to be approached with a certain humility and realism. There is a complexity in resettlement that is inherent in 'the interrelatedness of a range of factors of different orders: cultural, social, environmental, economic, institutional and political – all of which are taking place in the context of imposed space change and of local level responses and initiatives' (de Wet, 2006). An important goal for resettlement projects is to work out a system in which people can materially sustain themselves while they themselves begin the process of social reconstruction.

Conclusion

The combination of increasing coastal populations, population density, increasing poverty and occupation of more coastal lands has accentuated vulnerability to both sea level rise and coastal storms and increases the probability of forced migrations. Although the issue of environmental refugees has generated significant attention (and debate) over the last twenty years, appropriate policies pertaining to environmentally displaced peoples or other internally displaced populations have yet to attain legal status. Moreover, according to the International Federation of Red Cross and Red Crescent Societies, 'there are no well recognized and comprehensive legal instruments which identify internationally agreed rules, principles and standards for the protection and assistance of people affected by natural and technological disasters' (IFRC, 2004, p. 1). The UN's *Guiding Principles on Internal Displacement* (Deng, 1998), although widely recognized as an international standard, and certainly helpful in guiding NGOs and other aid organizations in assisting IDPs, have not been agreed upon in a binding covenant or treaty and have no legal standing. We must also recognize the very real potential for global climate change to generate displacements and migrations across international borders.

A policy-relevant research programme should clearly focus on documenting and analysing regionally and locally those combinations of exposure and vulnerability (sensitivity) that under expected sea level rise estimates can be projected to require significant adaptive efforts, including the potential displacement of communities. By the same token, research should also explore those adaptive strategies that permit continued occupation of lands affected by sea level rise. Policy goals should also include mitigation, vulnerability reduction and climate change stabilization. Research should also focus on deepening understanding of climate change-driven displacement. This research should inform the development of policy to improve resettlement practice with an emphasis on the education and training of resettlement project managers and personnel. It is both urgent and incumbent upon national and international actors and agencies to develop legally binding policies and informed practice to address the potentially massive displacement and resettlement that global climate change in general and sea level rise specifically are projected to cause.

References

Adger, W. N., Paavola, J., Huq, S. and Mace M. J. (eds). 2006. *Fairness in Adaptation to Climate Change*. Cambridge, Mass., MIT Press.

Bankoff, G. 2004. The historical geography of disaster: 'vulnerability' and 'local knowledge' in Western discourse. In: G. Bankoff, G. Frerks and D. Hilhorst (eds), *Mapping Vulnerability: Disasters, Development and People*. London, Earthscan, pp. 25–36.

Beever, L. B. 2009. Climate Ready Estuaries: Vulnerabilities and Adaptations, 15th Annual Public Interest Environment Conference, Levin College of Law,

University of Florida, 29 February. ftp:ftp.swfrpc.org (Accessed 3 March 2009.)

Bronen, R. 2009. Forced migration of Alaskan indigenous communities due to climate change: creating a human rights response. In: Oliver-Smith and Shen (eds), op. cit., pp. 68–73.

Button, G. and Peterson, K. 2009. Participatory action research: community partnership with social and physical scientists. In: Crate and Nuttall (eds), op. cit., pp. 209–17.

Cernea, M. 1996. *Eight Main Risks: Impoverishment and Social Justice in Resettlement.* Washington DC, World Bank Environment Department.

Crate, S. A. and Nuttall, M. (eds). 2009. *Anthropology and Climate Change: From Encounters to Actions.* Walnut Creek, Calif., Left Coast Press.

Deng, F. 1998. *Guiding Principles on Internal Displacement,* 11 February. (E/CN.4/1998/53/Add.2.)

Dasgupta, S., Laplante, B., Meisner, C., Wheeler, D. and Yan, J. 2007. *The Impact of Sea Level Rise on Developing Countries: A Comparative Analysis.* Washington DC, World Bank. (World Bank Policy Research Working Paper 4136.)

Dessai, S., O'Brien, K. and Hulme, M. 2007. Editorial: On uncertainty and climate change. *Global Environmental Change,* Vol. 17, No. 1, pp. 1–3.

de Wet, C. 2006. Risk, complexity and local initiative in involuntary resettlement outcomes. In: C. de Wet (ed.), *Towards Improving Outcomes in Development Induced Involuntary Resettlement Projects.* Oxford/New York, Berghahn Books.

Displacement Solutions. 2009. The Bougainville Resettlement Initiative Meeting Report (11 December 2008) Displacement Solutions. Canberra. http://www.displacementsolutions.org/files/documents/BougainvilleResettlementInitiativeMeetingReport.pdf

Dow, K., Kasperson, R. E. and Bohn, M. 2006. Exploring the social justice implications of adaptation and vulnerability. In: Adger et al. (eds), op. cit., pp. 79–96.

Dun, O. 2008. *Migration and Displacement Triggered by Flooding Events in the Mekong Delta, Vietnam.* Paper delivered at 3rd United Nations University Institute for Environment and Human Security-Munich Re Foundation Summer Academy, Schloss Hohenkammer, Munich, Germany, 28 July–1 August.

Ericson, J. P., Vorosmarty, C. J., Dingman, S. L., Ward, L. G. and Meybeck, M. 2006. Effective sea-level rise and deltas: causes of change and human dimension implications. *Global and Planetary Change,* Vol. 50, Nos. 1–2, pp. 63–82.

Finan, T. 2009. Storm warnings: the role of anthropology in adapting to sea level rise in southwestern Bangladesh. In: Crate and Nuttall (eds), op. cit.

Finan, T. J., West, C. T., McGuire, T. and Austin, D. 2002. Processes of adaptation to climate variability: a case study from the US Southwest. *Climate Research,* Vol. 21, No. 3, pp. 299–310.

Gemenne, F. and Shen, S. 2009. *Tuvalu and New Zealand Case Study Report #044468.* EACH-FOR (Environmental Change and Forced Migration Scenarios). Université de Liège, Belgium, CEDEM.

Hanna, E., Huybrechts, P., Janssens, I., Cappelen, J., Steffen, K. and Stephens, A. 2005. Runoff and mass balance of the Greenland Ice Sheet: 1968–2003. *Journal of Geophysical Research*, Vol. 110, D13108.

Hemming, D., Iowe, J., Biginton, M., Betts, R. and Ryall, D. 2007. Impacts of mean sea level rise based on current state-of-the-art modelling. Exeter, UK, Hadley Centre for Climate Prediction and Research.

Hugo, G. 1996. Environmental concerns and international migration. *International Migration Review*, Vol. 30, No. 1, pp. 105–31.

IFRC. 2004. *World Disasters Report.* International Federation of Red Cross and Red Crescent Societies, Bloomfield, Conn., Kumarian Press.

IPCC. 2001. *Climate Change 2001: Impacts, Adaptation and Vulnerability.* J. J. McCarthy, O. F. Canziani, N. A. Leary, D. J. Dokken and K. S White (eds). Cambridge, UK, Cambridge University Press.

—— 2007. Summary for Policymakers. In: *Climate Change 2007: The Physical Science Basis.* Contribution of Working Group I to the Fourth Assessment Report of the Intergovernmental Panel on Climate Change. Cambridge, UK/New York, Cambridge University Press.

Kintisch, E. 2009. Projections of climate change go from bad to worse, scientists report. *Science*, Vol. 323, 20 March, pp. 1546–47.

Lazrus, H. 2009*a.* The governance of vulnerability: climate change and agency in Tuvalu, South Pacific. In: Crate and Nuttall (eds), op. cit., pp. 240–49.

—— 2009*b.* Perspectives on vulnerability to climate change and migration in Tuvalu. In: Oliver-Smith and Shen (eds), op. cit., pp. 32–41.

Marino, E. 2009. Imminent threats, impossible moves and unlikely prestige: understanding the struggle for local control as a means toward sustainability. In: Oliver-Smith and Shen (eds), op. cit., pp. 42–50.

Marino, E. and Schweitzer, P. 2009. Talking and not talking about climate change in Northwestern Alaska. In: Crate and Nuttall (eds), op. cit., pp. 209–17.

McDowell, C. 2002. Involuntary resettlement, impoverishment risks, and sustainable livelihoods. *The Australasian Journal of Disaster and Trauma Studies*, Vol. 2002-2. http://www.massey.ac.nz/~trauma/issues/2002-2/mcdowell.htm (Accessed 15 October 2007.)

McGranahan, G., Balk, D. and Anderson, B. 2007. The rising tide: assessing the risks of climate change and human settlements in low elevation coastal zones. *Environment and Urbanization*, Vol. 19, No. 1, pp. 17–37.

Nakicenovic, N. and Swart, R. 2000. *Emissions Scenarios. Special Report of the Intergovernmental Panel on Climate Change.* Cambridge, UK, Cambridge University Press.

Nicholls, R. J. and Hoozemans, F. M. J. 2005. Global vulnerability analysis. In: M. Schwartz (ed.), *Encyclopedia of Coastal Science*. Dordrecht, Netherlands, Kluwer Academic Publishers, pp. 486–91.

Nicholls, R. J. and Tol, R. S. J. 2006. Impacts and responses to sea-level rise: a global analysis of the SRES scenarios over the twenty-first century. *Philosophical Transactions of the Royal Society A*, Vol. 364, No. 1841, pp. 1073–95.

Nicholls, R. J., Wong, P. P., Burkett, V. R., Codignotto, J. O., Hay, J. E., McLean, R. F., Ragoonaden, S. and Woodroffe, C. D. 2007. Coastal systems and low-lying areas. In: M. L. Parry, O. F. Canziani, J. P. Palutikof, P. J. van der Linden and C. E. Hanson (eds), *Climate Change 2007: Impacts, Adaptation and Vulnerability. Contribution of Working Group II to the Fourth Assessment Report of the Intergovernmental Panel on Climate Change*. Cambridge, UK, Cambridge University Press, pp. 315–56.

Oliver-Smith, A. 2004. Theorizing vulnerability in a globalized world: a political ecological perspective. In: G. Bankoff, G. Frerks and D. Hilhorst (eds), *Mapping Vulnerability: Disasters, Development and People*. London, Earthscan, pp. 10–24.

2009a. *Nature, Society and Population Displacement: Toward an Understanding of Environmental Migration and Social Vulnerability*. Bonn, United Nations University Institute for Environment and Human Security. (InterSecTions No. 8.)

2009b. *Sea Level Rise and the Vulnerability of Coastal Peoples: Responding to the Local Challenges of Global Climate Change in the 21st Century*. Bonn, United Nations University Institute for Environment and Human Security. (InterSecTions No. 7.)

2010. *Defying Displacement: Grassroots Resistance and the Critique of Development*. Austin, University of Texas Press.

Oliver-Smith, A. and Shen, X. (eds), 2009. *Linking Environmental Change, Migration and Social Vulnerability*. Bonn, United Nations University Institute for Environment and Human Security. (SOURCE No. 11.)

Pelling, M. and Uitto, J. I. 2001. Small island developing states: natural disaster vulnerability and global change. *Environmental Hazards*, Vol. 3, No. 2, pp. 49–62.

Pfeffer, W. T., Harper, J. T. and O'Neel S. 2008. Kinematic constraints on glacier contributions to 21st century sea-level rise. *Science*, Vol. 321, No. 5894, pp. 1340–43.

Poncelet, A. 2009. *Bangladesh Case Study Report, 'The Land of Mad Rivers'*. EACH-FOR (Environmental Change and Forced Migration Scenarios). Université de Liège, Belgium, CEDEM.

Ringot, E. and Kanagaratnam, P. 2006. Changes in the velocity structure of the Greenland Ice Sheet. *Science*, Vol. 311, No. 5768, pp. 986–90.

Scudder, T. 1981. What it means to be dammed: the anthropology of large-scale development projects in the tropics and subtropics. *Engineering & Science*, Vol. XLIV, No. 4, pp. 9–15.

Scudder, T. and Colson, E. 1982. From welfare to development: a conceptual framework for the analysis of dislocated people. In: A. Hansen and A. Oliver-Smith (eds), *Involuntary Migration and Resettlement*. Boulder, Col., Westview Press.

Tol, R. S. J., Bohn, M., Downing, T. E., Guillerminet, M. L., Hizsnyik, E., Kasperson, R., Lonsdale, K., Mays, C., and co-authors. 2006. Adaptation to five metres of sea-level rise. *Journal of Risk Research*, Vol. 9, No. 1, pp. 467–82.

Vaughan, D. G. and Spouge, J. R. 2002. Risk estimation of collapse of the West Antarctic ice sheet. *Climate Change*, Vol. 52, Nos. 1–2, pp. 65–91.

WBGU. 2009. *Solving the Climate Dilemma: The Budget Approach*. Berlin, German Advisory Council on Global Change. (Special Report.)

Wisner, B., Blaikie, P. Cannon, T. and Davis, I. 2004. *At Risk: Natural Hazards, People's Vulnerability and Disasters*, 2nd edn. New York, Routledge.

8

Environmental change and forced migration scenarios: methods and findings from the Nile Delta, Sahel, and Mekong Delta

KOKO WARNER, TAMER AFIFI, ALEX DE SHERBININ, SUSANA B. ADAMO AND CHARLES EHRHART

1. Introduction

Emerging empirical research indicates that environmental changes, including climate change, currently play a role in migration. As environmental changes increase, migration pressures related to these changes may also grow (IPCC 2007). Today, environmental change including climate change contributes to human mobility embedded in linked environmental and social processes. Social system characteristics including social networks play a mediating role in how environmental change affects whether people move away or stay at home. Migration can represent a response to changing environmental and economic conditions, such as farmers' choice to migrate due to failing crops and insecure livelihood prospects. Migration can also exacerbate environmental and economic problems in receiving areas. For example, urban areas attract migrants seeking better lives. High in-migration contributes to crowding and environmental/sanitation issues in slums.

This chapter explores how environmental shocks and stresses, especially those related to climate change and variability, can push people to leave their homes in search of better lives or mere survival. More erratic weather, rising sea level and other climate change impacts will exacerbate both migration pressures and environmental degradation. What is certain from empirical and theoretical research on environmentally-induced migration, in all its varieties, is that environmental change is one of many contributing factors.

This chapter shares the methods and fieldwork experiences of a first-time, multi-continent survey of environmental change and migration from the recently completed research project supported by the European Commission: Environmental Change and Forced Migration Scenarios (EACH-FOR; www.each-for.eu). The chapter also presents some of the empirical evidence generated during this project in the Nile Delta, the Sahel, and the Mekong Delta.

1.1   Aim of EACH-FOR project: Investigate environmental change and migration

In 2006, under the 6th Framework Program, the European Commission issued a call for proposals to support comparative research of factors underlying migration and refugee flows, including illegal immigration and trafficking in human beings. In response, seven organizations formed a consortium to investigate whether and how environmental change affects human migration.

The EACH-FOR Project was conceived as a multidisciplinary study that aimed to undertake original research through deskwork and case study research, complemented by statistical and other information sources. The project included researchers from many academic fields: migration specialists, economists, geographers, historians, environmental scientists, civil engineers, etc.

The research consortium created a set of broadly comparable studies using a unified methodological research approach. It was hoped that using one common approach across almost two dozen case studies would create a set of internally valid results, with some degree of external validity as well. Up until the time, there had been several studies of individual locations. But no study had been attempted using the same method across multiple case studies. EACH-FOR was the first global attempt to undertake field-based research on environmental change and migration at this scale. It also was apparent that the nature of the questions the project was asking would require a mixed methods approach. Given the parameters of the project that was in development in the autumn of 2006, EACH-FOR combined qualitative and quantitative approaches to first describe and then analyze the diversity of migration patterns it observed. Subsequent synthesis reports supported this approach (Boano et al., 2008; Brown, 2008; Kniveton et al., 2008; Piguet, 2008; McLeman and Smit, 2006).

The project objective was to better understand the role of environmental change in forced migration (both internal and international) and its related societal consequences, and provide plausible future scenarios of environmentally induced forced migration. The EACH-FOR project worked at the community level, interviewing experts and affected people or households. EACH-FOR worked in specific countries, but only in particular provinces expected to be more affected by both (documented) environmental change and migration. It was assumed that the household was the migration decision-making unit, and questions about the household unit are reflected in the questionnaires (understanding that migration decisions are made in combination with other risk management/ resource management decisions).

2. Definition of central concepts: environmental change and migration

Terms and concepts such as environmental or climate change migration, environmentally-induced or forced migration, ecological or environmental refugees, and climate change refugees continue to be used throughout the emerging literature, with no general agreement on precise definition (Hugo 1996; Dun and Gemenne 2008). Scholars have pointed out the challenge of isolating environmental factors from other migration drivers (Black 2001; Castles 2002; Boano et al. 2008).

Expert estimates range widely in part because no measurable definition exists. Estimates of the numbers of migrants and projections of future numbers are divergent and controversial. A middle-range estimate cited recently by IOM puts the figure at 200 million by 2050 (Brown 2008). The first controversy concerns the categorization of people made mobile by environmental factors including climate change. Some organisations refer to "environmental refugees" while others, following the position of UNHCR, stress that the word "refugee" has a specific legal meaning in the context of the 1951 Geneva Convention Relating to the Status of Refugees.

It was recognized from the beginning of the EACH-FOR project that it might be difficult to interpret research results, as the literature has established that migration outcomes have multiple causal factors. This has led some "minimalists" to claim that environmental factors are contextual issues in migration decisions, while "maximalists" make stronger claims about the causal driving role of environmental change in migration and displacement. The existing data on migration are uneven, with much of the information based on international migration figures from census data that do not necessarily capture temporal or geographic dynamics of human movement (Kniveton et al., 2008; Afifi and Warner, 2008). Institutions have not gathered systematic and comparable time series data about migration, with the exception of international migration (measured as stocks in national census, although migration processes are flows in reality). This presented a situation in which it would be challenging to measure any difference between migration in the absence of the independent variable and migration in the presence of the independent variable (environmental change). Pragmatic budgetary and time constraints would prevent the project from filling these gaps by undertaking long-term observations to assess environmental change processes, and how these might affect migration through time.

EACH-FOR used the working definition of environmentally induced migrants proposed by the IOM: "Environmentally induced migrants are persons or groups of persons who, for compelling reasons of sudden or progressive changes in the environment that adversely affect their lives or living conditions, are obliged to leave their habitual homes, or choose to do so, either temporarily or permanently, and who move either within their country or abroad" (IOM, 2007, p1).

This working definition is comprehensive, and identifies environmental degradation as an important push factor triggering migration. Its limitations include that it does not distinguish between temporal or permanent migration, the particular environmental circumstances people migrated under, whether return is possible or not, etc. The working definition does not contribute to isolating the independent variable or understanding the interaction between it and the dependent variable of interest. New typologies are being developed to address these issues (Renaud et al. 2007).

3.   Project design for investigating environmental changes and human migration

The project method included the following phases: questions and hypotheses, literature review and general overview studies in each region of investigation, case study selection, fieldwork method design, fieldwork, analysis of findings and reporting. This section focuses on the project's design[1] including hypothesis formation, literature review, case study selection, and fieldwork.

The EACH-FOR project faced many considerations about how to design the research approach. These considerations stemmed from the multidisciplinary nature of the research question, but especially the omnipresence and characteristics of the independent variable environmental change. Table 8.1 outlines the research design process of the EACH-FOR Project. This section describes and analyses how the project addressed the challenges involved in investigating links between environmental change and migration.

3.1   Hypothesis and research questions about environmental change and migration

EACH-FOR was conceived as an initial study, at the community level, upon which further extensive research would be built. EACH-FOR came up with a set of questions to help gather observations in the field that could test the central hypothesis of the project. The project chose the dependent variable migration (including a range from internal to international migration), and a range of environmental change factors as the independent variable. The EACH-FOR Project started its inquiry by forming a hypothesis to test in desk and field research: There is a discernible environmental signal in human migration patterns today. The project considered this general hypothesis to hold true if fieldwork found qualitative and quantitative evidence that migration occurred, in part, due to environmental factors. The failure to find migrants in whose mobility pattern environmental causes were negligible or played no role would negate the central hypothesis.

The project defined eight central questions to guide interviews during the collection of data in desk study and fieldwork. It was hoped that the answers to these questions would aid researchers in determining the validity of the hypothesis, which is crucial given the challenge of defining meaningful control groups (discussed below). These questions tried to avoid drawing a deterministic relationship between environmental degradation and migration, while helping to identify cases where environment plays an important role as a contributor to population movement.

---

[1] Other works have presented the findings and next steps for research, and are not discussed in this chapter. For the interested reader the following references may be useful: Jäger et al. 2009; Warner et al. 2009a; Warner and Laczko 2008.

Table 8.1 *EACH-FOR project research steps and design issues*

Step	Description	Design issue
1. Hypothesis	Discernable environmental signal in migration today. Null hypothesis: no discernible environmental signal in migration today.	How to establish whether the environmental signal is discernible in migration patterns? How to assess or measure environmental signals?
2. Variables	Independent variable of interest: environmental change. Dependent variable: migration.	How to isolate the independent variable of environmental change? How to determine that presence of independent variable caused dependent variable?
3. Intervention group and control group	Intervention group made up of people who will experience environmental change. Control group made up of people who will not experience environmental change.	How to isolate control group that does not experience environmental change (independent variable)?
4. Introduce intervention	Environmental change.	Impossible to control environmental change; need to carefully select case study countries.
5. Measure dependent variables in intervention group and control group	Did migration occur when environment changed?	How to prove that migration would not have occurred in the absence of environmental change?

1. Who is migrating away from situations of environmental degradation/change?
2. Where are environmentally induced migrants coming from and where are they going to?

3. Why have people migrated (i.e., what role has environmental degradation or change played)?
4. How does environmental degradation interplay with other social, economic and political factors in decisions relating to migration?
5. What might prevent people from migrating when they are faced with environmental degradation (i.e., what assistance was needed, what was lacking)?
6. Why do some people remain in areas of environmental degradation/change while others migrate (i.e., what are their coping/adaptation strategies and capacities)?
7. How does environmentally induced migration occur (e.g., choice of destination, networks used)?
8. What is the role of people's perception of environmental degradation in triggering them to move?

These eight questions provided a basis upon which individual case studies could build additional hypotheses about the particular relationships between environmental factors in specific areas and migration trends there. Its case studies were intended to provide insights into the many possible hypotheses that could subsequently be formulated and tested.

Defining a control group to test hypothesis

Taking environmental change as an independent variable was associated with at least three issues. First, the set of environmental variables that make up 'environmental change' are difficult to isolate from other factors driving migration. The limitation of this hypothesis lies in determining a measure for 'discernible' and environmental 'signals' and in defining a null hypothesis. The latter would require isolating variables so one could establish that the environment played no role in migration. The EACH-FOR Project treated the independent variable as certain types of environmental change: it employed a multi-case study approach to examine types of environmental change and how these environmental factors affect migration. Similarly, the control of the independent variable in this kind of research attempt is challenging: there are likely few cases of migration where research can fully exclude the environmental variables of interest. Environmental processes are often ongoing and present as a major background factor in all migration or non-migration situations. This presents a need to devise a methodology that can test the impact of environmental change on migration.

The ideal design would allow for the isolation of intervention groups and control groups of migrants and non-migrants in every case study area, so that the hypothesis could be established or rejected in each case. EACH-FOR faced two practical difficulties: The researcher could not randomly assign individuals into the two groups; and it was unclear how to isolate a control group in each case that does not experience environmental change. Some case study areas offered conditions in which some parts of the country experienced a particular kind of environmental change, while other areas remained intact[2].

Measuring the intervention and control group

EACH-FOR researchers tried to find ways to establish whether migration would not have occurred in the absence of environmental change. To test whether there was indeed an impact on migration when the environment became less hospitable, the project had a three-step procedure: first, desk research was undertaken to examine historical patterns of both environmental change and migration; secondly, expert interviews were conducted to help capture the dynamics of environmental change and how this might have affected human mobility in the past and current situation in a given case study; and, finally, a questionnaire was given to migrants and non-migrants who had stayed behind in areas with documented cases of environmental degradation.

This latter comparison of migrants and non-migrants was hoped to reveal answers to the central question of the project: what role has environmental degradation or change played in people's decision to migrate or not to migrate? For those individuals that remained behind, the project was keen to understand what factors intervened to keep people from migrating, even when they faced environmental problems. It was hoped that this set of answers would help researchers verify or reject the hypothesis, and also help measure the environmental signal in migration patterns (especially in relation to other factors).

3.2 Literature review and general overview studies

At the time, this type of investigation had not been done, so researchers first conducted an analysis of literature and methodologies (see Vag

---

[2] A pre-test (assessment of the migration situation before environmental change was introduced) was not possible, due to limited time and budget, and the more important fact that it is impossible to control the independent variable.

et al., 2007). This review showed a number of individual case studies that mentioned migration or environmental change. Yet existing studies lacked consistent and comparable data on migration related to environmental change. The project then conducted general overview studies of each major region to be investigated. The studies were divided into sections that looked at environmental issues in the region, migration and displacement in the region, and interactions between environmental factors and human mobility in the region.

Research on environmentally induced migration falls into the realm of complex interactions of human and natural systems. These interactions include livelihood impacts, conflict, and a variety of human-induced environmental change including climate change. Lonergan described migration as "an extremely varied and complex manifestation and component of equally complex economic, social, cultural, demographic, and political processes operating at the local, regional, national, and international levels" (Lonergan, 1998). Environmental factors are in most cases not solely responsible for driving migration, so understanding the phenomena becomes a task of defining causes and consequences of environmentally induced migration. It remains difficult to isolate these contributing factors from one another to determine their relative importance (Black, 2001; Castles, 2002; Biermann, 2007; Boano et al., 2008). Quantifying the numbers of environmentally induced migration is problematic, in part because no measurable definition exists (Döös 1997).

3.3   Case study selection

The project design aimed to observe cases where both independent and dependent variables were present, in order to determine whether there was a discernible environmental signal in migration patterns (i.e., whether the independent variable affected migration). The case studies were conducted in areas within countries where it was thought that both independent and dependent variables were present. Figure 8.1 indicates the countries which were selected as case studies, focusing on developing and transition countries because of their particular exposure to extreme events and climate change-related stressors (IPCC 2007).

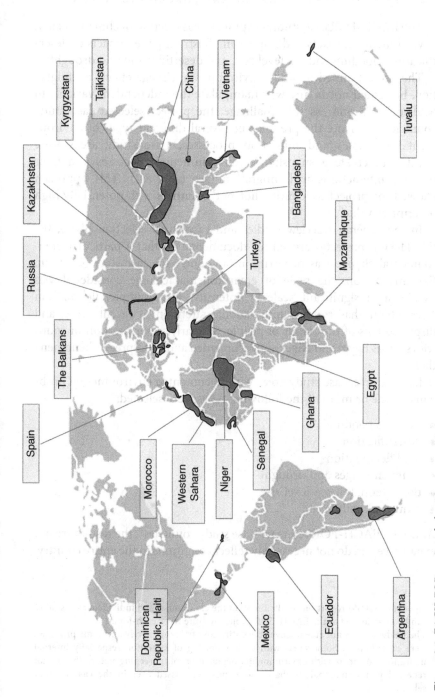

Figure 8.1 EACH-FOR case study locations

The EACH-FOR case studies explore a cross section of climate-related environmental impacts and migration, illustrating the impacts of desertification, flooding and sea level rise, and desertification and drought[3].

The project aimed to study environmental change effects on migration, but could not in any way manipulate the independent variable in fieldwork. To address this challenge, the project selected case study countries to ensure the presence of several different types of environmental processes and migration processes. Case study areas were selected to create a snapshot of environmental processes and their possible interactions with migration (see table 8.2). This approach allowed the project to identify 'hotspot' countries with potentially high descriptive value[4].

In the general overview studies and case study selection phase, the EACH-FOR project searched for documentation that a particular environmental change was occurring in a country for case study selection. For this reason, the scale of environmental change considered was usually at a significant scale or intensity. For example, flooding in Mozambique has reached international attention in recent years, and displaced tens of thousands of people. Similarly, desertification in countries such as Egypt and Niger are documented and widespread phenomena there.

For example, case study areas with documented environmental problems of one or more of the following types were selected:

- extreme flooding
- desertification
- land degradation
- water shortages and drought
- the potential of sea-level rise
- industrial pollution.

Although EACH-FOR refers to "case study countries", the results are not exhaustive and do not necessarily reflect conditions in the entire country.

---

[3] Jäger et al. (2009) report on results for all of the case studies and individual case studies can be downloaded at the EACH-FOR project website www.each-for.eu.

[4] The fieldwork was not able to cover the entire country exhaustively and some processes were cross-border. Some areas with underreporting of migration (especially internal migration), or areas with environmental degradation of a creeping nature that is not reflected in international databases, were possibly passed over in the case country selection.

Table 8.2 *EACH-FOR case studies overview (regions/countries; environmental issues addressed and case study sites)*

Case study region/ countries	Environmental issues addressed	Case study sites
Asia-Pacific		
Bangladesh	Sea-level rise; cyclones	Coastal Regions of Bangladesh (South West), chars (moving islands) on Jamuna River (North West), Dhaka
China	(a) Dam construction	(a) Three Gorges Dam, affected regions: Shangdong Province, Jiangsu Province, Chongming Island of Shanghai. Zhejiang Province
	(b) Desertification	(b) Erenhot, Inner Mongolia Autonomous Region
Tuvalu	Sea-level rise, erosion, waste disposal, water stress	Funafuti Atoll, Tuvalu and Auckland, New Zealand
Viet Nam	Flooding	Mekong Delta, particularly An Giang Province
Central Asia		
Kazakhstan	(a) Desertification and water stress	(a) Aral Sea region, Almaty
	(b) Nuclear testing	(b) Semipalatinsk
Kyrgyzstan	Soil pollution, waste disposal, landslides, earthquakes	Whole country, with particular focus on the Ferghana Valley
Tajikistan	Soil pollution, degradation and erosion, mud flow, landslides, floods, earthquakes	Whole country, with particular focus on the Ferghana Valley
Europe and Russia		
Spain	Water shortage and desertification	South-eastern regions of Spain – Murcia and Almeria
Turkey	(a) Dam construction	(a) South-east Turkey (Adiyaman – Samsat District; Urfa city centre); West of Turkey (Didim-Yalikoy village; Izmir – Torbali)
	(b) Water destruction	(b) South-east Turkey (Urfa – Suruc District); Istanbul

Table 8.2 (*cont.*)

Case study region/countries	Environmental issues addressed	Case study sites
Balkans	Unavailable at time of printing	Danube Basin
Russia	Unavailable at time of printing	Unavailable at time of printing
Latin America and Caribbean		
Argentina	(a) Floods, increase of rain – water excesses with periods of abnormal droughts	(a) Pampa Arenosa and Depresión del Salado north-west of the Province of Buenos Aires
	(b) Droughts – decrease in water availability, melting of glaciers	(b) Pre-Andean region (Comahue and the city of Jáchal, San Juan)
	(c) Droughts – decrease in water availability.	(c) Yungas in the Salta Province
Dominican Republic, Haiti	Deforestation (and its consequences during tropical storms)	Province of Independencia, Dominican Republic; Port-au-Prince, Haiti
Ecuador	Water quality and availability; soil degradation; climate issues (ENSO and its consequences);	Guayas, El Oro, Pichincha, Manabi, Imbabura, Bolívar, Tunguruha, Azuay eta Quevedo provinces
Mexico	(a) Tropical storms, landslides, flooding;	(a) Soconusco/Chiapas, south-eastern Mexico;
	(b) Desertification, soil degradation	(b) Western Tlaxcala (approx 60km east of Mexico City)
Middle East and North Africa		
Egypt	Water shortage	Newly reclaimed desert lands (Western Cairo), Cairo slums, Nile Valley and Nile Delta, Upper Egypt (Southern Egypt).
Morocco	Water shortage, desertification and the impact of other environmental challenges on rural villages in arid areas	Desert fringe villages in south-east Morocco: the two most southern Oases of the Draâ river valley: Mhamid and Tagounite (Province of Zagora)

Western Sahara	Desertification and water shortage	Algeria: Interviews with refugees from Western Sahara in refugee camps in Algeria (Tindouf region) under the control of the *Frente Polisario* government-in-exile
Sub-Saharan Africa		
Ghana	Unreliable rainfall, poor soil fertility	Source area: Upper West Region; Destination area: Brong Ahafo Region
Mozambique	Flooding, droughts	Central Mozambique - Zambezi River Valley
Niger	Droughts, deforestation, overgrazing, land degradation, Niger River problems and Lake Chad drying out	Niamey, Tilabéri
Senegal	Desertification, drought and water management	Fatick and Kaolack (the Peanut Basin) in Central Senegal and the Sénégal River Valley in Northern Senegal

Rather, "hotspot" locations within each country were examined and analysed[5]. The case studies, however, do provide a good overview of patterns which are now subsequently under more profound investigation.

Additionally, as the number of questionnaires in each case study area was limited and sampling techniques not exhaustive, EACH-FOR results do not lend themselves to robust statistical analysis. The general trends do provide insights, however, about decisions to migrate (or not) related to environmental processes.

The EACH-FOR Project involved multiple case studies, research teams and field workers, and it created a comparable questionnaire for both migrants and non-migrants, as well as guidelines for semi-structured expert interviews.[6] The questionnaires were pre-tested in an early case study and then adjusted and revised before all other case study work began. All investigators received field guidelines about how to work with participants/interviewees, record their results, and interpret the results (to ensure interpreter reliability).

3.4   Design of fieldwork approach

The next step in the method was to design the fieldwork approach. To explore the questions outlined above, the EACH-FOR project asked experts, migrants and non-migrants about their perceptions of environmental factors and whether these factors had any relationship with the decision to migrate or not to migrate. Where participants answered positively, this was considered as evidence that environmental factors were perceived as having played a role in migration (discernible).

The approach was first tested in early case studies, then discussed by the project and refined. Each case study gathered 30 or more migrant and non-migrant interviews per case study, in addition to 15 or more expert interviews. On average, field researchers gathered about 65 observations per case study area, and some case studies gathered substantially more observations. Over the full project, about 1500 observations/surveys were gathered from migrants and non-migrants who indicated whether environmental factors had played a role in their decision to migrate.

---

[5] The reports Jaeger at al (2009) and Warner et al. (2009a) more specifically delineate the areas within countries (based on provincial boundaries) where fieldwork was conducted.

[6] The migrant and non-migrant questionnaires can be found online at: http://www.each-for.eu/index.php?module=project_outline.

Confounding factors in fieldwork

Even in the preparation phase for fieldwork, the project identified several limitations of field questionnaires (one developed for migrants, and one for non-migrants) and interviews. These general limitations of the EACH-FOR project in establishing the relationship between the independent variables (types of environmental degradation) and the dependent variable (migration) included limits in fieldwork budgets. This constrained time spent in case study areas and placed constraints on gathering extensive field observations. Further, the reality of fieldwork inevitably led some researchers to change wording to meet local conditions. Language translation further exacerbated the instrumentation confound, and complicated the interpretation and comparison of results from one case study to another. Similar to the issues mentioned above, this approach created locally-specific and useful case studies with a moderate degree of comparability.

To address language and local context challenges, EACH-FOR worked in many case study countries with the International Organization for Migration (IOM) local in-country offices, and in all case study countries with local researchers and experts. IOM and local experts, to the extent possible, were engaged as partners prior to the fieldwork commencing. They played a crucial role as locally-based partners in identifying and establishing initial contact with relevant experts, arranging logistics and assisting with translation and implementation of the EACH-FOR questionnaire. A key contact point within IOM was established as the main interpreter and assistant for the duration of the field research in all locations.

Controlling for threats to validity of project findings

The project considered several designs to help control for threats to validity and increase the internal and external validity of results. An 'ex-post facto' design and a 'static group comparison design' were particularly considered. In the end the 'ex-post facto' design was chosen. The reasons for this choice are discussed below.

The ex-post facto design is used when a single group of people is measured on some dependent variable (migration) after an intervention (environmental change) has taken place. This research design is often used when it is impossible to manipulate the dependent variable (migration). The researcher tries to evaluate the experiment by interviewing people (observation) and assessing the impact of the intervention. This design involves no pre-test or control group – two characteristics that fit the situation of the EACH-FOR Project.

The project considered an alternative research design: the two-group post-test design. In this design, the researcher has no control over assignment of participants. This leaves the static-group comparison design open to irresolvable validity threat. There is no way of telling whether the two groups were comparable at time 1, before the intervention, even with a comparison of observations 1 and 3. The researchers can only guess whether the intervention caused any differences in the groups in time 2. The short nature of the project (24 months) weighed against the nature of the independent variable (environmental change, which may happen abruptly or gradually or not at all in a discernible way in 24 months) and did not offer significant advantages over the ex-post facto research design. Yet this research design makes it difficult to be sure that the observations from fieldwork are the result of some particular intervention (environmental change). In spite of this weakness, the ex-post facto design has the potential to produce powerful intuitive results based on numerous migrant responses about environmental factors that contributed to household migration decisions.

3.5   Applying the EACH-FOR project design in fieldwork

Following the design stage, the EACH-FOR Project undertook fieldwork in 23 areas across the world to try and apply the methods described above. The aim was, to the extent possible, to produce a set of globally comparable case studies. This section examines some of the fieldwork experiences using this methodology in the case study countries portrayed in Figure 8.1 above.

Non-probability sampling technique in fieldwork

A challenge was the possibility of selection bias related to limitations around the control group, random assignment and control over assignment of participants to groups. The project limitations did not allow for a random sample of a large population of people. Field researchers were looking for people exposed to environmental problems in order to ask whether those problems affected the participants' decisions about whether or not to migrate. In most cases, researchers were only able to interview people from a limited number of areas, due to time and budget constraints. Expert interviews and desk studies were used to help balance the sampling biases that would emerge in migrant and non-migrant questionnaires.

A non-probability sampling method was chosen because it was suitable for research during which the population of interest is not fully visible and where accurately defining the population of interest is problematic. This method fits the nature of the problem: to better understand the impact of a little-understood variable (environmental change) on decisions relating to migration. The snowball, or chain-referral, sampling method was used in the project. Researchers identified an initial set of relevant respondents in pre-fieldwork preparations. During field interviews, researchers requested that participants suggest other potential subjects who shared similar characteristics or had relevance in some way to the object of study. This second set of subjects was then interviewed, and also requested to supply names of other potential interview subjects. The process continued until the individual researcher felt that the sample was large enough for the purposes of the study. The limited amount of time for each case study – an average of seven weeks – prevented an exhaustive sampling.

The researcher for each case study was directly involved in developing and managing the initiation and progress of the sample. Each researcher sought to ensure that the chain of referrals remained within boundaries that were relevant to the study. Researchers were instructed to ensure that the initial set of respondents was sufficiently diverse so that the sample was not skewed excessively in any one particular direction (Tansey, 2006).

Perception of environmental change

From the outset, the project sought a set of methodologies that would allow for an ex-post observation (observation after the fact) of whether or not environmental variables affected migration and, if so, how this occurred. One implication of this was that an important part of the EACH-FOR Project's work was to gather information about how people *perceive* the influence of environmental factors on their decision to migrate.

Because the EACH-FOR Project placed such importance on an ex-post methodology, the ability of subjects to perceive change required them to be at a boundary where change could be observed – either a physical boundary, such as the desert noticeably advancing onto a subject's field, for example, or a noticeable time boundary, such as a violent storm or an exceptionally dry period (in a time scale relevant to human memory).

Timing issues: Gap(s) between environmental events and site visits

Researchers found that the more recently a particular locality had experienced the environmental problem or issue under investigation, the more people in that locality were aware of the situation. Therefore, they were able to discuss their recent experience as a factor leading to livelihood impacts and possible migration or resettlement. For example, in the EACH-FOR Mozambique case study, communities in flood-prone areas had experienced the 2007 flooding event six months prior to the researcher's visit. On the other hand, in the Vietnam case study, the last major disaster flood in the Mekong Delta occurred in 2000 (Dun, 2009). Considering that the research was conducted seven years later, it was more difficult to pinpoint the exact migratory impacts of that particular event, despite the fact that people could clearly remember the event.

In addition, the personal experience of a migrant or non-migrant with the environmental event in the past also played a significant role in shaping research findings. For example, in Mozambique, interviews were carried out with people who had experienced multiple flooding events. This revealed a change in attitude towards migration or resettlement from temporary evacuation to permanent resettlement. The affected people accepted the fact that they should not move back to their places of origin. Similarly, in Vietnam, the research found that the impacts of repeated flooding (as opposed to a unique flooding event) could play a crucial role in prompting people to migrate. In Niger, gradual environmental change is a constant part of life, the time gap between environmental events and the site visit did not play a significant role in influencing the results of migrant interviews (Afifi 2009a). Even in the cases of the severe droughts of 1973 and 1984, farmers and cattle herders have continued to suffer from the cumulative impacts of those events.

Another timing issue that had to be taken into account was the day-time and seasonal timing, especially for rapid research, for interviewing migrants/non-migrants in the field location. For example, the researchers in Mozambique and Niger found that the farmers and cattle herders were only available in the villages at certain times of the day and the year (Stal 2009; Afifi 2009a). Likewise, when the farmers living along the

Zambezi River Valley were busy seeding their almost inaccessible fields, it was mainly the elders and the children who were interviewed. In the case of Niger, the researcher visited the country in the dry season (January/February 2008), during which many farmers wander with their cattle to other regions that were not covered in the field visit. Therefore, the researcher had to rely on interviews with people who stayed in the Tilabéri region, as well as people in the capital.

Additional research challenges

While there was one overall field research goal (that of investigating whether there were linkages between the environment and migration), there were several field research objectives. The focus on whether there was a link between environment and migration meant that the field research did not strike a balance in terms of exploring the range of other reasons why people migrated. The project attempted to address each objective with the questionnaire, semi-structured migrant and non-migrant interviews and interviews with experts. Yet each of the field objectives could require a specific type of methodology.

The fieldwork was a scoping exercise, but one that did not allow for repeated visits to follow up on information gathered. Researchers conducted both expert interviews, migrant and non-migrant interviews and questionnaires within the field research phase. There was little time for analysis of expert interviews before the researchers interviewed migrants and non-migrants. At times, for example, the researchers obtained crucial pieces of information from experts, following interviews with migrants – information that would have been useful to know before interviewing migrants.

The project recognized the difficulty of attempting to explain the patterns and trends of migration, both international and internal, using only one approach or academic discipline. For example, push factors were frequently mentioned in fieldwork, as migrants sometimes mentioned declining livelihoods from farming at home, due to land degradation or erosion, and the sense that a combination of environmental and economic factors contributed to migration.

At the outset of each field study, field researchers visited experts and institutions in the case study country (usually in capital cities). It was hoped that these experts could assist in "ground truthing" background preparation and offer more specific insights about the overall country situation, the kinds of policies that affect environmental degradation and migration, and feedback about identification of locations where migrant questionnaires could be conducted fruitfully within the countries. The experts sought were usually in government ministries of agriculture, environment, justice (dealing with border issues, migration), and disaster risk management and humanitarian assistance. NGOs, international organizations, and local academics provided further inputs at this stage of field research.

The EACH-FOR Project also had several assets to address some of the practical limitations of completing an ambitious two-year project on a complex research topic. The project had access to good statistical data sources and geo-information, as well as a partnership with numerous local partners, research organizations, and international organizations. In many of the field studies, the International Organization for Migration (IOM) played a critical role in linking EACH-FOR researchers with these local experts. Where possible and appropriate, IOM accompanied researchers in several case studies and helped provide input about the local context. Field researchers used sampling techniques such as the snowball method to identify migrants who had departed a situation of *sudden* or *gradual* environmental change. Field researchers also tried to assess the situation of potential migration, by looking for contacts with people or groups who had not departed the same situation of sudden or gradual environmental change (control) for particular reasons but who might migrate in the future. The project relied on the relative importance that interviewees placed on environmental factors to begin to isolate the relevance of environmental change variables in the migration choice. Second, 'environmental change' is comprised of many different phenomena, spanning different geographical and temporal scales.

3.6   Analysis of project findings and reporting

In summary, given these methodological issues, some scholars expressed caution in a 'driver-focused' framing of environmental factors and migration (Black, 2001). However, the EACH-FOR consortium recognized the need to address the knowledge gaps, particularly in light of findings from the IPCC fourth assessment report indicating that the impacts of climate change are expected to increase in the future (IPCC 2007), which coincided with the project's start. The EACH-FOR project was given the timely opportunity to gather information from the field and report its findings back to a wider academic and policy-centered community. The EACH-FOR Project accepted these limitations and shaped itself as a scoping study that would contribute to the building of a basis upon which more rigorous scientific studies about the relationships between environmental change, intervening variables like livelihood security, and human mobility could be undertaken. At the time of publication, a new generation of research is underway to build upon EACH-FOR's research contributions.

We now turn to a summary of three case studies to illustrate the kinds of results yielded by the EACH-FOR approach.

4. Selected results related to migration, sea level rise and desertification, and flooding

As noted, the EACH-FOR Project undertook fieldwork in 23 areas across the world to try and apply the methods described above. The aim was, to the extent possible, to produce a set of globally comparable case studies. This section examines some of the fieldwork experiences using this methodology.

In the subsections which follow, this paper examines the qualitative results of EACH-FOR case studies in Egypt (the Nile Delta), Vietnam (the Mekong Delta), and Senegal, Niger, and Ghana (Western Africa). These case studies were selected for this chapter because they represent a cross section of climate-related environmental impacts and migration, illustrating the impacts of desertification, sea level rise, flooding, desertification and drought.

4.1   Nile Delta: Between desertification and sea level rise[7]

In Egypt slow-onset events like sea level rise and desertification affect the Nile Delta (Jäger et al. 2009). The total area of the Arab Republic of Egypt is about one million km², most of which has an arid and hyper-arid climate. The most productive zones in Egypt are the Nile Delta and Nile Valley (3 percent of the total land). Projected increases in sea levels will pressure a quickly growing population into more concentrated areas. Desertification and soil degradation claim large swaths of land on the

---

[7] This section is based on Afifi, 2009a.

Eastern and Western Nile Delta. Vast areas of land may be rendered unusable by the dual climate change-related forces of desertification and sea level rise. In the future, sea level rise could affect an additional 16 percent of the population (ibid).

The overall area influenced by the active encroachment of sand and sand dunes is estimated to be roughly 800,000 hectares (Hegazi and Bagouri, 2002). Land productivity has diminished by about 25 percent compared to its original productivity (ACSAD, 2000). The annual erosion rate has been estimated between 0.8 and 5.3 ton/ha/year (Editorial Board, 2005). Desertification and land degradation drive some people to migrate internally in search of livelihoods.

The government of Egypt combats desertification through an internal migration scheme related to the Mobarak National Project in the Western and Eastern Delta. The program was initially designed to alleviate environmental problems but also unemployment, poverty, and overpopulation in Cairo, Beheira, Kafr El-Sheikh, and Qalioubia. This project aimed to create an internal urban-to-rural migration flow towards the edges of the Delta.

People who were resettled in the *Eastern* Delta were mainly unemployed young men from urban slums. In contrast, the people who moved to the *Western* Delta were mainly farmers affected by a law that favored land owners who could easily drive away share croppers from desirable agricultural areas. After eviction, the share croppers were moved by the government to the Western Delta.

The program allocated each sharecropper/farmer in the Eastern and Western Delta a land parcel of 10,500m^2, and often additional migrants came to work as peasants in these areas. Soon, however, reclaimed areas began to manifest soil and water salinity problems. When it became too expensive to dig new wells for groundwater, many landowners sold their land and evicted the migrant peasants (Afifi 2009a). The new immigrants received shelter and agricultural extension and veterinary services from the government and NGOs. Government funding provided migrants with pesticides and artificial crop pollination. Yet initial investments and incentives to encourage poor people to migrate to new areas tapered off with time. The Western and Eastern Delta lack access to potable water, proper infrastructure, public facilities, schools, health care, and well-functioning sewage systems. Consequently, many migrants did not stay and others are expected to leave either to other regions or to return to their original regions. Today, only half of designated resettlement land has been utilized.

4.2   The Sahel: Pressure on livelihoods and creeping onward migration[8]

Land degradation, desertification, and deforestation are factors that potentially result in mobility as a household adaptation strategy (de Sherbinin et al 2007). Land degradation, as defined by Article 1 of the Convention to Combat Desertification, is defined as a "reduction or loss of biological or economic productivity of ecosystems resulting from climatic variations, land uses and a combination of processes such as: soil erosion, deterioration of soil properties and long-term vegetation loss." Thus, losses of land productivity are inextricably linked to climate change.

Although precise estimates of the land affected by degradation are difficult to obtain, some estimates suggest that more than one-third of drylands are affected by land degradation (Clarke and Noin, 1998). Land degradation is a major concern in West Africa, where about 65 percent of the cultivable lands have degraded. From 2000–2005, West and Central Africa lost 1.36m ha of forest cover per year, or a total of 67,800 sq km (FAO, 2005). More than 300 million people in Africa already live with water scarcity, and areas experiencing water shortages are likely to increase by almost a third by 2050.

West Africa is made up of a diversity of ecosystems, ranging from more tropical humid in the South to arid in the North. While climate change projections of seasonal or annual precipitation are uncertain, the projected increase in intensity of rainfall events, superimposed on the region's already high climate variability, is likely going to lead to increased frequency of droughts and floods. Water shortage and land degradation affect large areas of the Sahel, a region south of the Sahara and north of the humid zone that spans west to east across nine countries from Mauritania and Senegal to Sudan. In the Sahelian zone of Western Africa, two different drought events – a large drought from 1968–74 and a slightly less intense one from 1982–84 – were among the worst on record (Hulme, 2001). During the first drought, more than 100,000 people died, most of whom were children (Bryson and Paddock 2003). By 1974, more than 750,000 people in Mali, Niger and Mauritania were totally dependent on food aid (Wijkman and Timberlake, 1984). These droughts and consequent land degradation are now understood to have been caused in part by a pattern of warming of the tropical oceans which

---

[8] This section is based on Afifi, 2009a.

itself may have been driven by anthropogenic climate change (Giannini et al, 2008). Such environmental pressures could grow in the future with climate change.

Forty-four percent of West Africa's population works in the agricultural sector, most of them at a subsistence level (FAO 2005). Despite the high dependence on agriculture in this climatically variable region, the actual areas under irrigation are among the lowest on a per-area basis for any region in the world. For example, in Senegal in 2005, only 67,000 ha were irrigated out of 8.8m ha, or less than 1 percent of the total (ibid). Although the Sahel has seen a "greening" since the mid-1980s drought, at 2.6 percent the region still has the second highest population growth rate in the world (after Central Africa). This population growth combined with climatic trends and land degradation could lead to:

- declining per capita production for the agriculture, including animal husbandry
- shortage of fuelwood
- declining rainfall in some regions with consequences for rain-fed and irrigated agriculture
- food shortages and famines in drought years
- movement to urban areas or to more fertile farming areas, such as recently opened areas in the Savannah zone owing to the eradication of river blindness (UNEP, 2008).

Migration, particularly circular mobility, is a traditional coping mechanism in the region, representing a livelihood diversification strategy (Cordell and Piché, 1996; Rain, 1999). But in some areas these traditional patterns have changed in recent decades (Henry et al., 2003; Dietz and Veldhuizen, 2004; Guilmoto, 1998). Each location has its specific characteristics, but migration and pressures on water and land systems are common denominators. A significant proportion of environmental migrants are displaced due to land degradation and drought in the Sahel, though drought-induced migration is often only temporary. Generally, there is a large migration movement to the coastal and urban agglomerations, and to the coastal states (Cour, 2001; Raynaut, 2001).

One study of the impact of climate change on drylands in West Africa noted that between 1960 and 2000, deteriorating situations due to rainfall decreases, land degradation, and violence in the arid and semi-arid areas of Senegal, Mali, Burkina Faso and Niger resulted in a rapid intra-country migration southward and a swelling of big cities like Dakar, Bamako, Ouagadougou, Niamey and Kano (Dietz and Veldhuizen, 2004). Estimates for Burkina Faso suggest that close to half of the adult population born there has moved for at least part of the year to coastal states like Ivory Coast and Ghana (ibid).

Even those not directly dependent on natural resources for their live-lihoods can be affected by desertification and motivated to migrate (Afifi, 2009a). Traditionally pastoralism has represented an important mecha-nism for adjusting to climate variability, since pastoralists can move their herds along with the rainfall (Bascom, 1995; Suliman, 1994). A symbiotic relationship often formed between herders and agriculturalists, with agriculturalists receiving animal manure to fertilize their crops in return for allowing livestock to graze on plant stubble. However, as the Sahel has become more densely settled, increasingly severe conflicts over land and water resources have erupted between pastoralists and sedentary farmers (Tonah, 2003).

In Senegal, fieldwork revealed that environmental changes negatively affect agricultural livelihoods, and contribute to migration through different mechanisms. For areas where irrigated agriculture is possible, farmers living close to the Senegal River expect that their way of life will continue to be possible and therefore do not intend to migrate in the future. But in areas like the Peanut Basin, where land degradation is severe, interviewees said they plan to move away if agricultural live-lihoods do not improve. Most migrants who already migrated said they would return home to the countryside if agricultural livelihoods improved. In Senegal experts observe an increasing movement of people back to the countryside due to the global economic crisis. However, that coping mechanism is running into counter-pressures because areas people are returning to are in many cases already degraded. Conflict over access to land seems to be increasing.

Rather than returning after migrating, the trend goes in the opposite direction. People increasingly migrate step-by-step in pursuit of envi-ronments that will support them. The village Caré in the Tilabéri region of Niger is now home to migrants from another village called Farka where soil degradation has made crop cultivation impossible. In another study in Burkina Faso, researchers found that people from drier regions are more likely to migrate temporarily and to a lesser extent permanently to other rural areas (rural–rural migration), compared with people from wetter areas. A rainfall deficit increases the rural-rural migration but decreases migration to abroad. No rush to cities has been observed during periods of drought. Studies in other regions support this finding, and suggest that environmental conditions often play a more direct role in short-term moves rather than long-term ones (Massey et al. 2007). And yet, if environmental changes render "home" unlivable, short-term migration can develop into a pattern of creeping onward movement.

4.3   Vietnam: Living with floods and resettlement

Environmental degradation, particularly impacts caused by flooding, is a contributing factor to rural out migration and displacement in the Mekong Delta of Vietnam. The Vietnamese portion of the Mekong Delta is home to 18 million people, or 22 percent of Vietnam's population. It provides 40 percent of Vietnam's cultivated land surface and produces more than a quarter of the country's GDP. Half of Vietnam's rice is produced in the Mekong Delta, 60 percent of its fish-shrimp harvest, and 80 percent of Vietnam's fruit crop. Ninety percent of Vietnam's total national rice export comes from the Mekong.

Flooding plays an important role in the economy and culture of the area. People live with and depend on flood cycles, but within certain bounds. For example, flood depths of between half a meter up to three meters are considered part of the normal flood regime upon which livelihoods depend. These are so-called "nice floods" [ngập nông] by Vietnamese living in the delta, such as upstream in the An Giang Province. Flood depths beyond this such as between three and four meters [ngập vừa], however, challenge resilience capacities of affected people and often have harrowing effects on livelihoods.

Floods exceeding the four meter mark, called "ngập sâu" for severe flooding, have increased in magnitude and frequency in Vietnam in recent decades. Fieldwork from the EACH-FOR project indicated that lack of alternative livelihoods, deteriorating ability to make a living in the face of flooding, together with mounting debt, can contribute to the migration "decisions" in the Mekong Delta. People directly dependent on agriculture for their livelihood (such as rice farmers) are especially vulnerable when successive flooding events destroy crops. This can trigger a decision to migrate elsewhere in search of an alternative livelihood. During the flooding season, people under-take seasonal labor migration and movement towards urban centers to bolster livelihoods. As an extreme coping mechanism, anecdotal information from fieldwork pointed to human trafficking as one strategy adopted by some families who have suffered from water-related stressors.

The government in Vietnam has a program known as "living with floods". This program may become more important as the impacts of climate change become more pronounced. The government, as part of this flood management strategy, is currently resettling people living in vulnerable zones along river banks in the An Giang province (Le et al, 2007). Almost 20,000 landless and poor households in this province are targeted for relocation by 2020. Households are selected for resettlement based on a number of factors related to the environment, such as living in an area at risk of natural calamities (flooding, landslides) or river bank erosion. These resettlement programs allow families to take up a five year interest free loan to enable them to purchase a housing plot and basic house frame. Households then often need a further loan to complete building the house (PCAGP, 2006). The clusters provide few infrastructure services like access to schools, health, or water and sewage treatment facilities (Dun, 2009). People planned for relocation are usually the landless who have nowhere else to move if their houses collapse and are often too poor to move to urban areas. For these people, social networks provide the link to livelihoods – most rely on day-to-day employment as laborers. Although the "residential clusters" are usually located only 1–2 kilometers away from the former residence, moving people out of established social networks threatens their livelihoods and contributes to a sense of isolation. The resettlement clusters are not yet planned in a way that allows participation of potential residents.

The Vietnamese strategy of "living with floods" will combine resettlement, shifting livelihoods (i.e. from rice to fishery-based jobs), and some migration. In the future one out of every ten Vietnamese may face displacement by sea level rise in the Mekong Delta (Dasgupta et al. 2007).

5. Conclusions

In sum, even given its limitations, the EACH-FOR project represented a milestone in field-based research on the topic of environmental change and migration. It significantly expanded the empirical base of investigation, spawned greater discussions of methodological and conceptual development, and analyses of policy implications. Today several prominent projects in different phases of completion continue to expand the knowledge base and provide a more refined understanding of how environmental factors, including current weather extremes and the potential for significant longer term changes in climatic systems, affect migration and displacement. This growing knowledge base will be reflected in the upcoming 5th Assessment Report of the Intergovernmental Panel on Climate Change (IPCC) which is the authoritative summary of science relevant to the impacts of climate change on human and natural systems[9]. The topic will also be addressed in several other chapters, particularly regional chapters, as a cross cutting issue.

The EACH-FOR Project represented the first major, global attempt to explore a set of hypotheses in fieldwork, and provided a valuable point of departure for further research. Some of the most significant results of the project were that it created an overview of patterns of environmental change and migration in different types of ecosystems worldwide – from drylands and small island developing states, to deltas, mountain areas, and flood-prone areas (Warner et al. 2009a). The 23 EACH-FOR case studies provided insights about ways that environmental factors affect human mobility – from sudden natural hazards, such as flooding and storms, to gradual phenomena, such as desertification, sea-level rise and other forms of land and water degradation. What is certain from field-based and theoretical research on the issue, in all its varieties, is that environmental change is one of many contributing factors to human mobility. For many areas in the world, more erratic weather, rising sea levels and other climate change impacts will motivate resettlement, forced migration, or other forms of human mobility (Bogardi and Warner 2008). Emerging empirical research such as that described here underscores that environmental change plays a role in human mobility, but more research is necessary to understand the dynamics and policy implications.

EACH-FOR fieldwork suggested that as environmental quality declines, vulnerable people may have fewer options for managing related risks (especially to livelihoods). This narrowing of risk management alternatives may change the role of human mobility in coping with a variety of stressors including environmental change.

---

[9] Chapter 13, 2nd working group.

In every EACH-FOR case study, scenarios in the future, and literature reviews underlined that changes in the environment do affect decisions about migration. Migration is a significant – and growing – response to a spectrum of environmental changes. Environmentally induced migration in different areas in the world is not as "simple" as climate and environmental changes causing people to leave as a direct causal effect. Instead, the main question is about alternatives for managing environmental stressors, with a mix of mobility, livelihood options, and social and other forms of capital for affected areas.

The debate about environmental change and migration is still emerging; it is not yet clearly framed. But as the dialogue on the topic increases, there will be a struggle to define the debate as one of security, of humanitarian action, of development, of disaster management, or of climate change. While the parameters are still undefined, more field-based research and data will be needed in order to better understand the interactions between environmental factors and human mobility, as well as the intervening factors that affect migration or non-migration when environmental changes occur. These questions are of paramount policy importance in the current context of climate change, and will continue to grow in importance in coming decades. Empirical research – and its underlying methods of investigation – has a role to play in helping to support and shape effective policy.

References

Afifi, T. 2009a. Case study report on Egypt for the Environmental Change and Forced Migration Scenarios Project. http://www.each-for.eu/documents/CSR_Egypt_090130.pdf.

Afifi, T. 2009. Niger Case Study Report for the Environmental Change and Forced Migration Scenarios Project, www.each-for.eu

Afifi, T., and K. Warner. 2008. The impact of environmental degradation on migration flows across countries. Working Paper No.5/2008, UNU-EHS Working Paper Series, United Nations University, Institute for Environment and Human Security. Bonn. http://www.ehs.unu.edu/article:476?menu=94

Arab Center for the Studies of Arid Zones and Dry Lands (ACSAD). 2000. Regional Report on Desertification in the Arab World. Damascus.

Bascom, J. 1995. The new nomads. An overview of involuntary migration in Africa. In Baker, J., T. A. Aina, eds, The Migration Experience in Africa.

Uppsala, Sweden: Nordiska Afrikainstitutet [Scandinavian Institute of African Studies], 197–219.

Biermann, F. 2007. "Earth system governance as a crosscutting theme of global change research". *Global Environmental Change*, 17: 326–337.

Black, R. 2001. "Environmental refugees: myth or reality?" In *New Issues in Refugee Research*. Working Paper No. 34, University of Sussex, Brighton. ISSN 1020–7473 http://www.jha.ac/articles/u034.pdf

Boano, C. et al. 2008. *Environmentally displaced people: Understanding the linkages between environmental change, livelihoods and forced migration. Forced Migration Policy Briefing 1*. Refugee Studies Centre, Oxford Department of International Development, University of Oxford.

Bogardi J., Warner, K. 2008. Here comes the flood. Nature. Nature reports Climate Change. Published online 11 December 2008. doi.10.1038/climate.2008.138

Brown, O. 2008. *Migration and Climate Change*. IOM Migration Research Series No. 31. International Organization for Migration, Geneva.

Bryson, R., C. Paddock. On the climates of history. In Rotberg, R. and T. Rabb, eds., Climate and History: Studies in Interdisciplinary History. Princeton: Princeton University Press, 3–4.

Castles, S. 2002. "Environmental change and forced migration: making sense of the debate". In *New Issues in Refugee Research*. Working Paper No. 70. United Nations High Commissioner for Refugees (UNHCR), Geneva.

Center for International Earth Science Information Network (CIESINb), Columbia University; International Food Policy Research Institute (IFPRI); The World Bank; and Centro Internacional de Agricultura Tropical (CIAT). 2009b. Global Rural-Urban Mapping Project (GRUMP), Beta Version: Population Density Grids. Palisades, NY: Socioeconomic Data and Applications Center (SEDAC), Columbia University. Soon to be available at http://sedac.ciesin.columbia.edu/gpw

Center for International Earth Science Information Network (CIESINc), Columbia University; International Food Policy Research Institute (IFPRI); The World Bank; and Centro Internacional de Agricultura Tropical (CIAT). 2009c. Global Rural-Urban Mapping Project (GRUMP), Beta Version: Population Grids. Palisades, NY: Socioeconomic Data and Applications Center (SEDAC), Columbia University. Soon to be available at http://sedac.ciesin.columbia.edu/gpw

Clarke, J., and D. Noin. 1998. Introduction. In J. Clarke and D. Noin, eds., Population and Environment in Arid Regions. Paris: UNESCO / Partenon Publishing Group, 1–18.

Cordell, D., J. Gregory and V. Piché. 1996. Hoe and Wage: A Social History of a Circular Migration System in West Africa. Boulder: Westview Press.

Cour, J.-M. 2001. The Sahel in West Africa: Countries in transition to a full market economy. Global Environmental Change 11: 31–47.

Dasgupta, S., B. Laplante, C. Meisner, D. Wheeler, and J. Yan. 2007. The impact of sea level rise on developing countries: A comparative analysis. World Bank Policy Research Working Paper 4136 (WPS4136), World Bank, Washington.

de Sherbinin, A., L. VanWey, K. McSweeney, R. Aggarwal, A. Barbieri, S. Henry, L. Hunter, W. Twine, and R. Walker. 2007. Household demographics, livelihoods and the environment. Global Environmental Change 18: 38–53.

Dietz, T. and E. Veldhuizen. 2004. Population dynamics. An important intervening variable. In Dietz, A., R. Ruben, and A. Verhagen, eds., The Impact of Climate Change on Drylands. With a Focus on West Africa. Dordrecht: Kluwer Academic Publishers.

Döös, B. 1997. "Can large-scale environmental migrations be predicted?" In Global Environmental Change, 7(1):41–61.

Dun, O. 2009. Linkages between flooding, migration and resettlement. Case study report on Vietnam for the Environmental Change and Forced Migration Scenarios Project, 17. http://www.each-for.eu/documents/CSR_Vietnam_090212.pdf Pp. 17.

Dun, O. and Gemenne, F. 2008. "Defining Environmental Migration", Forced Migration Review, 31: 10–11.

Editorial board. 2005. Egyptian National Action Program to Combat Desertification. Arab Republic of Egypt Ministry of Agriculture and Land Reclamation, UNCCD, and Desert Research Center, Cairo. http://www.unccd.int/actionprogrammes/africa/national/2005/egypt-eng.pdf.

Food and Agriculture Organization of the United Nations (FAO). 2005. Forest Resources Assessment. Rome: FAO.

Giannini, A., M. Biasutti and M. Verstraete. 2008. A climate model-based review of drought in the Sahel: Desertification, the re-greening and climate change. Global Planetary Change 64: 119–128. DOI: 10.1016/j.gloplacha.2008.05.004.

Guilmoto, C. 1998. Institutions and migrations. Short-term versus long-term moves in rural West Africa. Population Studies 52 (1): 85–103.

Hegazi, A., M. and I. H. El Bagouri. 2002. Arab Republic of Egypt National Action Plan for Combating Desertification (Provisional). Cairo: Arab Republic of Egypt.

Henry, S., P. Boyle, and E. Lambin. 2003. Modeling inter-provincial migration in Burkina Faso, West Africa: the role of socio-demographic and environmental factors. Applied Geography 23: 115–136.

Hugo, G. 1996. Environmental concerns and international migration. International Migration Review 30: 105–131.

Hulme, M. S. 2001. Climatic perspectives on Sahelian desiccation:1973–1998. Global Environmental Change 11:19–29.

International Organization for Migration (IOM) 2007. Discussion Note: Migration and the Environment (MC/INF/288 – 1 November 2007 – 94th

Session), International Organization for Migration, Geneva. 14 February 2008.

IPCC (Intergovernmental Panel on Climate Change). 2007 Climate Change 2007: Impacts, Adaptation and Vulnerability. Contribution of Working Group II to the Fourth Assessment Report of the Intergovernmental Panel on Climate Change, Parry, M. L., et al. (eds), Cambridge University Press, Cambridge.

Jäger, J., J. Frühmann, S. Grünberger, and A. Vag. 2009. D.3.4 Synthesis Report. Environmental Change and Forced Migration Scenarios Project, 64–66. http://www.each-for.eu/documents/EACH-FOR_Synthesis_Report_090515.pdf.

Kniveton, D. et al. 2008. *Climate change and migration: Improving methodologies to estimate flows.* IOM Migration Research Series, No. 33. International Organization for Migration, Geneva.

Le, T., H. Nguyen, H. Nhan, E. Wolanski, T. Tran, and H. Shigeko. 2007. The combined impact on the flooding in Vietnam's Mekong River delta of local man-made structures, sea level rise and dams upstream in the river catchment. Estuarine, Coastal and Shelf Sciences 71: 110–116.

Lonergan, S. 1998. The Role of Environmental Degradation in Population Displacement. Global Environmental Change and Security Project Report, Research Report 1, July 1998 (2nd edition), University of Victoria, Canada.

Massey, D., W. Axinn, and D. Ghimire. 2007. Environmental change and out-migration: Evidence from Nepal. Population Studies Center Research Report 07–615 (January).

McLeman, R. and Smit, B. 2006. "Migration as an adaptation to climate change". *Climatic Change* 76: 31–53.

PCAGP. People's Committee of An Giang Province. 2006. Project: Removal of Canal Houses to Secure Environmental Sanitation of An Giang Province from now to 2020 (English translation). An Giang: People's Committee of AnGiang Province.

Piguet E. 2008 Climate change and forced migration. Geneva, United Nations High Commissioner for Refugees. (New Issues in Refugee Research, No. 153.)

Rain, D. 1999. Eaters of the Dry Season: Circular Labor Migration in the West African Sahel. Boulder: Westview Press.

Raynaut, C. 2001. Societies and nature in the Sahel: Ecological diversity and social dynamics. Global Environmental Change 11:9–18.

Renaud, F. G., Bogardi, J. J., Dun, O., Warner, K. 2007. *Control, Adapt or Flee: How to Face Environmental Migration?* United Nations University Institute for Environment and Human Security, Bonn.

Suliman, M. 1994. The predicament of displaced people inside the Sudan. Environmental degradation and migration in Africa. In Bächler, G., ed.,

Umweltflüchtlinge: das Konfliktpotential von morgen? [Environmental Refugees: A Potential of Future Conflicts?]. Münster, Germany: agenda Verlag GmbH & Co, 111–132.

Tansey, O. (2006). Process Tracing and Elite Interviewing: A Case for Non-probability Sampling. http://www.nuff.ox.ac.uk/politics/papers/2006/tansey.pdf

Tonah, S. 2003. Integration or exclusion of Fulbe pastoralists in West Africa: A comparative analysis of interethnic relations, state and local policies in Ghana and Cote d'Ivoire. Journal of Modern African Studies 41 (1): 91–114.

UNHCR (United Nations High Commissioner for Refugees). 2006. Convention and Protocol Relating to the Status of Refugees: Text of the 1951 Convention Relating to the Status of Refugees, Text of the 1967 Protocol Relating to the Status of Refugees, and Resolution 2198 (XXI) adopted by the United Nations General Assembly, UNHCR, Geneva, Available at: http://www.unhcr.org/protect/PROTECTION/3b66c2aa10.pdf, 22 February 2007.

Vag, A., Faist, T., Enzinger, H, Jäger, J., Bogardi, J. 2007. *Research Guidelines Book 1, Version 3.* Working document. EACH-FOR Work Package 3: Methodology & Synthesis. www.each-for.eu

Warner, K.; Stal, M.; Dun, O.; Afifi, T. 2009a. "Researching Environmental Change and Migration: Evaluation of EACH-FOR Methodology and Application in 23 Case Studies Worldwide" in: Lazcko, F., Aghazarm, C. (Eds.) *Migration, Environment and Climate Change: Assessing the Evidence.* IOM, Geneva.

Warner, K, Laczko, F. 2008b. Migration, Environment and Development: New Directions for Research. In *International Migration and Development, Continuing the Dialogue: Legal and Policy Perspectives.* Joseph Chamie and Lucca Dall'Oglio (eds.). International Organization for Migration and Center for Migration Studies (CMS). New York and Geneva.

Wijkman, A., and L. Timberlake. 1984. Natural disasters. *Acts of God or acts of man?* London: Earthscan.

PART 2

Policy responses, normative issues and critical
perspectives

9

How they became the human face of climate change. Research and policy interactions in the birth of the 'environmental migration' concept

FRANÇOIS GEMENNE

Introduction

The literature on the nexus between the environment and migration is relatively recent, mainly dating back to the mid-1980s, a period characterized by asylum crises and major natural disasters. The nexus has been explored in a variety of ways, but its two components have mainly been associated in a causal relationship. A few studies have focused on the 'impacts of refugee movements on the environment, whereas more recent studies have primarily addressed the impacts of environmental changes on migration flows.

Overall, four themes permeate the literature on the nexus: that research is (a) impeded by a lack of empirical studies; (b) driven by a climate change-dominated agenda; (c) abundantly supplemented by 'grey' literature; and (d) marked by disciplinary divides.

The lack of empirical research was already evident at the fifth meeting of the International Research and Advisory Panel on Forced Migration, held in 1996, when a 'disappointingly small number of papers' on the topic were presented. During the keynote address of the meeting, Gaim Kibreab stressed that 'research on refugees [had] been largely environmentally-blind . . ., and that in the absence of a body of empirical research, a number of myths and misperceptions still predominate[d]' (Koser, 1996). Similar comments still held valid more than a decade later, as identified by Brown (2008) and Kniveton et al. (2008). Some progress in this area has been made thanks to the EACH-FOR project, whose conclusions and findings are described in Chapter 8 of this volume, but much remains to be done.

The risk of migration flows associated with climate change was highlighted in the *First Assessment Report* of the IPCC (McTegart et al.,

1990), and the impacts of climate change have since increasingly overshadowed other types of environmental change as migration drivers. The
shift of focus has been so evident that it has led some authors to fear that
people displaced by environmental disruptions not related to climate
change may be forgotten by future studies and policies (Lassailly-Jacob,
2006). Indeed, the majority of recent works and conferences on the topic
focus on climate change, and do not address other environmental
changes as root causes of migration (Piguet, 2008; Biermann and Boas,
2010; Brown, 2008; Meze-Hausken, 2004; McLeman and Smit, 2006;
Kniveton et al., 2008). Furthermore, many of these works – but not
all – contain the implicit assumption that conclusions reached with
regard to climate change hold true for other kinds of environmental
disruptions, largely because the impacts of global warming, such as
droughts or floods, do not seem to be fundamentally different in nature
from other environmental disruptions.

Numerous reports and papers on the topic are part of a growing body of
'grey' literature, which forms a significant part of research on the nexus, and
often drives the research agenda. This trend is also apparent for migration
research in general, as 'many of the information producers are governments,
international agencies, and non-governmental organisations', which produce
literature with a 'practical orientation' (Mason, 1999). Grey literature has been
particularly influential in supporting alarmist forecasts of future migration
flows, sometimes based on inflated estimates (Kniveton et al., 2008).

Finally, literature on the nexus derives from a variety of academic
disciplines that offer different, and sometimes conflicting, viewpoints on
the topic. Indeed, the study of environmental migration is multidisciplinary by nature: whereas the study of environmental change usually
draws on the natural sciences for its evidentiary basis, the study of
migration is typically the preserve of the social sciences. This chapter
argues that the literature on the subject is split between an alarmist and a
sceptical perspective, rooted in the disciplinary divide between natural
and social sciences. The alarmist perspective, often championed by
environmental scholars, the media and civil society, claims that environmental disruptions, among which the impacts of climate change in
particular, will induce massive population displacements. On the contrary, the sceptical perspective, voiced by migration scholars, insists that
migration is multicausal by nature, and that environmental drivers
should not be set apart from other migration drivers.

I contend that these conflicting viewpoints have shaped the policy
debate on the responses to what has been labelled 'environmental

migration'. Alarmists called for the development of new policy instruments to better protect those displaced by environmental changes and natural disasters, whereas sceptics argued that the development of such instruments was unnecessary, as 'environmental migrants' did not constitute a new category of migrants.

This chapter is based on the fundamental assumption that researchers are not policy neutral: especially in this area, they should be considered as policy entrepreneurs, whose perception of their research object is shaped by a series of policy objectives and fundamental values. Ideas are an important input of the policy process, and researchers are obviously prime providers of ideas. My goal is therefore, here, to examine the mutual interactions of science and policy in the conceptualization of environmental migration.

The point of my argument is to show that not only does research influence the policy process, but that this process also informs research, and shapes the conceptualization of environmental migration. I therefore consider the concept of environmental migration as a political construct, shaped by both ideational linkages and policy responses.

A key argument is that the opposition between the alarmist and the sceptical perspectives has played a major role in the definition of policy responses that addressed – or failed to address – population displacements associated with environmental changes. A first section therefore reviews the emergence of the ideational linkages between environmental changes and migration in the literature, and attempts to show how the opposition between the alarmist and sceptical perspectives has structured the literature on the topic. The second section assesses the influence of these discursive linkages on the development of policy responses in two areas that are at the frontline of environmental migration: environmental and migration policies. This section seeks in particular to explain how different research perspectives can explain the emergence of different policy outcomes. Finally, a concluding section examines what these different policy outcomes reveal about the conceptualization of environmental migration as a political construct.

How discursive linkages between environment and migration emerged in the literature

Early texts

The issue of ecological refuge was mentioned in 1948 (Vogt, 1948), but the first use of the term 'environmental refugee' in the literature is uncertain: Kibreab (1997) detects its first occurrence in 1984 in a briefing document from the International Institute for Environment and Development, while Black (2001) traces its origins to speeches and reports by environmentalist Lester Brown of the WorldWatch Institute in the 1970s. There seems to be universal agreement, however, attributing the first official use of the term to El-Hinnawi (1985) in a UNEP report entitled *Environmental Refugees*.

In 1988, a working paper by Jodi Jacobson from the WorldWatch Institute attempted to systematize the study of this new category of forced migrants. Jacobson proposed a typology similar to that put forward by El-Hinnawi, distinguishing between temporary displacements associated with temporary environmental stress, permanent displacements associated with permanent environmental stress, and temporary or permanent displacement due to progressive environmental change (Jacobson, 1988). She contended that the term 'environmental refugee' was first used in reference to Haitian boat people, arguing that land degradation in Haiti created these desperate people and their dangerous journey to south Florida.

Both El-Hinnawi's and Jacobson's reports were received with great interest in the field of environmental studies, and attracted harsh criticism in the field of refugee studies: they had a 'short-lived shock-effect on the public debate but were rejected as unserious by scholars' (Suhrke, 1993). At the times of publication, El-Hinnawi was working for the United Nations Environment Programme, while Jacobson was a member of the WorldWatch Institute, an environmental think-tank: the reports were therefore perceived as an attempt to use forced migration to draw attention to environmental problems. Irrespective of its legal meaning, the use of the word 'refugee' was criticized. Suhrke and Visentin (1991) stated that the definition provided by El-Hinnawi was

> So wide as to render the concept virtually meaningless ... Uncritical definitions and inflated numbers lead to inappropriate solutions and compassion fatigue. We should not, however, reject outright the concept of environmental refugees. Instead we should formulate a definition that is more narrow but more precise.

Likewise, McGregor argued that 'the category "environmental refugee" confuses rather than clarifies the position of such forced migrants, as it lacks both a conceptual and a legal basis', contending that the category involved a 'false separation between overlapping and interrelated categories' (1993). McGregor's criticism was actually aimed at the very concept of 'environmental refugee', prefiguring the academic controversies that were soon to appear with regard to the conceptualization of environmental migration.

Richmond, in *Global Apartheid*, devoted a chapter to 'environmental refugees', outlining his theoretical framework surrounding environmental migration, and attempting for the first time to situate it within migration system and theories. He proposes a multivariate model of environmentally related population movements that acknowledges the mingling of environmental factors (constraints or facilitators) with social, economic, political and technological factors (1994). The model replicates the continuum between proactive and reactive migration he had developed in an earlier work (Richmond, 1993), suggesting that this continuum should replace the traditional dichotomy between voluntary and forced migration. Richmond also proposes a typology of environmentally related disasters, classified in different categories according to their origin: natural, technological, economic, political or social (1994). Outlining many of the challenges facing the study of environmental migration, he notes that the scale of this kind of migration is 'difficult to estimate', depending greatly on

> whether past, present, or possible future movements are considered; whether worldwide migration is considered or only that occurring in developing countries is considered; whether internal as well as external migrations are taken into account; and whether environmental degradation is considered in isolation or in conjunction with other political, economic, and social determinants of population movement (Richmond, 1994).

Nevertheless, he goes on to suggest some policy ramifications: the need for a new instrument of international law to address the 'humanitarian needs of all those displaced from their homes', the need for a system of humanitarian priorities, the importance of more effective coordination of the work of UN agencies, and finally the need to integrate population movements in the concept of sustainable development.

At the time of publication of *Global Apartheid*, the first conference on the nexus between environment and migration was organized in Nyon (Switzerland), jointly sponsored by the Swiss Department of Foreign Affairs, the International Organization for Migration (IOM), and the Refugee Policy Group (RPG). The background paper of the conference aimed to synthesize the burgeoning academic debates on the nexus, with an emphasis on migration induced by environmental changes. The paper classified the most important causes and dynamics of environmental

migration into six categories: elemental, biological, slow-onset, acciden-
tal, development-induced disruptions, and environmental warfare. The
classification proved short-lived and of little practical use, but did
acknowledge other intervening factors, and the multicausality of displacement. Debates on conceptualization were soon to crystallize around
the two conflicting perspectives of alarmists and sceptics.

Conceptualizing the nexus: alarmists and sceptics

The 1992 conference in Nyon invited further research on the conceptualization of the nexus, following early endeavours by El-Hinnawi and
Jacobson. This conceptualization addressed both aspects of the nexus:
the impact of environmental changes on migration, as well as the
environmental impacts of migration, though the latter aspect was
addressed by fewer researchers. A clear divide quickly emerged between
those who adopted a maximalist perspective and those with a minimalist
perspective: the former insisted on strong causal relationships between
both sides of the nexus, whereas the latter stressed the multicausality of
the nexus and other intervening factors. Logically, the maximalists forecasted waves of 'environmental refugees' and pinpointed environmental
factors as a major driving force of migration, whereas the minimalists
adopted a more sceptical stance vis-à-vis the empirical reality of such
migration flows, insisting on the complexity of the migration process.
The former are henceforth described as 'alarmists' and the latter as
'sceptics'. These perspectives initially formed around scholars from
different disciplines: alarmists were mostly scholars from the natural
sciences, and security experts; whereas sceptics were found among social
scientists, and migration scholars in particular. NGOs and interest
groups usually sided with alarmists, and the grey literature also tends
to adopt a maximalist perspective.

This debate emerged soon after the coining of the expression 'environmental refugees', and has been ongoing since. Already in 1993, Suhrke
had noted that

> While literature on environmental change and population movement is
> quite limited, two different and opposing perspectives can be discerned.
> One – which I call the minimalist view – sees environmental change as a
> contextual variable that can contribute to migration, but warns that we
> lack sufficient knowledge about the process to draw firm conclusions.
> The other perspective sets out a maximalist view, arguing that

environmental degradation has already displaced millions of people, and more displacement is on the way (Suhrke, 1993).

Today, this debate continues in pretty much the same terms.

The alarmist perspective

The taxonomy established by El-Hinnawi (1985) and Jacobson (1988) paved the way for an alarmist perspective that was used to forecast impressive migration flows related to a wide variety of environmental changes. Many scholars who adopted this perspective were initially interested in the environment–security nexus (Westing, 1989; Homer-Dixon, 1991; Swain, 1996b) – out of concern for the linkage between environmental disruption and conflicts – and deployed refugee flows as an exploratory variable to justify a causal relationship between environmental change and conflict.

Homer-Dixon took the debate a step further, contending that environmental change would lead to acute armed conflicts (Homer-Dixon, 1991; 1994); with a distinctly Malthusian air he opined that 'waves of environmental refugees that spill across borders with destabilizing effects on the recipient's domestic order and on international stability' would be a key consequence of environmental change (1991). Homer-Dixon, however, also invoked other factors, such as vulnerability, more acute in the South than in the North. He developed his research agenda further in a subsequent paper, in which he used three hypotheses to link six types of environmental change with violent conflict. The second of these hypotheses holds that 'large population movements caused by environmental stress [will] induce "group identity" conflicts, especially ethnic clashes' (Homer-Dixon 1994). He tests this hypothesis with empirical evidence from Bangladesh, where significant numbers of migrants have fled to the adjacent Indian states.

Migration flows from Bangladesh to India are also cited by Swain (1996a) as empirical evidence of conflicts induced by environmental disruption, through migration flows. Swain's thesis is that 'population migration transports . . . the conflict from the environmentally affected regions to the migrant receiving areas' (Swain, 1996b). He contends that environmental migration poses important security challenges to developing countries, and should therefore be at the top of the global political agenda.

To summarize the general approach of these works, the initial alarmist approach to the nexus assumed that environmental disruptions were major contributors to insecurity. Migration was conceptualized both as a consequence of environmentally induced conflicts and as a trigger of future conflicts over natural resources. The theories were deeply rooted in a neo-Malthusian perspective, and gained authority with the commonly held perception that climate change was a threat to the world's security. Climate change prompted a deep questioning over the notion of security, and alarmist theories were quick to make their way into the policy realm.

From the mid-2000s onwards, different governments commissioned or were recipients of reports warning of the threat that climate change posed to national or international security. The first report of this kind – and the one portraying the most doom and gloom-laden scenario – was commissioned by the US Department of Defense, and reportedly censored by the White House. The report evokes an apocalyptic scenario in which brutal change of weather conditions, induced by the crossing of a climate threshold, triggers massive flows of migrants worldwide, who compete for resources and ultimately threaten US and international security (Schwartz and Randall, 2003). The report warns that such a scenario is plausible, yet not the most likely, and reveals its political agenda by urging the United States Government – which, notoriously, did not ratify the Kyoto Protocol – to take climate change more seriously.

A report on the same topic was submitted to the Canadian Security Intelligence Service the following year (McLeman and Smit, 2004). The authors noted that 'the consequent displacement [due to the impacts of climate change] of large numbers of people causes substantial disruption in the source area, but also places stress on areas that receive the unexpected migrants' and concluded that 'security implications are a combination of those in the source area and the receiving one' (2004).

Another report by the German Advisory Council on Global Change (WBGU, 2008), submitted to the German Government and endorsed by UNEP, also addresses climate change as a security risk. The report warns that climate change amplifies the mechanisms that lead to insecurity and violence, such as political instability and weak governance structures. Among threats to international stability and security, the triggering and intensification of migration is mentioned as one of the potential major fields of conflict in international politics. The authors assert that there is a 'particularly significant risk of environmental migration occurring and increasing in scale' in developing countries (WBGU, 2008). The report goes on to recommend a reform of the UN Security Council and UNEP in order to address the challenge, as well as increased cooperation among migration management agencies, including a new, ad hoc convention to protect environmental migrants.

Finally, a report recently prepared for the European Council adopts a similar stance, and warns that 'Europe must expect substantially increased migratory pressure', especially from Africa (European Commission and the Secretary-General/High Representative, 2008). The report recommends that a comprehensive European migration policy take into account environmentally triggered additional migratory stress, but does not further elaborate on this.

The ongoing conflict in Darfur is often cited as an empirical evidence of neo-Malthusian theories: it is a case in which environmental change and resource scarcity have induced migration, leading to violent conflicts. The UN Secretary-General Ban Ki-moon has endorsed such assumptions, and argued in an Op-Ed article in *The Washington Post* that 'amid the diverse social and political causes, the Darfur conflict began as an ecological crisis, arising at least in part from climate change' (Ki-moon, 2007). The UN chief added that the repercussions reached far beyond Darfur, contending that conflicts in Somalia, Côte d'Ivoire and Burkina Faso were rooted in similar ecological crises.

Overall, the alarmist perspective views environmentally induced migration as a security threat, a threat that has been exacerbated and brought to policy level by climate change. Attention to the linkages between climate change and security is rapidly gaining currency, and the recent award of the Nobel Peace Prize to Al Gore and the IPCC can be interpreted as an acknowledgement of such linkages. In April 2007, the question of the linkages between climate change and security was discussed for the first time in the UN Security Council, at the request of the British chair of the Council.

Building upon this initial approach to environment and security, some scholars tried to examine the linkages between environmental disruptions and migration, and forecast future migration flows. The priority of their research was no longer security threats, but the risks facing the environment and the consequences of environmental disruptions. They were led by environmentalist Norman Myers, without doubt the most prominent whistleblower in the field. Myers wrote extensively on the topic and dared to forecast precise estimates, which were broadcast widely in the media (1997, 2002, 1993). Myers was himself inspired by the works of Westing (1992), who was among the first to conceptualize what he saw as a new form of displacement. In 1995, Myers published jointly with Kent a book entitled *Environmental Exodus*, whose impact has remained considerable: it is one of the most-cited sources on the subject, for good reason. The study was the first to offer a forecast of future flows, as well as to identify hot spots at country level. Myers and Kent fed the well-known media appetite for numbers, sometimes at the risk of oversimplifying a complex situation. In particular, they insisted that demographic growth, sea level rise and natural disasters could be used as explanatory variables for future flows of environmental migrants.

Myers updated his estimates in 2002, forecasting that 'when global warming takes hold, there could be as many as 200 million people overtaken by sea-level rise and coastal flooding, by disruption of monsoon systems and other rainfall regimes, and by droughts of unprecedented severity and duration' (Myers, 2002). Through frequent repetitions, this latest figure, though highly speculative and questionable, has become taken as empirical evidence, and has been frequently cited in media reports and other studies, most notably in the Stern review on the economics of climate change (Stern, 2007).

Myers can certainly be credited with drawing worldwide attention to the topic of environmentally induced migration. However, his work is largely based on speculative common sense rather than on actual figures and estimates – a point that has been vigorously criticized by scholars adopting a more sceptical perspective. Other scholars and many NGOs, however, followed the path forged by Myers.

This deterministic perspective is a common feature of works written from an alarmist viewpoint. Bates (2002) assumes a direct causal relationship between environmental changes and migration, and attempts to provide a typology of these changes. Her classification is based on three binary criteria related to the environmental disruptions causing the migration: their origin (natural or man-made), their duration (acute or gradual), and whether migration was an intentional outcome of the disruption or not. Byravan and Rajan (2006) focus on sea level rise and insist on the 'inevitability' of the displacement of people living in coastal areas and small islands by 2050, because of sea level rise. In line with Myers, they estimate that about 200 million people will be at risk with a 1 m rise in sea level, representing a land loss of about 212,000 km².

Numerous reports from NGOs have also contributed to the alarmist perspective, and provided additional estimates and forecasts. The Red Cross stressed in 2001 that more people were forced to leave their homes because of environmental disasters than war (IFRC, 2001), while Friends of the Earth Australia (2004) emphasized Myers' predictions and urged the Australian Government to take action against climate change. Lester Brown, who was among the first to use the term in the 1970s, noted that flows of 'environmental refugees' were just beginning and were 'yet another indicator that our modern civilization is out of sync with the Earth's natural support systems' (Brown, 2008). In a much-debated report, the NGO Christian Aid dramatically revised Myers' forecasts, and predicted that up to 1 billion people could be displaced by environmental disruption by 2050 (Christian Aid, 2007). Even though the report acknowledged that 600 million of the predicted 1 billion would actually be displaced because of development projects, rather than actual environmental change, the estimate was still significantly higher than those made previously.

Even though environmental degradation has been recognized as a root cause of refugee flows by the UNHCR since 1993 (Ogata, 1993), nonetheless many scholars claim that no protection has evolved to help people displaced by environmental change. Accordingly, they initiated a debate at the policy level, where the alarmist perspective translated into a case for the development of new policies and instruments to fill in what was perceived as a protection gap.

Hermsmeyer (2005) has claimed that the inadequacies of the international refugee regime have denied the humanitarian rights of environmentally displaced people. Her work reviews some causes of environmental migration, and suggests a series of policy measures, for example, the creation of a specialized agency, and development of effective strategies for prevention, response and recovery. King (2006) has also proposed the creation of a dedicated mechanism to address environmental migration – an International Coordination Mechanism for Environmental Displacement (ICMED) – whose role would be to coordinate the activities of the different agencies involved in providing relief to the displaced. Conisbee and Simms (2003) have made yet another case for a radical policy change, and pleaded for the inclusion of 'environmental refugees' in the population protected by the Geneva Convention, arguing that environmental degradation is, in many ways, a form of political persecution. If their amendments were adopted, people uprooted by environmental change would be entitled to the same protection as those uprooted by war and violence. Finally, Bell's contribution to the debate (2004) took a more original approach, as his is one of the few papers to frame the issue in terms of global environmental justice, concluding that a new framework for global justice is required.

Such claims intersected directly with those put forward by another perspective that had developed in reaction to Myers' and Homer-Dixon's theories. Led by two senior figures of migration studies, Gaim Kibreab and Richard Black, the sceptical approach contended that the concept of environmental migration made little sense, and that environmental impacts of migration flows were grossly overestimated.

The sceptical perspective

Kibreab (1997; 1994) was the first to attack the alarmist approach, which was increasingly accepted as 'scientific truth'. According to him, the concept of environmental migration served to depoliticize the causes of displacement, allowing states to derogate their obligations to provide asylum, as 'environmental conditions (did) not constitute a basis for international protection' (1997); he perceived the concept as a threat to refugee protection, and an excuse for governments to justify restrictive asylum policies. He argued that the concatenated causal relationship between environmental degradation, population displacement and insecurity needed to be reversed, insecurity being the cause – and not the consequence – of population displacement. Though Kibreab's argument might be perceived as radical, it is important to note that he does not challenge the significance of environmental change in relation to migration, but rather the nature of its relationship with insecurity.

Black adopted a similar sceptical approach in his milestone book, *Refugees, Environment and Development*, published in 1998, which asks the pertinent question, 'What is the reason for the linkage between forced migration and environmental change?' Black notes that 'one reason frequently given in recent years to explain mass population displacement is the growth of environmental problems', and wonders 'in whose interest it is that environment degradation should be seen as a possible cause of mass displacement?' (Black, 1998). However, he refutes Kibreab's argument that the concept is just an excuse for Northern governments to enforce tighter asylum policies, given that most of the literature on the topic calls for an extension of the current refugee regime, rather than a restriction. Yet he agrees with Kibreab that most of the 'alarmist' literature 'serves only to differentiate a single cause of migration' (1998) and therefore may lead to restrictions in asylum policies in the North, which is important because most of these migration flows occur in developing countries. Black notes however that the origin of the concept lies in the environmentalist and 'conflict studies' literature rather than in asylum literature, and (rightly) suspects that Myers' concern is not migration, but the threat associated with climate change.

In a subsequent working paper for the UNHCR's *New Issues in Refugee Research* series, Black further questions the very notion of 'environmental refugees', saying that the 'linkages between environmental change, conflict and refugees remain to be proven'. He asserts that current statistics and case studies on 'environmental refugees' are not 'encouraging in terms of staking out a new area of academic study or public policy', and that these so-called refugees may be no more than a myth. However, he concedes that environmental changes can be factors behind large-scale migration, but raises doubts about the possibility of defining these migrants adequately. He concludes with some reflections on a possible international protection regime for 'environmental refugees', asking if such a regime would rather help or hinder 'the battle to focus the world's attention on pressing environmental problems' (Black, 2001).

Wood also agrees that there is no simple relationship between environmental causes and societal effects, and argues, along with Black, that the debate on 'environmental refugees' has been driven by 'simplistic generalisations rather than solid empirical research' (2001). He insists that it is impossible to separate economic and environmental factors as root causes for population displacement. The impossibility of isolating environmental from other factors is also emphasized by Lonergan and Swain (1999), who stress that generalizations about the environment–migration nexus mask 'a great deal of the complexity that characterizes migration decision-making'.

The coexistence of two such antagonistic positions as the alarmist and sceptical ones in the literature struck Castles, who ventured to compare both perspectives through Myers' and Black's arguments, trying to 'make sense of the debate', as the title of his paper suggests (Castles, 2002). He notes that the two approaches are difficult to reconcile, for example, in Black rejecting the 'apocalyptic vision' put forward by Myers. He underlines a methodological difference between the conflicting perspectives: Myers uses wide-ranging estimates in a deductive perspective, whereas Black uses empirical studies at the national and local levels. In particular, Castles stresses that Myers does not provide figures on actual displacements, but only on potential displacements. The disagreement, however, is far from being purely methodological, and Castles points out that 'general forecasts and common sense linkages do little to further understanding', and that it is crucial to look at specific cases, and strengthen empirical research (2002). Castles then abandons his neutral stance and reaches the same conclusion as Kibreab and Black: 'the notion of the "environmental refugee" is misleading and does little to help us understand the complex processes at work in specific situations of impoverishment, conflict and displacement'. He later writes that 'emphasis on environmental factors is a distraction from central issues of development, inequality, and conflict resolution' (Castles, 2007). He further sides with Black in arguing that the term 'environmental refugee' is 'simplistic, one-sided and misleading, (and) implies a mono-causality that very rarely exists in practice' (Castles, 2002). Revealing his policy agenda, he invites scholars to do their utmost to defend the Geneva Convention, but also to call for an improved international legal regime and institutions to protect other types of migrants – agendas that might appear mutually contradictory.

Suhrke (1993) also reflects on the divide between the alarmist and sceptical perspectives, and affirms that the distinction between migrants and refugees must be restored in order to produce more accurate estimates of people forcibly displaced by environmental changes. She sees environmental degradation as a 'proximate cause of migration' that interacts with demography and political economy, but wishes to escape the 'trap of environmental determinism'. She asserts that 'it is difficult to argue that environmental degradation produces particular forms of out-migration except in one respect: the appearance of distress migrations occasioned by sudden or extreme environmental degradation': only in the latter case is the term 'environmental refugees' appropriate, she argues. Suhrke's reflection on the divide between what she calls 'maximalist' and 'minimalist' perspectives does not absolutely reject the alarmist views in the same way that Black, Kibreab and Castles do. Instead, her proposal transcends the divide and emphasizes the multi-causality of migration. In doing so, she opened the way for a more moderate approach to the nexus in which the importance of environmental factors was acknowledged, as was the intertwining of these factors with other migration drivers. This stance is adopted by Gonin and Lassailly-Jacob (2002), Kliot (2004) and Hugo (1996).

The demarcation between the alarmist and sceptical approaches is principally their stance towards the importance of environmental factors as (sole) migration drivers. The controversy between the two is connected with the very conceptualization of environmental migration, and impacts upon the debate on policy implications – alarmists favouring reform and sceptics the status quo. This debate was first addressed by legal scholars, then by a new, revived stream of research on the topic, prompted by natural disasters in the mid-2000s and worldwide concern with climate change.

How these perspectives shaped the policy debates

Proposed policy developments

Contemporaneously with debates over the conceptualization of environmental migration, scholars started to enquire into the regime protecting those uprooted by environmental changes. Sceptics argued that the term 'environmental refugee' was misleading from a sociological viewpoint, because it presumed a single causality in the migration decision; legal scholars asserted that the term was equally misleading from a legal perspective, as it did not fall within the scope of the 1951 Geneva Convention. This latter objection to the term, however, reinforced the alarmist perspective: most scholars argued that a new protection regime was needed to address what were perceived as migration flows neglected by policy-makers. The development of legal studies in the field was mainly conducted by environmental law scholars, rather than by refugee law scholars; an overwhelming majority of papers addressing this topic were published in environmental law journals and reviews.

The first thorough study on the topic of migration in the law was conducted by Magniny (1999), who raised the hypothesis of creating a special status in international law for 'environmental refugees'. Starting with a discussion of the origins and founding principles of refugee law, as well as the limits of the Geneva Convention, she elaborated on the controversial concept of environmental damage, and proposed considering victims of the damage in a collective dimension. This laid the basis for a discussion of a specific legal status for 'environmental refugees', as well as of the organizations that might implement such a status. Magniny's work opened the gates for a flood of legal reflections on the status of 'environmental refugees', particularly in France, which culminated in a conference addressing the issue held in Limoges on 23 June 2005,[1] followed in 2008 by

---

[1] The conference, entitled Les Réfugiés Ecologiques (Ecological Refugees), was convened by the Centre de Recherches Interdisciplinaires en Droit de l'Environnement, de l'Aménagement et de l'Urbanisme (CRIDEAU – CNRS/INRA), in collaboration with the Observatoire des Mutations Institutionnelles et Juridiques (OMIJ) and the Centre International de Droit Comparé de l'Environnement (CIDCE).

the proposal of a new international convention on environmental displacement (Prieur et al., 2008).

Along with Magniny, Cournil (2006) underlined several inadequacies of the current regime, pointing to the weaknesses of specific legal texts (such as the Geneva Convention or the Addis Ababa Convention of the Organization of African Unity). She stressed that human rights were not guaranteed for 'ecological migrants', and that a case-by-case, inconsistent approach prevailed in the treatment of asylum claims on an environmental basis. In subsequent papers written in collaboration with Mazzega (Cournil and Mazzega, 2006; 2007), the insufficiencies of international norms and national protections were discussed further, including an assessment of the pros and cons of some innovative protection mechanisms, including ecological intervention[2] and environmental asylum. These two authors agree that an international protection regime for 'environmental refugees' is currently unrealistic, and suggest instead reinforcing the protection of internally-displaced persons (IDPs). Were an international regime to be developed, however, they favour protection with a collective dimension and variable duration, according to the circumstances (if there is the possibility of return or not), and rights for those displaced.

Michelot (2006) introduces the issue of environmental responsibility to the debate. In her analysis, the figure of 'environmental refugee' embodies the threats facing humanity: accordingly, the recognition of a status for those displaced might imply the acknowledgement of global responsibility and implementation of the principles of environmental justice. Chemillier-Gendreau (2006) also reflects on the possibility of creating an international status for 'ecological refugees'. Claiming that it is humanity's duty to express its solidarity through adequate international instruments, she proposes extending the definition of refugee to 'all people whose life is threatened or whose normal living conditions are made impossible due to an environmental degradation, whether its origin is natural or the result of human actions perpetrated by states or private agents' (2006), adding that the granting of refugee status could be based on class action, rather than on individual procedures.

Some legal scholars have argued that the refugee definition should be extended to include 'environmental refugees'. Cooper suggests a new definition of refugee that would include any person who,

---

[2] This term refers to the French doctrine of '*droit d'ingérence*', by which a state intervenes in another state's internal affairs with the mandate of a supranational authority, usually for humanitarian reasons.

owing to degraded environmental conditions threatening his life, health, means of subsistence, or use of natural resources, is outside the country of his nationality and is unable or ... unwilling to avail himself of the protection of that country (Cooper, 1997).

Most of her fellow scholars, however, contended that broadening the definition of a refugee was not the way to go, and that environmental law offered alternative solutions.

An early work by McCue judges that 'the refugee system was never intended to address the problem of environmental refugees' (1993). Given the reluctance of states to expand the current refugee regime, McCue suggests that advocacy efforts to protect 'environmental refugees' should be directed at securing a multilateral convention in environmental law, rather than at extending the refugee protection regime. He argues for the adoption of basic principles of international environmental law, such as the duties to prevent environmental disasters, to notify and provide information about them when they occur, and to develop contingency plans.[3] The convention would go further and include a general duty to compensate the victims, in what the author admits is the 'most difficult political leap' posed by his proposal.

Falstrom (2001) also makes the case for a new convention to address both the causes of the problem (environmental degradation) and its consequences (displacements). She notes that most protection mechanisms focus on the consequences of the problem, but fail to deal with its root causes. Hence she advocates a convention that would address displacements, but also contain obligations for states to prevent environmental damages from occurring. She suggests that such a convention could be modelled on the United Nations Convention Against Torture, drafted in 1984: as she notes, this convention applies to all persons fearing torture, whether acts of torture have been perpetrated or not, and whether the person is prosecuted because of belonging to a specific group or not.

The idea of a convention similar to the Convention Against Torture is enthusiastically endorsed by Lopez (2007), who details the insufficiencies of the current protection regimes in Europe and the United States, and observes that neither European directives on temporary and subsidiary

---

[3] Such contingency plans, applied to environmental migration, would 'extend the principle of *non-refoulement* in the case of an environmental event causing trans-border migration in an amount recognized by the convention' (McCue, 1993).

protection nor the American temporary protected status[4] can offer adequate protection to 'environmentally displaced persons'. Thus she considers Falstrom's proposal the most concrete framework to deal with the issue.

Finally, a recent paper by Biermann and Boas (2010) advocates a new *sui generis* protection regime, to be added as a protocol to the UN Framework Convention on Climate Change (UNFCCC). The authors argue that 'climate refugees' need international recognition and protection, as well as financial assistance, but that current mechanisms are insufficient to deal with the issue. The additional protocol they suggest would best fit within the framework of the UNFCCC, and would thus benefit from the wide support of parties to the climate convention. They suggest the establishment of a committee on recognition, protection and resettlement of 'climate refugees', which would determine the populations needing relocation due to climate change. They acknowledge that such a proposal is likely to create some friction with the Geneva Convention, but contend that there is no reason to reserve the term 'refugee' for people fleeing persecution and not apply it to people fleeing climate change impacts. They also propose founding a new, specific funding mechanism aimed at fully reimbursing the incremental costs incurred by the resettlement of 'climate change refugees', although they remain vague about how the fund would be financed.

Although these studies seem unconnected from debates on the very concept of environmental migration and the meaning of the environment–migration nexus, they also contributed to the conceptualization of the issue. A majority were published in environmental law journals, further contributing to the divide between migration and environment scholars. Far from being confined to academic spheres, this divide is also apparent in the policy process, and can explain how and why environmental and migration policies followed different paths of evolution, as we shall now see.

Actual policy developments

Two policy areas are particularly concerned with environmental migration: migration policies and environmental policies. While environmental policies, at the international level, have increasingly accounted for migration flows triggered by environmental changes, international migration

---

[4] For example, this status was provided to the people of Montserrat displaced by a volcanic eruption in 1997, and to the people of Honduras and Nicaragua displaced by Hurricane Mitch in 1998.

and asylum policies have remained largely blind to environmental migration.

Another major difference between the two policy areas is their relationship with science. Environmental policies tend to rely heavily upon the natural sciences. Rosenbaum (2005) notes that 'what often distinguishes environmental policy making from other policy domains is the extraordinary importance of science, and scientific controversy, in the policy process'. Science, indeed, is often at the centre of policy debates, and its impartiality and objectivity can sometimes be tested. Linkages between natural sciences and environmental policies are particularly apparent in the case of natural disasters and of course climate change, where an ad hoc scientific body, the Intergovernmental Panel on Climate Change (IPCC), has been established by two United Nations agencies (United Nations Environment Programme and World Meteorological Organization) to review scientific evidence relating to global warming, and to assess the associated risks and impacts. The relationship between science and policy is significantly different in the area of migration policies, and the limited impact of social sciences on policies (and migration policies in particular) is often lamented (Weiss, 1978). Why is this the case? Florence and Martiniello (2005) note that academic discourse on migration faces increasing competition from the media and other non-academic discourses, which are often preferred to the former; academic sources may be disregarded for being too complex and critical, and if they are used, it is merely to legitimate a policy action. Furthermore, research findings on migration often conflict with politicized preconceptions of migration, both in public opinion and among policy-makers themselves. Bearing this in mind, I now look at the different directions taken by environmental and migration policies with regard to environmental migration. Only the international level is considered here, despite the obvious importance of the local and national dimensions of policy-making in these areas.

Environmental policies

Before states were concerned with slow-onset environmental changes, they were concerned with natural disasters. These are probably the most obvious cause of environmentally induced forced displacements. Policies implemented to deal with disasters evolved considerably throughout history (Haddow and Bullock, 2004). From the late 1970s, the sharp increase in disaster casualties and damages prompted a deep questioning of disaster policies, which at the time were focused almost exclusively on emergency relief and humanitarian aid. It was realized that a solely humanitarian approach could not suffice to deal with disasters. The chimerical pursuit of absolute security and risk eradication was abandoned, and so was the deterministic approach to disasters: the role of vulnerability was increasingly recognized. Soon states attempted to mutualize the risk through the creation of national agencies for disaster management, as well as the payment of damage compensation for victims of disasters.

As disaster management became internationalized, a number of other actors emerged and started playing an expanding role. Besides the United Nations, NGOs, experts and the international media gained a wider role at the scene of disaster relief. Regarding population displacements, both the United Nations High Commissioner for Refugees (UNHCR) and the International Organization for Migration (IOM) were increasingly involved in disaster relief following the tsunami of 2004 and other large-scale catastrophes of the mid-2000s.

Although formal cooperation mechanisms for disaster management were not implemented until the 1970s, the deployment of international aid in the face of disasters is not a recent idea. The interwar period saw the first and only attempt to provide disaster relief assistance in the form of a treaty, the International Relief Union (IRU) (Macalister-Smith, 2007). Founded in the aftermath of the Messina earthquake of 1908, the IRU was the civil equivalent of a military alliance: members pledged to come to each other's aid in case they were hit by a disaster due to *force majeure*. International assistance was no longer a matter of goodwill and charity, but a matter of common responsibility guaranteed by a treaty. The IRU proved short-lived however: following the foundation of the United Nations, international disaster assistance was split between different UN agencies, such as UNICEF or the World Health Organization. The major disasters that occurred in the 1970s and 1980s increased the need for a coordinated response, and provided the background for general agreement that the role of the UN in disaster relief should be expanded. An international disaster regime started to emerge in the 1970s, with the creation of the UN Disaster Relief Office (UNDRO) in 1971, replaced by the Office for Coordination of Humanitarian Affairs (OCHA) in 1991. One of the motivations for establishing OCHA was the deep concern about 'flows of refugees' and the 'mass displacement of people'. This concern materialized in the creation of an Inter-Agency Standing Committee (IASC), a platform of exchange and policy-making involving diverse UN agencies and other organizations, including the UNHCR and IOM.

This myriad of actors constitutes the embryo of a governance system for natural disasters, but this system remains highly fragmented, increasingly specialized, and marred by institutional rivalries. Thus international normative frameworks on disaster management remain disparate, but have increasingly addressed the issue of disaster-induced displacement. The most ambitious document is undoubtedly the Hyōgo Framework for Action 2005–2015, which is the main outcome of the World Conference on Natural Disaster Reduction, held in Kobe (Japan) in 2005. The Hyōgo Framework is supported by the Global Facility for Disaster Reduction and Recovery (GFDRR), and fund-managed by the World Bank. Although this is not explicitly stated, the Fund can be used to facilitate the return of those displaced by disasters and to aid the reconstruction of their homes.

Another key document is the *Operational Guidelines on Human Rights and Natural Disasters*. In the absence of binding normative frameworks, international organizations such as the UNHCR and IOM have stepped up to provide operational responses to natural disasters. The assessment of these on-the-ground responses translated into a policy document edited by the IASC (2006) and drafted in the aftermath of the 2004 tsunami and the 2005 Hurricane Katrina. The document outlines four types of protection that ought to be guaranteed to people affected by disasters:

- the protection of life, security of the person, physical integrity and dignity;
- the protection of rights related to basic necessities of life;
- the protection of other economic, social and cultural rights; and
- the protection of other civil and political rights.

These guidelines are undoubtedly the most advanced effort to address the issues of people displaced by disasters directly.

The policy approach to natural disasters is now more international and institutionalized than it has ever been. Yet, despite the greater acknowledgement of social vulnerabilities and the prevention imperative, disaster management remains largely focused on emergency operations, with little consideration of the long-term needs of those displaced. Displacement is often viewed as temporary, and few contingency plans exist for long-term displacement and possible eventual relocation. Disaster management continues to construct environmental migration as a temporary displacement, rather than a long-term or permanent relocation. Hence, despite some recent progress, it continues to limit itself to an emergency response for displaced people, and fundamentally biases the conceptualization of the displacement as short term, whereas empirical evidence has often proved otherwise.

Natural disasters are best understood as a special case of environmental change, addressed by specific machinery. Over time, climate change has come to embody all facets of environmental hazards, and the debate on migration induced by climate change has consistently overshadowed the debate on migration triggered by other kinds of environmental disruptions. Four key reasons account for this myopic focus on migration triggered by climate change:

- First, climate change encompasses a wide range of environmental changes, making it difficult to distinguish between the events related to climate change and those that are not. Hence any migration linked to environmental change – with few exceptions[5] – can be described as a consequence of climate change.
- Second, climate change is expected to dramatically increase the number of environmental migrants, both because many environments will become increasingly degraded, but also because other environments are expected to become more favourable.[6] There is, therefore, a new dimension to the scale of the problem.
- Third, negotiations concerning the future climate regime are currently ongoing. Authors who make the case for the development of new normative frameworks often consider the possibility of including an additional protocol to the UNFCCC (Biermann and Boas, 2010). There is therefore an incentive for policy-oriented research to focus on climate change-induced migration.
- Fourth, most of the funding possibilities for research in the field, as well as media requests, are currently related to projects on the impacts of climate change. This creates a strong incentive to orient research in this direction.

Since the signature of the United Nations Framework Convention on Climate Change in 1992, a considerable number of climate policies, institutions and instruments have emerged to tackle global warming. These form what is commonly called the 'international climate change

---

[5] Such exceptions include earthquakes, tsunamis and volcanic eruptions.
[6] This is the case, in particular, of Siberia and northern Canada, which may become new destination areas for migration in the near future.

regime'. The fight against climate change has taken two directions, which have at times been opposed to each other. One was concerned with mitigation, that is, the reduction of greenhouse gas emissions, while the other dealt with adaptation to the impacts of climate change. Adaptation to climate change has long been considered a failure of mitigation, a 'hypothetical possibility best kept in the background lest it reduce the felt urgency of mitigation' (Wilbanks et al., 2003; Biermann and Boas, 2010), and mitigation has consistently been prioritized over adaptation – a situation that continues today, to a large extent. In recent years, as it has become progressively apparent that a number of climate change impacts cannot be avoided, many voices, from developing countries in particular, have called for a greater emphasis on adaptation. At the same time, adaptation was perceived as an effective strategy to reduce the negative impacts of climate change – Tol et al. (1998) remark that some authors would even see it as better response strategy than mitigation – and also as a way to achieve climate justice, and thus convince developing countries to accept a mandatory reduction of their carbon emissions (Roberts and Parks, 2007). Environmental migration is most concerned with adaptation, whether a failure of adaptation or a successful adaptation strategy.

Overall, the international climate change regime remains largely focused on mitigation. Although migration, as both a consequence of climate change and an adaptation strategy, is mentioned in the IPCC reports and the Stern review on the economics of climate change, this concern has not been translated into the UNFCCC, the Kyoto Protocol, or any instrument or mechanism relating to these agreements. Yet many scholars and NGOs place high hopes in international talks on climate for the development of new mechanisms to address climate-related migration. At this stage, however, and despite a short-lived inclusion in the negotiating text of the 2009 Copenhagen Conference, there is no sign that migration will be placed on the agenda of climate negotiations in the near future.

Environmental changes as a whole are now overshadowed by global warming. Climate change is addressed through a complex international regime, made up of binding laws, funds and institutions. This regime has mainly focused on the source of the problem, the emission of greenhouse gases, rather than on the different adaptive strategies that could be developed to cope with the impacts of the change. Migration is increasingly recognized as one of these possible strategies, but this aspect remains overshadowed by the view of migration as a failure of adaptation. Although migration appears to be a core aspect of global climate change, it is not yet on the agenda of ongoing talks on the future climate change regime.

Migration policies

Institutional frameworks dealing with forced migration have undergone rapid development over the second half of the twentieth century, a development that accelerated from the asylum crisis of the 1990s. A common but simplistic argument often raised in conferences and workshops is that environmental 'refugees' are actually not refugees, because they do not meet the criteria of the Geneva Convention. As Burson has pointed out (2008), this view is over-simplistic: 'Caution needs to be exercised before rigid and immutable distinctions are drawn between environmentally displaced persons and those to whom the Convention's protection regime can extend.' Indeed, although people displaced by environmental changes do not fit the criteria for being recognized as 'refugees' under the Geneva Convention, there are some instances in which the Convention can be helpful in addressing their plight: this is the case when environmental changes are associated with conflicts, or when people are affected by environmental changes as a result of their belonging to a specific group.

At the same time, the regime associated with the Geneva Convention has repeatedly been called into question, particularly since the asylum crisis that started in the late 1980s (Garvey, 1985; Dacyl, 1995; Hathaway, 1997). Critiques of the regime have argued that reforms are necessary to make the regime relevant again, and that it is not able, in its current state, to accommodate the large numbers of forcibly displaced people who are not covered by the current refugee status.

The adoption of the Geneva Convention on the status of refugees in 1951 was the outcome of a long maturation process, and followed numerous international arrangements to deal with refugees throughout the world. Along with its additional protocol of 1967, the Convention is still considered today as the cornerstone of international refugee protection. Its goal was not only to formalize a refugee definition, but also to define the precise rights of refugees and the obligations of states towards them.

Over time, the Geneva Convention has acquired the status of a human rights treaty (McAdam, 2006), and some of its core principles, including that of finding asylum from persecution, are to be found in the Universal Declaration of Human Rights.

Studies and initiatives on environmental migration, however, have repeatedly called for a modification of the 1951 Convention, or an additional protocol, in order to grant appropriate protection to people displaced by environmental changes. These calls have come from a wide range of sources: scholars, NGOs, but also policy-makers. While it is impossible to draw up a comprehensive list of these initiatives, it is possible to classify them into two categories:

- Those who argue that environmental degradation is a form of political persecution, and that environmental displacees therefore have a legitimate right to claim refugee status (Conisbee and Simms, 2003);
- Those who contend that the refugee definition no longer matches the realities of forced migration, and must be revised in order to incorporate environmental displacees (Cooper, 1997; Chemillier-Gendreau, 2006).

These interventions have not been well received by scholars working on refugee rights, who were the key members of the sceptic perspective. Most were afraid that these calls were an excuse to limit governmental responsibility for forced displacement (Kibreab, 1997; 1994), or that they would result in a watering down of the protection mechanisms in place for current refugees.

Some initiatives went a step further and called officially for recognition of 'environmental refugees' within the Geneva Convention. In recent years the Belgian Parliament has been the most active of all legislative bodies on this issue. In 2006, the Senate passed a resolution urging the Belgian Government to work towards better protection of 'environmental refugees' in the Geneva Convention (Mahoux, 2006). The governments of Finland and Sweden both decided in the mid-2000s to grant asylum to those displaced by natural disasters.

Refugee law, which revolves around the Geneva Convention, deliberately dismisses factors other than persecution when defining who may qualify as a refugee. A core function of the Convention is to define a refugee, and yet it does so in a performative way: refugees are those who fit the criteria of the Convention; migrants in refugee-like situations that do not fit these criteria cannot be labelled refugees. However, study of the preparatory works for the Convention reveals how the refugee definition was based on political motives, and not on scientific grounds. Furthermore, the work of the UNHCR depends, to a great extent, upon the range of this definition. This certainly sheds some light on the conceptualization of refugee as a political construct, and can illuminate debates on the conceptualization of environmental migration or 'environmental refugees'.

With regard to protection, international refugee law can do very little for those uprooted by environmental changes, except in specific cases such as environmental racism. On the operational level, the UNHCR mandate has been considerably stretched in recent years. The agency has increasingly been involved with internal displacement and situations of natural disasters, to the extent that the protection of officially recognized refugees now represents a diminishing share of its operational activities. As the UNHCR seems to be evolving towards a global protection agency, it could take a bigger role in dealing with environmental displacement. This would be, however, a further stretch of its mandate.

Overall, unlike the areas of natural disasters and global warming, the international refugee regime remains dominated by a sceptical approach, which can explain why the regime has not been significantly modified since the 1960s. In particular, a common feature of the different evolutions of the regime is their reactive character: it tried to adapt to some new realities of forced migration, but never did try to anticipate these new realities. This is a major difference between this policy area and the areas considered above, which take a more proactive and prospective approach.

Empirical evidence, however, shows that most migration movements linked to environmental stressors are internal displacements. Although the UNHCR has been increasingly concerned with internally displaced persons, this type of displacement does not fall within the scope of the international refugee regime. Concerns for IDPs began in the early 1990s, when the United Nations first came to deal with the issue. International concerns grew with the nomination of Francis Deng as Representative of the Secretary-General on the Human Rights of Internally Displaced Persons, and culminated with the adoption of the *Guiding Principles on Internal Displacement* by the UN General Assembly in 1998 (Deng, 1998).

Internally displaced persons were first defined negatively: they were people who had fled their homes, but did not qualify as refugees (Phuong, 2004). The conceptualization of IDPs is indeed deeply linked to the refugee problem, and takes place in the context of containment efforts by receiving states aiming to restrict asylum and keep displaced people within the borders of their own country (Dubernet, 2001). In the early 1990s, the international community perceived the need to define IDPs in order to provide them with adequate assistance. The UNHCR favoured an approach whereby IDPs would be defined as people who would have been refugees had they left their country. The definition that was finally adopted in 1998 departed from this approach and chose to focus on the causes of displacement, with a view to the prevention of forced displacement. Phuong (2004) notes that this focus on the causes permitted a discussion of some root causes of displacement which had not been considered with regard to refugee movements, such as natural disasters or development projects, despite the reluctance of some authors to consider people displaced by disasters as IDPs. The definition includes people displaced by natural disasters, but not those uprooted by slow-onset environmental changes. Some impacts of climate change, such as sea level rise, are likely to bring huge challenges to this definition. Walter Kälin, former Representative of the Secretary-General on the Human Rights of Internally Displaced Persons, recently identified three types of slow-onset environmental changes brought upon by global warming which have the potential to trigger internal displacements (Kälin, 2008):

- Governments will need to identify areas at high risk from environmental change and displace – perhaps by force – people from these zones;
- Environmental degradation and slow-onset disasters, for which 'we need criteria to better determine where to draw the line between voluntary movement and forced displacement'; and finally
- The case of 'sinking' small island states, which currently lies in legal limbo.

These three areas have the potential not only to induce forced migration, but also to call into question the root causes for displacement addressed by the definition.

The *Guiding Principles of Internal Displacement* consist of thirty recommendations upholding the human rights of IDPs. The *Guiding Principles* are 'soft law', which states are free to apply or disregard. Despite their non-binding character, the *Guiding Principles* have been the basis of many programmes on internal displacement, and have been widely used by UN agencies, NGOs, and some governments.

Although some norms and policies do exist, important protection gaps remain, especially with regard to displacements induced by slow-onset environmental changes. Another significant policy gap which is often unmentioned relates to migration management. It is increasingly recognized that voluntary migration can be an adaptation strategy to environmental changes, yet no consistent international framework exists for a global, proactive governance of migration. Forced migration is only one dimension of environmental migration, one that can often (but not always) be avoided through voluntary migration. A significant policy gap lies therefore in the absence of mechanisms aimed at facilitating migration. Overall, even more so than in the case of forced migration, migration policies remain the exclusive privilege of states, and international cooperation in this field has been extremely limited. As Hannah Arendt (1973) put it, 'sovereignty is nowhere more absolute than in matters of emigration, naturalization, nationality, and expulsion'.

Conclusion – How 'climate refugees' became the face of climate change, or what policy responses reveal about our conceptualization of environmental migration

Environmental and migration policies have evolved along different lines, at different speeds, over recent decades. Both sets of policies aspire to develop mechanisms of global governance within their respective fields, but the desire of states to maintain sovereign control over these policy areas has impeded the development of comprehensive policies. Environmental migration, lying at the crossroads of these policy areas, has thus been addressed in a fragmented fashion by different levels of governance.

I explain this fragmented approach by the lack of connections between environmental and migration policies and the influence of different research perspectives in these policy areas: the alarmist perspective weighs heavily on the development of environmental policies, whereas migration policies remain under the influence of a more sceptical perspective.

This chapter has tried to show the divergent ways in which environmental migration has been addressed by different policy areas. Displacements as a result of environmental disruption have been a matter of increasing concern in the field of disaster reduction and adaptation to climate change, whereas asylum and migration policies have remained timid in addressing the issue, despite creating some subsidiary mechanisms of protection at regional level.

Overall, the policy process relating to environmental migration is dominated by an alarmist perspective. This perspective is fuelled by a set of values, interests and strategies aimed at fostering the protection of the environment, and the fight against climate change in particular. As 'climate refugees' were portrayed as the human face of climate change, climate negotiations became the central focal point of the policy subsystem of environmental migration, and it is expected that normative frameworks to deal with the phenomenon of environmental migration will develop primarily in international climate policies.

This shapes the current conceptualization of environmental migration in different ways:

- First, it denies the role played by migration policies in determining the size and patterns of migration flows, by implying that flows are primarily dependent on the extent of climate change impacts and the adaptation strategies that will be implemented to mitigate these impacts. Thus the embedding of policy debates within the sphere of climate change reflects a deterministic perspective that assumes that environmental displacements depend primarily on the impacts of climate change, when in fact these impacts can be mitigated by adaptation policies.
- Second, it narrows cases of environmental migration to those displacements induced by climate change, and thus excludes displacements associated with other environmental disruptions. Furthermore, it has a strong bias towards future displacements and considers environmental migration to be a new issue, despite the existence of past and current migrations associated with environmental disruptions, including the effects of climate change.
- Third, environmental migration is considered as a forced movement, and the voluntary dimension of some migration flows is not addressed. Therefore, policy proposals are often exclusively geared towards protection and compensation, rather than governance and facilitation of migration.
- Finally, and most importantly, it tends to conceptualize environmental migration as a specific type of migration that must be addressed in environmental forums rather than in debates over migration policies and governance. Undoubtedly, this is the clearest sign of the dominance of the alarmist perspective in the policy subsystem.

Overall, policy debates have cemented a concept of environmental migration as a forced migration caused primarily by the impacts of climate change, a humanitarian disaster in the making. Yet this deterministic view is not supported by empirical evidence, as numerous studies show that migration is only one of some possible reactions to environmental changes, determined by a wide series of factors, other than just the nature or extent of environmental degradation.

Ideas and representations matter in the policy process. Environmental migration has been constructed as a catastrophic human impact of climate change, a failure to cope with environmental disruptions. Such a political construction is primarily due to the influence of an alarmist perspective in the policy process, and bears significant policy implications, as if migration flows associated with environmental changes were distinct from other forms of mobility, requiring specific policy responses.

This deterministic perspective serves many worthwhile goals, and highlights in particular the need to fight against climate change. Thus portraying 'climate refugees' as the human face of climate change is helpful in many ways, but certainly not to accurately describe the realities of migration associated with environmental changes.

References

Arendt, H. 1973. *The Origins of Totalitarianism*. New York, Harvest Books.

Bates, D. C. 2002. Environmental refugees? Classifying human migrations caused by environmental change. *Population and Environment*, Vol. 23, No. 5, pp. 465–77.

Bell, D. 2004. Environmental refugees: What rights? Which duties? *Res Publica*, Vol. 10, pp. 135–52.

Biermann, F. and Boas, I. 2010. Preparing for a warmer world. Towards a global governance system to protect climate refugees. *Global Environmental Politics*, Vol. 10, No. 1, pp. 60–88.

Black, R. 1998. *Refugees, Environment and Development*. Harlow, UK, Addison Wesley Longman.

2001. *Environmental Refugees: Myth or Reality?* Geneva, United Nations High Commissioner for Refugees. (*New Issues in Refugee Research*, No. 34.)

Brown, L. R. 2008. *Troubling New Flows of Environmental Refugees*. Earth Policy Institute 2004 [cited 17 January 2008]. http://earth-policy.org/Updates/Update33.htm

Brown, O. 2008. *Migration and Climate Change*. Geneva, International Organization for Migration. (Migration Research Series No. 31.)

Burson, B. 2008. *Environmentally Induced Displacement and the 1951 Refugee Convention: Pathways to Recognition*. Paper presented at Environment, Forced Migration and Social Vulnerability conference, 9–11 October, Bonn.

Byravan, S. and Rajan, S. C. 2006. Providing new homes for climate change exiles. *Climate Policy*, Vol. 6, No. 2, pp. 247–52.

Castles, S. 2002. *Environmental Change and Forced Migration: Making Sense of the Debate*. Geneva, United Nations High Commissioner for Refugees. (*New Issues in Refugee Research*, No. 70.)

2007. *Confronting the Realities of Forced Migration*. Oxford, UK, Migration Policy Institute 2004 [cited 18 March 2007]. http://www.migrationinformation.org/Feature/display.cfm?ID=222

Chemillier-Gendreau, M. 2006. Faut-il un statut international de réfugié écologique? *Revue européenne de droit de l'environnement*, No. 4, pp. 446–52.

Christian Aid. 2007. *Human Tide: The Real Migration Crisis*. May, London.

Conisbee, M. and Simms, A. 2003. *Environmental Refugees. The Case for Recognition*. London, New Economics Foundation.

Cooper, J. B. 1997. Environmental refugees: meeting the requirements of the refugee definition. *New York University Environmental Law Journal*, Vol. 6, No. 2, pp. 480–503.

Cournil, C. 2006. Vers une reconnaissance des 'réfugiés écologiques'? Quelle(s) protection(s), quel(s) statut(s)? *Revue du droit public*, No. 4, pp. 1035–66.

Cournil, C. and Mazzega, P. 2006. Catastrophes écologiques et flux migratoires: comment protéger les 'réfugiés écologiques'? *Revue européenne de droit de l'environnement*, No. 4, pp. 417–27.

— 2007. Réflexions prospectives sur une protection juridique des réfugiés écologiques. *Revue européenne des migrations internationales*, Vol. 23, No. 1, pp. 7–34.

Dacyl, J. W. 1995. Europe needs a new protection system for 'non-convention' refugees. *International Journal of Refugee Law*, Vol. 7, No. 5, pp. 579–605.

Deng, F. 1998. *Guiding Principles on Internal Displacement*, 11 February. (E/CN.4/1998/53/Add.2.)

Dubernet, C. 2001. *The International Containment of Displaced Persons. Humanitarian Spaces Without Exit*. Aldershot, UK, Ashgate.

El-Hinnawi, E. 1985. *Environmental Refugees*. Nairobi, United Nations Environment Programme.

European Commission and the Secretary-General/High Representative. 2008. *Climate Change and International Security*. Brussels, Council of the European Union.

Falstrom, D. Z. 2001. Stemming the flow of environmental displacement: creating a convention to protect persons and the environment. *Colorado Journal of International Environmental Law & Policy Yearbook*, Vol. 6, No. 1, pp. 2–32.

Florence, E. and Martiniello, M. 2005. The links between academic research and public policies in the field of migration and ethnic relations: selected national case studies – thematic introduction. *International Journal on Multicultural Societies*, Vol. 7, No. 1, pp. 3–10.

Friends of the Earth Australia. 2004. *A Citizen's Guide to Climate Refugees*. Fitzroy, Australia.

Garvey, J. 1985. Toward a reformulation of international refugee law. *Harvard International Law Journal*, Vol. 26, No. 2, pp. 483–502.

Gonin, P. and Lassailly-Jacob, V. 2002. Les réfugiés de l'environnement. Une nouvelle catégorie de migrants forcés? *Revue européenne des migrations internationales*, Vol. 18, No. 2, pp. 139–60.

Haddow, G. and Bullock, J. A. 2004. *Introduction to Emergency Management*. Amsterdam, Butterworth-Heinemann.

Hathaway, J. C. (ed.). 1997. *Reconceiving International Refugee Law*. The Hague, Netherlands, Martinus Nijhoff.

Hermsmeyer, H. A. 2005. *Environmental Refugees: A Denial of Rights*. San Diego, Calif., Center for Comparative Immigration Studies, University of California. (Contemporary Topics in Forced Migration.)

Homer-Dixon, T. 1991. On the threshold. Environmental changes as causes of acute conflict. *International Security*, Vol. 16, No. 2, pp. 76–116.

—— 1994. Environmental scarcities and violent conflict. Evidence from cases. *International Security*, Vol. 19, No. 1, pp. 5–40.

Hugo, G. 1996. Environmental Concerns and International Migration. *International Migration Review*, Vol. 30, No. 1, pp. 105–31.

IASC. 2006. *Protecting Persons Affected by Natural Disasters*. IASC Operational Guidelines on Human Rights and Natural Disasters. Washington DC/Geneva, Brookings-Bern Project on Internal Displacement/Inter-Agency Standing Committee.

IFRC. 2001. *World Disasters Report. Focus on Recovery*. Geneva, International Federation of Red Cross and Red Crescent Societies.

Jacobson, J. 1988. *Environmental Refugees: A Yardstick of Habitability*. Washington DC, World Watch Institute. (World Watch Paper 86.)

Kälin, W. 2008. *The Climate Change–Displacement Nexus*. Paper presented at ECOSOC Panel on Disaster Risk Reduction and Preparedness: Addressing the Humanitarian Consequences of Natural Disasters, Geneva.

Kibreab, G. 1994. Migration, environment and refugeehood. In: B. Zaba and J. Clarke (eds), *Environment and Population Change*. Liège, Belgium, International Union for the Scientific Study of Population.

—— 1997. Environmental causes and impact of refugee movements: a critique of the current debate. *Disasters*, Vol. 21, No. 1, pp. 20–38.

Ki-moon, B. 2007. A climate culprit in Darfur. *Washington Post*, 16 June, A15.

King, T. 2006. Environmental displacement: coordinating efforts to find solutions. *Georgetown International Environmental Law Review*, Vol. 18, No. 3, pp. 543–66.

Kliot, N. 2004. Environmentally induced population movements: their complex sources and consequences. A critical review. In: J. D. Unruh, M. S. Krol and N. Kliot (eds), *Environmental Change and Its Implications for Population Migration*. Dordrecht, Netherlands, Kluwer Academic Publishers.

Kniveton, D., Schmidt-Verkerk, K., Smith, C. and Black, R. 2008. *Climate Change and Migration: Improving Methodologies to Estimate Flows*. Geneva, International Organization for Migration. (Migration Research Series No. 33.) www.iom.int/jahia/webdav/site/myjahiasite/shared/shared/mainsite/published_docs/serial_-publications/MRS-33.pdf

Koser, K. 1996. Changing Agendas in the study of forced migration: a report on the Fifth International Research and Advisory Panel Meeting, April 1996. *Journal of Refugee Studies*, Vol. 9, No. 4, pp. 53–66.

Lassailly-Jacob, V. 2006. Une nouvelle catégorie de réfugiés en débat. *Revue européenne de droit de l'environnement*, No. 4, pp. 374–80.

Lonergan, S. and Swain, A. 1999. *Environmental Degradation and Population Displacement*. Victoria, BC, Global Environmental Change and Human Security (GECHS). (AVISO Working Papers.)

Lopez, A. 2007. The protection of environmentally-displaced persons in international law. *Environmental Law*, Vol. 37, No. 2, pp. 365–409.

Macalister-Smith, P. 2007. The International Relief Union of 1932. *Disasters*, Vol. 5, No. 2, pp. 147–54.

Magniny, V. 1999. Les réfugiés de l'environnement. Hypothèse juridique à propos d'une menace écologique. Doctoral dissertation. Paris, University Paris I Panthéon-Sorbonne, Department of Law.

Mahoux, P. 2006. Proposition de résolution visant à la reconnaissance dans les conventions internationales du statut de réfugié environnemental. In: B. Senate (ed.), *3–1556*. Brussels, Belgian Senate.

Mason, E. 1999. Researching refugee and forced migration studies: an introduction to the field and reference literature. *Behavioral & Social Sciences Librarian*, Vol. 18, No. 1, pp. 1–20.

McAdam, J. 2006. *The Refugee Convention as a Rights Blueprint for Persons in Need of International Protection*. Geneva, United Nations High Commissioner for Refugees. (*New Issues in Refugee Research*, No. 125.)

McCue, G. S. 1993. Environmental refugees: applying international environmental law to involuntary migration. *Georgetown International Environmental Law Review*, Vol. 6, No. 1, pp. 151–90.

McGregor, J. 1993. Refugees and the environment. In: R. Black and V. Robinson (eds), *Geography of Refugees: Patterns and Processes of Change*. London, Belhaven.

McLeman, R. and Smit, B. 2004. Climate change, migration and security. *Commentary*. Ottawa, Canadian Security Intelligence Service.

2006. Migration as an adaptation to climate change. *Climatic Change*, Vol. 76, Nos. 1–2, pp. 31–53.

McTegart, W. J., Sheldon, G. W. and Griffiths, D. C. (eds). 1990. *Impacts Assessment of Climate Change. Report of Working Group II. Intergovernmental Panel on Climate Change*. Canberra, Australian Government Publishing Service.

Meze-Hausken, E. 2004. Migration caused by climate change: how vulnerable are people in dryland areas? *Mitigation and Adaptation Strategies for Global Change*, Vol. 5, No. 4, pp. 379–406.

Michelot, A. 2006. Enjeux de la reconnaissance du statut de réfugié écologique pour la construction d'une nouvelle responsabilité environnementale. *Revue européenne de droit de l'environnement*, No. 4, pp. 428–45.

Myers, N. 1993. Environmental refugees in a globally warmed world. *BioScience*, Vol. 43, No. 11, pp. 752–61.

1997. Environmental refugees. *Population and Environment*, Vol. 19, No. 2, pp. 167–82.

2002. Environmental refugees: a growing phenomenon of the 21st century. *Philosophical Transactions of the Royal Society of London B*, Vol. 357, No. 1420, pp. 609–13.

Myers, N. and Kent, J. 1995. *Environmental Exodus: an Emergent Crisis in the Global Arena*. Washington DC, Climate Institute.

Ogata, S. 1993. *The State of the World's Refugees 1993*. New York, Penguin.

Phuong, C. 2004. *The International Protection of Internally Displaced Persons*. Cambridge, UK, Cambridge University Press.

Piguet, E. 2008. *Climate Change and Forced Migration*. Geneva, United Nations High Commissioner for Refugees. (*New Issues in Refugee Research*, No. 153.)

Prieur, M., Marguénaud, J.-P., Monédiaire, G., Bétaille, J., Drobenko, B., Gouguet, J.-J., Lavieille, J.-M., Nadaud, S. and Roets, D. 2008. Projet de convention relative au statut international des déplacés environnementaux. *Revue européenne de droit de l'environnement*, No. 4, pp. 81–93.

Richmond, A. 1993. Reactive migration: sociological perspectives on refugee movements. *Journal of Refugee Studies*, Vol. 6, No. 1, pp. 7–24.

1994. *Global Apartheid. Refugees, Racism, and the New World Order*. Toronto, Oxford University Press.

Roberts, J. T. and Parks, B. C. 2007. *A Climate of Injustice. Global Inequality, North-South Politics, and Climate Policy*. Cambridge, Mass., MIT Press.

Rosenbaum, W. A. 2005. *Environmental Politics and Policy*. 6th ed. Washington DC, CQ Press.

Schwartz, P. and Randall, D. 2003. *An Abrupt Climate Change Scenario and Its Implications for United States National Security*. Report commissioned for the US Department of Defense. San Francisco, Calif., Global Business Network.

Stern, N. 2007. *The Economics of Climate Change: The Stern Review*. Cambridge, UK, Cambridge University Press.

Suhrke, A. 1993. *Pressure Points: Environmental Degradation, Migration and Conflict*. Cambridge, Mass., American Academy of Art and Science.

Suhrke, A. and Visentin, A. 1991. The environmental refugee: a new approach. *Ecodecision*, Vol. 2, September, pp. 73–84.

Swain, A. 1996*a*. Displacing the conflict: environmental destruction in Bangladesh and ethnic conflict in India. *Journal of Peace Research*, Vol. 33, No. 2, pp. 189–204.

1996*b*. Environmental migration and conflict dynamics: focus on developing regions. *Third World Quarterly*, Vol. 17, No. 5, pp. 959–73.

Tol, R. S. J., Fankhauser, S. and Smith, J. B. 1998. The scope for adaptation to climate change: what can we learn from the impact literature? *Global Environmental Change*, Vol. 8, No. 2, pp. 109–23.

Vogt, W. 1948. *Road to Survival*. New York, William Sloane Associates.

WBGU. 2008. *Climate Change as a Security Risk*. Berlin/London, German Advisory Council on Global Change/Earthscan.

Weiss, C. H. 1978. Improving the linkage between social research and public policy. In: L. E. Lynn (ed.), *Knowledge and Policy: The Uncertain Connection*. Washington DC, National Academy of Sciences.

Westing, A. H. 1989. The environmental component of comprehensive security. *Bulletin of Peace Proposals*, Vol. 20, No. 2, pp. 129–34.

1992. Environmental refugees: a growing category of displaced persons. *Environmental Conservation*, Vol. 19, No. 3, pp. 201–07.

Wilbanks, T. J., Kane, S. M., Leiby, P. N., Perlack, R. D., Settle, C., Shogren, J. and Smith, J. B. 2003. Integrating mitigation and adaptation. Possible responses to global climate change. *Environment*, Vol. 45, No. 5, pp. 29–38.

Wood, W. 2001. Ecomigration: linkages between environmental change and migration. In: A. R. Zolberg and P. Benda (eds), *Global Migrants, Global Refugees*. New York, Berghahn Books.

Lessons from past forced resettlement for climate change migration

GRAEME HUGO

Introduction

In the recent burgeoning of interest in the relationship between climate change and migration, there is a danger that such migration will be considered as being separate from other types of mobility. In fact climate change should be seen as a new and increasingly significant driver among a constellation of several dynamic forces impinging on mobility with which it interacts. Considering environment-related migration separately from other mobility is dangerous not only because most climate change-related movement is driven by multiple causes but also because there is a substantial body of knowledge of migration which is of relevance to better understanding the complex climate change – migration relationship (Hugo, 2010). Separating climate change-related migration from existing knowledge of migration theory and practice would significantly delay progress in improving understanding and could lead to inappropriate, ineffective and inequitable policy intervention.

This chapter seeks to make a contribution by focusing on the extent to which existing knowledge is of relevance to one of the most discussed and controversial dimensions of the climate change – migration relationship – forced displacement of individuals, families and communities from areas impacted by climate change and the resettlement of these people. It begins with a brief consideration of the potential for climate change to displace significant numbers of people. This is a highly contested area but one where there has been much uninformed speculation and a lack of appreciation that the impact of climate change will not always be resettlement. Too often it is assumed that severe environmental impact *must* result in displacement but the adaptation process is much more complex. The discussion below relates to the existing knowledge of the nature and scale of forced displacement and resettlement, especially in low-income countries, and their relevance to climate change-related displacement. It is argued that while there are several types of such displacement, those associated with large-scale infrastructure projects and resettlement schemes have perhaps the greatest relevance to climate change displacement. The major part of the discussion then attempts to draw out the key lessons that can be drawn from existing research relating to the process of displacement and resettlement which are likely to be of relevance to the impacts of climate change. Finally, some implications for climate change-related migration policy are considered.

Climate change and population displacement

Most scientifically robust projections of climate change indicate that there will be significant changes in communities' ability to earn a livelihood in hot spot areas which will experience the greatest impact. While much of the response will involve *in situ* adjustment, some mobility is likely to result (Hugo et al., 2009). It needs to be recognized, however, that migration responses to climate change are of two types:

- Migration of some people out of areas influenced by climate change on a *temporary* or *permanent* basis can enhance the capacity of those left behind to adapt to climate change.
- In extreme cases where climate change makes it impossible for communities to remain in their home areas, *displacement migration and resettlement* elsewhere offers a last resort.

Some of the literature exaggerates the second type of migration response but it is only one among the array of *in situ* and migration responses to climate change and even then is usually the last resort when other adaptation mechanisms have been exhausted and community resilience broken down. In this paper, however, we focus on this displacement and resettlement process.

A key distinction also has to be made between: (a) the sudden onset of cataclysmic events which destroy or rapidly change livelihoods or displace population on a permanent or temporary basis; and (b) the more long-term, slow-onset processes which see an incremental decline in the ability of an area to provide a livelihood for its resident population. In both cases mobility based and non-mobility based strategies, including displacement and resettlement, can be initiated. However, the institutions and strategies which are needed to cope effectively often differ across the two types of impact.

One of the challenges is the need to convince policy-makers to take action regarding responses to *both* sudden and slow-onset impacts. There is a considerable body of experience with respect to disaster response, refugees and internally displaced persons (IDPs) which is of relevance to responding to climate change. While creating policies, actions and governance systems to cope with cataclysmic changes is by no means trivial, the existence of a substantial body of relevant experience and the sudden and extreme nature of its impact mean that policy-makers have responded with a degree of urgency. On the other hand more long-term, incremental impacts have relatively less immediacy so there is the danger that policy-makers will defer action. It is of critical importance for policy-makers to recognize that while the full impact of these incremental processes will generally not be fully evident for several decades, the interventions needed to offset or ameliorate them are often of such large scale and complexity that they will need to be operationalized over decades. The crucial point is that the need for action on *both* sudden-impact and slow-impact hazards is urgent.

Our focus here is upon the more or less permanent displacement and resettlement associated with climate change impacts and especially those associated with slow-impact, gradual deterioration of the environment due to climate change. The more dramatic, sudden displacement of people as a result of an extreme environmental event tends to dominate the discourse on this topic but the 'silent violence' (Spitz, 1978) of unrelenting gradual impacts also needs attention for the following reasons:

- Taking action on them can prevent suffering and loss of human life *before* the situation becomes extreme.
- The numbers impacted in this way by climate change will be substantially greater than those displaced by sudden extreme events.
- If dealt with in a timely way the displacement and resettlement processes can be planned to maximize the chances that those displaced can at least maintain, and hopefully enhance, their level of living.

However, responding to the slow-onset effects of climate change holds a number of challenges:

- As it lacks the immediacy of extreme events it is more difficult to attract the attention of policy-makers and planners.

- The long time period of the impact results in policy-makers deferring intervention in order to deal with what are seen as more pressing concerns.
- Local resilience and adaptation can mask the inevitability of at least some of the local community affected by climate change having to eventually be resettled.

While our main concern here is with slow-onset impacts of climate change on resettlement, much of what is discussed is also of relevance to the resettlement of those displaced by sudden impacts of extreme events. Hence before considering the gradual impacts of climate change on resettlement we briefly discuss displacement by extreme events associated with climate change.

Displacement by extreme events

One of the distinctive features of climate change is the increased incidence of extreme environmental events that will produce sudden displacements of population. These displacements are often temporary and people are able to, and/or want to, return to their former home area. It is often the assumption that such displaced persons all wish to return but it has been the experience of recent disasters that significant numbers do not wish to return and prefer to settle elsewhere. Following Hurricane Katrina in New Orleans in 2005 many displacees chose not to return, although their former community was reconstructed (Grier, 2005; Sastry and Gregory, 2009). One survey (Page, 2005, p. 1) conducted seven weeks after the disaster found that nearly four in ten displaced persons did not plan to move back to New Orleans. Accordingly the onset of sudden disasters associated with climate change can create two types of need for resettlement: (a) where the devastation means that the displaced persons are unable to return to their home area; (b) where the displaced persons no longer wish to live in that area.

The suddenness of such events can place great pressure on regional and national authorities to provide both immediate shelter and longer-term resettlement for those affected.

There are already substantial international initiatives regarding disaster response and management and most countries have in place structures and institutions to respond to disasters. Climate change, however, is likely to result in more frequent and widespread extreme events:

> ... of the type associated with disasters, such as heatwaves, changes in weather patterns, longer and more intense drought, more intense rainfalls, and more frequent coastal and inland flooding (Basher, 2008, p. 35).

Moreover, these sudden disasters associated with climate change are likely to occur most frequently in climate change hot spots – low-lying coastal areas, heavily populated delta areas, desertified low rainfall areas, low-lying islands and atolls, etc. – where communities are already highly vulnerable to climate-related hazards.

An important point here is that there are already mechanisms in place both internationally and nationally regarding disaster preparedness, response and management. Displacements associated with climate change-related extreme events need to be dealt with within this existing framework rather than seeking to create a new structure. However, that existing framework does need to be modified to explicitly recognize that climate change is likely to lead to hazard events and produce changes to natural resource conditions on an unprecedented scale and frequency. Hence a more comprehensive integration of disaster preparedness and climate change strategies is essential. A starting point here is the Hyōgo Framework for Action (www.unisdr.org/eng/hfa/hfa.htm) developed in response to the 2004 Indian Ocean tsunami. It elaborates five priorities for action based on a review of past successes and failures (Basher, 2008, p. 36):

- Prioritize disaster risk reduction nationally and locally with a strong institutional basis for implementation.
- Identify, access and monitor disaster risks and enhance early warning.
- Use knowledge, innovation and education to build a culture of safety and resistance at all levels.
- Reduce the underlying risk factors.
- Strengthen disaster preparedness for effective response at all levels.

There are clear overlaps here with an agenda for dealing with the impacts of climate change. At the national level there is a need to integrate planning for dealing with climate change population displacement with disaster preparedness. However, as Basher (2008, p. 35) points out:

> This convergence is easier said than done, as the two issues of disaster risk and climate change are usually dealt with as separate policy processes and by different government departments.

There is particular convergence on the need for tackling the root causes of disasters but especially the crucial task of providing a humanitarian response to accommodate the needs of displaced people. While national preparedness to deal with displaced populations is critical, the international community often will be required to play a crucial role. The experience associated with the 2004 Asian tsunami has provided a wealth of understanding of what will be required to cope with sudden massive population displacements associated with climate change (Laczko and Collett, 2005; UNHCR, 2006).

Considering that the vast bulk of displacement due to extreme events will occur within low-income countries and that the resources which poor countries have to mobilize are limited, the assistance of the international community is often crucial. In the Asian tsunami this was demonstrated in the massive mobilization of international resources involving the pledging of US$6.8 billion worth of assistance, the involvement of sixteen United Nations agencies, eighteen Red Cross Response Teams, more than 160 international NGOs, hundreds of private civil society groups and thirty-five armed forces (Laczko and Collett, 2005). It is apparent that there are significant problems of coordination in the delivery of this assistance. The disorganized response to Hurricane Katrina in New Orleans provides a stark example of the difficulty of coping effectively with massive displacement from sudden events, even in wealthier nations. There have been significant and ongoing improvements in the international disaster preparedness and emergency response systems but it is clear that much is still to be done. In 2006 the United Nations set up the UN Central Emergency Response Fund to collect funding to deal with emergency responses and disburse it to countries affected. There is a need to incorporate climate change impacts into these institutional structures.

Systems for coping with displacements caused by disasters, induced by climate change as well as other causes, are needed at national, regional and international levels. It is important that these initiatives are properly funded and resourced and that they are carefully coordinated and integrated so that the response can be immediate and effective. Part of the response will need to relate to providing and supporting a range of settlement options for the temporarily displaced populations which will include, among others, resettlement.

At present, international organizations such as the UNHCR and IOM are playing an important role in providing assistance to groups displaced by extreme events such as the Asian tsunami and cyclone Nargis in Myanmar. There have been discussions as to whether these organizations, or others, should take primary international responsibility for coordinating the response to sudden displacements by extreme events such as those associated with climate change. However, the mandates of these organizations are quite specific and while they can make, and have made, notable contributions in this space, they are probably not the right vehicle for coordinating efforts. The most appropriate policy approach is to strengthen and enhance the United Nations system's disaster response strategies and framework (Warner et al., 2008a; 2008b). There is a crucial need for building up capacity in the areas of disaster preparedness, managing the evacuation of large numbers of people displaced by sudden events, organizing their accommodation in temporary settlement and ensuring their eventual repatriation or resettlement. The latter need to take into account the lessons drawn from other resettlement experience discussed below.

Permanent displacement due to climate change

A distinction is often drawn between environmental risks which involve a gradual decline in the quality or quantity of natural resources and the increased risk of sudden environmental hazards including typhoons, storm surges, flooding, droughts and associated socio-ecological events including famines, disease outbreaks, etc. In the discussion on climate change and migration, much attention has focused on population displacement as a result of linear or gradual deterioration of local environments due to climate change, either in relation to the decline in natural resource condition or the gradual increase in the likelihood or severity of hazards. The underlying idea is that at some point a region ceases to be able to sustain the livelihood of its resident population, or there is a perception of such a condition, and a tipping point or threshold of tolerance is reached after which there is a non-linear increase in the number of people forced to emigrate (Meze-Hausken, 2008; Hugo et al., 2009).

Heine and Petersen (2008, p. 50) argue: 'Generally, the international community tends to regard migration as an adaptation failure.' Yet, migration need not be conceptualized purely in this manner. While it is of course very important that there are policies and programmes in place that give local people the choice to adapt *in situ* to climate change, equally it needs to be recognized that in some cases resettlement will become necessary, when all other avenues have been exhausted. Boano et al. (2008, p. 19) maintain that migration is often less a function of immediate stress undertaken upon the onset of disaster but is more frequently a proactive diversification strategy taken in anticipation of such events in the future or to cope with long-term declines in livelihood. Certainly at times there will also be ongoing or sudden impacts of climate change that lead to thresholds being reached after which large numbers of people are forced to migrate as it becomes untenable for a community to remain (Lonergan and Swain, 1999). However, migration needs to be seen as both a proactive and reactive response to the onset of climate change impacts or the expectation of them.

While much of the discussion on climate change and migration to date has exaggerated the likely extent of displacement and resettlement due to climate change, the risk is real and considerable. Despite the enormous uncertainty, there are significant risks if systems are not established to manage new orders of magnitude of internal or external displacement. The Precautionary Principle should apply in this case (defined under Art. 15 of the Rio Declaration, 1992). It is important to put in place mechanisms at national and regional levels which have the capacity to facilitate the process of resettlement of individuals, families and in some cases entire communities displaced by climate change.

It is important also to recognize that the bulk of climate change-induced forced migration will involve poor people as they will be the most vulnerable to climate change impacts and will probably have been further impoverished by the deteriorating local situation. Moreover, most of the resettlement will occur within poor low-income countries. One of the clear lessons from decades of resettlement experience (e.g. Cernea and McDowell, 2000) is that successful resettlement is not cheap. If resettlement is not to lead to further impoverishment, there will need to be a significant investment of resources made to successfully establish those displaced in new locations and provide them with security and sustainable livelihood opportunities. It is imperative, therefore, that the lessons learned from experience with resettling large populations be incorporated into planning for resettlement of people displaced by climate change impacts. Not to do so will risk repeating the mistakes of the past and wasting the scarce resources that will be available for resettlement.

Types of resettlement policies and programmes in low-income countries

Over recent decades there have been many attempts in low-income countries to resettle significant numbers of people within their national boundaries, either in response to an event which renders living in an area impossible or to achieve a change in the spatial distribution of the national population for economic, social or political reasons. In this section the major types of such policies and programmes are briefly outlined and their relevance to climate change displacement discussed. The focus is on resettlement programmes which have had government involvement. This is deliberate since it is apparent that governments need to play a key role in climate change-induced resettlement. At the outset a number of common features of these programmes should be noted:

- Resettlement has overwhelmingly occurred *within* countries and this is also likely to be the case with climate change-related displacement and resettlement.
- They predominantly have involved rural-agricultural populations resettled in rural-agricultural destination communities. This is an issue of some debate in the context of climate change-related displacement as in many contexts it could be argued that the most efficacious resettlement would be likely to involve rural to urban migration.
- They have often involved poor, vulnerable and powerless groups, and while there is considerable variation between individual cases there is usually a degree of force or necessity in the migration and choices about whether to stay or move are constrained.
- There is often strong central government involvement and a lack of consultation with local government or communities is also characteristic.
- They are often under-funded so that those resettled are less well-off than they were in their now untenable original locations.

While the types of resettlement programme discussed below are common to low-income regions of the world, most examples are drawn from Asia where the writer has most first-hand experience.

Land resettlement

Several countries have sought to redistribute their national populations in order to achieve a better match between resources and population and relieve population pressure in regions of high population density. There is a particular focus on resettling agricultural families in rural areas and the heyday of the schemes was in the 1960s, 1970s and 1980s (Oberai, 1988; MacAndrews, 1979; Chambers, 1969; Nelson, 1973; Bahrin, 1981; World Bank, 1994). One of the most substantial has been Indonesia's Transmigration Program. Figure 10.1 shows the number of families resettled from the densely populated islands of Java, Bali and Lombok to the less densely settled islands of Sumatra, Sulawesi, Kalimantan and West Papua (Arndt, 1983; Hardjono, 1977; 1986; World Bank, 1988). Some have argued

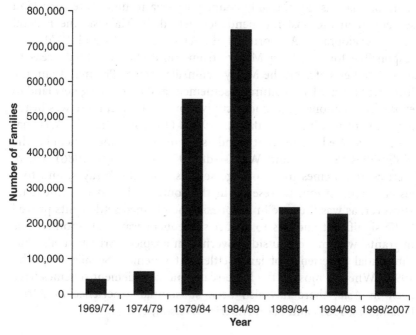

Figure 10.1 Number of families moved under Indonesia's Transmigration Program (1969–2007)

that the transmigration scheme had a political motive in part to 'Javanize' Indonesia's Outer Islands with the largest group in the nation (the Javanese) being resettled in areas where other groups predominate (Tirtosudarmo, 2001). In most areas rural-based agricultural families were resettled on land in the Outer Islands under government auspices. While the nature of the programme and the resources provided by the government varied considerably over time, it involved selection of candidates (families with relatively young working males with agricultural experience, etc.), government preparation of land and community services and utilities in the destination and resources to maintain families in the early period of settlement, which often involved clearing land. In some cases pre-existing communities in destination areas were culturally different to the newcomers or felt discriminated against because of the resources and services made available to the resettlers. While the origins of the transmigration scheme date back to colonial times (Pelzer, 1948) under post-Independence governments a separate ministry and a significant under-bureaucracy were established to plan and operationalize the policy and programmes.

Indonesia was not the only country in Asia to undertake such land settlement in the post-Independence period. In Malaysia the Federal Land Development Authority (FELDA) in the 1960s and 1970s was responsible for resettling Malays from more closely settled areas to more frontier states on the Malay Peninsula such as Pahang. Again the focus was on rural agricultural resettlement and on clearing new land to expand the national area under cultivation. Moreover, there was heavy government involvement (Bahrin, 1988). Other countries in Asia that have developed significant land settlement schemes include the Philippines (Simkins and Wernstedt, 1971; Paderanga, 1986), where there were schemes to encourage settlers from the Visayas, and to a lesser extent Luzon, to resettle in the southern island of Mindanao. However, as was the case in Indonesia, the planned settlements proved to be significant anchors to attract subsequent waves of 'spontaneous migrants' who moved outside government auspices. Sri Lanka also has substantial experience of land settlement schemes (Senaka-Arachchi, 1995). Whereas most of the large-scale land resettlement schemes have been in Asia, there has also been substantial experience in Africa (Chambers, 1969) and Latin America (Nelson, 1973).

Urban resettlement

While there has been a huge investment in resettling displaced persons in low-income countries in agriculture-based rural communities, little attention has been given to resettlement in urban areas. This is despite the fact that rural to urban migration has been the dominant spontaneous internal migration process in those countries (UN, 2007). It could be argued, however, that among populations displaced from both rural and urban areas as a result of climate change it is likely that resettlement in urban areas will be more important (Hugo et al., 2009). This is because of the structural changes in the economies of low-income countries which will result in the proportion of their populations living in urban areas increasing from 43.8% in 2007 to 53.2% in 2025 and 67.0% in 2050, while the population living in rural areas will decline from 56.2% in 2007 to 46.8% in 2025 and 33.0% in 2050 (UN, 2007).

There has been little experience in resettling displaced persons, especially those from rural areas, in urban localities. Tan (2008) explains that 58.6% of the 1.3 million people displaced by the Three Gorges Dam were resettled in urban areas but the focus of research is on rural settlers. In fact, as Simmons (1981) outlines, in several low-income countries significant efforts have been made to slow down or reverse rural to urban migration. In addition, others have sought to divert rural to urban migrants from large cities towards smaller and intermediate centres (Hansen, 1981). In more recent decades, however, there has been a move away from such 'anti-urban' policies but the dominant attitude has been to accommodate migrants in cities rather than to facilitate and encourage rural to urban migration. The resettlement of persons displaced by climate change impacts in urban areas, however, will be of substantial significance over the next four decades.

Internally displaced persons (IDPs)

The United Nations High Commissioner for Refugees (UNHCR) has in recent years identified persons who are displaced by similar forces to those experienced by mandated refugees but who move within national boundaries. Displaced persons are defined as:

> persons or group of persons who have been forced to flee or to leave their homes or places of habitual residence, in particular, as a result of, or in order to avoid the effects of, armed conflict, internal strife, systematic violations of human rights, or natural or man-made disasters, and who have not crossed an internationally recognized state border (OCHA, 2004).

In 2008 it was estimated that there were 26 million IDPs worldwide (UNHCR, 2009), significantly more than the number of mandated refugees (15.2 million). Experience has been gained in resettling IDPs in several countries but again Indonesia provides an excellent example (Hugo, 2002) where at one stage in the early 2000s the number of official IDPs exceeded 1.3 million. While many IDPs can be repatriated to their home area once security has been restored, others need to be resettled because of long-term difficulties in the origin. Unlike those moving under land settlement schemes, IDPs usually are suddenly uprooted from their homes and forced to move without being able to take most of their assets with them and are similar in many respects to refugees. In the Indonesian case many IDPs are drawn from minority ethnic or religious groups so that the resettlement options for them need to be sensitive to those factors.

Resettlement associated with environmental disasters

There have unfortunately been a number of recent disasters in Asia which have displaced very large numbers of people, such as the 2004

Indian Ocean tsunami (between more than 1 million [UNHCR, 2006, p. 21] and over 2 million [AidWatch, 2006] displaced), Cyclone Nargis in Myanmar (800,000 displaced [UNICEF, 2009]) and the Szechuan earthquake in China (362,000 displaced) (Chinese Academy of Science, 2008). These disasters have produced massive displacements of people and although much of the displacement is temporary they have also involved significant resettlement. As with the previous group, there has been an abrupt displacement of whole communities which have been forced to move without planning and without most of their assets. The experience of the Indian Ocean tsunami has been very well documented (Laczko and Collett, 2005) and has involved a major resettlement effort both within the area directly influenced by the tsunami and elsewhere.

It is not only in cases of the effects of sudden events that have led to displacement and resettlement in Asia but also where there has been resettlement attempted in anticipation of such an event. Lucardie (1981), for example, has studied the case of the island of Makian in Eastern Indonesia where a volcanic eruption was expected to occur, showing that successive volcanic eruptions on Makian have made cultivation increasingly difficult because of the fresh overlays of lava and volcanic rock and increasing scarcity of water. The Makianese adapted to this deterioration of their ability to earn a livelihood through agriculture by adopting a range of mobility coping strategies. Each village developed its own constellation of temporary and permanent migration strategies to cope with the environmental deterioration:

> For the Makianese, geographical mobility was the only alternative way to make a living outside the poor resources of their island. Various forms of geographical mobility of the Makianese have been reported as early as the middle of the nineteenth century, but probably are dating back to the period after the first great eruption of the Kie Besi in the year 1646 (Lucardie, 1981, p. 3).

Given this tradition of adapting to gradual environmental deterioration by adopting circular migration strategies of supplementing their local livelihood with earnings elsewhere, it would perhaps be anticipated that the Makianese would be amenable to a planned resettlement scheme. However, Lucardie (1981) found that this was not the case. He argues that the resistance to resettlement among the Makianese is only partly explained by the traditional problems reported in the Indonesian Transmigration Program (Hardjono, 1977) such as: (a) insufficient funding of the move; (b) inappropriate land at destination; (c) conflicts with local population at destination; (d) insufficient compensation; (e) insufficient preparation of resettlement areas.

He argues that there are also important emotional considerations related to the Makianese's attachment to their home community:

> Resettlement of the whole population of Makian in a Malifut transmigration project simply would mean the end of the country's old migratory traditions with Makian as the pivot. Depopulation of Makian certainly will lead to a complete disturbance of their mental map (Lucardie, 1981, p. 11).

This study has pointed to a number of issues of relevance to the present study:

- Communities can readily adapt to severe environmental deterioration through adopting permanent and temporary migration strategies that allow them to supplement their declining livelihood at home by working outside the area.
- Comprehensive preparation and planning and adequate funding and compensation are crucial to the success of resettlement.
- Emotional bonds to areas are very strong so that permanent resettlement should not be seen as the only way to cope with severe environmental deterioration.

Displacement and resettlement associated with mega projects

Over the last half century there have been massive displacements of population in Asia caused by the construction of large-scale infrastructure projects, especially dams, which have led to the inundation of large areas of agricultural land and entire settlements. In the 1980s alone it has been estimated that between 80 million and 90 million people were displaced and resettled due to the construction of dams and transportation (World Bank, 1994, p. i). The construction of the Three Gorges Dam on the Yangtze River in China led to the displacement and resettlement of more than 1.3 million people in the decade up to 2008 (Tan, 2008). In several respects it is the displacement and resettlement of communities, families and persons associated with dam construction that will provide the most relevant experience to guide climate change-induced displacement and settlement policy and practice. There are several reasons for this:

- Displacement from gradually rising water levels of dams is analogous to the slow-onset effects of climate change-induced sea level rise and desertification, so that there is substantial lead time to plan the displacement process and prepare the resettlement destination.

- In both cases there is a strong inevitability of the eventual destruction of the living environment and associated with this the powerlessness of the bulk of the populations affected.
- While both urban and rural communities have been influenced by infrastructure construction, it is especially rural agriculturalists losing their source of livelihood who are impacted.
- In both cases it is the poor in the affected areas who are most powerless and least able to make plans to move.
- There is a substantial literature on the experience of displacement and resettlement associated with mega projects because evaluation has been a major element in such projects in recent years (Cernea and McDowell, 2000; Cernea, 1997; Tan, 2008).

Lessons from the displacement-resettlement literature

In distilling the lessons from the experience of displacement and resettlement it is important to note that there is no single magic recipe for initiating successful resettlement schemes, as circumstances vary considerably from place to place. Nevertheless, there are clearly a number of issues that recur in the literature. It also needs to be said that while it is possible to identify 'best practice' in displacement and resettlement (World Bank, 1994), many of the lessons come from failure rather than success and few areas of public policy have a more sustained record of failure (Cernea, 1997). Ultimately, the key indicator of success in displacement and resettlement must be that those displaced need to be established at the destination with minimally the same level of living they enjoyed at the origin but desirably an improved standard of living. As Cernea has pointed out:

> Impoverishment of displaced people is the central risk in . . . involuntary population resettlement. To counter this central risk, protecting and reconstructing displaced peoples' livelihoods is the central requirement of equitable resettlement programs (Cernea, 1997, p. 1569).

He quotes the case of India where by the early 1990s some 20 million people had been displaced by development programmes and three-quarters had not been 'rehabilitated'. What, then, are the lessons from over five decades of experience of displacement and resettlement which need to be heeded in coping with the anticipated forced displacements of people caused by the impact of climate change in low-income countries?

1. Provision of sufficient and properly allocated funding

One of the chronic problems of resettlement programmes has been a failure to provide the necessary level of funding for the displacement and resettlement processes and, equally significant, the misallocation of those funds through corruption and poor planning. Too often the fund-, ing allocation is based purely on the resources made available by government rather than a careful analysis of what actual costs will be incurred by relocating from the origin and re-establishing livelihoods at the destination. Tan (2008), for example, found in an analysis of resettlement associated with China's Three Gorges Dam that the compensation paid for assets lost by families in most cases was insufficient for them to re-establish themselves at an equivalent level at the destination. Cernea (1995) argues that one of the problems in past resettlement programmes has been that all that has been involved is payment of a compensation amount for assets lost rather than a comprehensive funding of the process of dislocation and resettlement at the destination. The latter is a *sine qua non* of successful resettlement programmes.

To resettle those families and communities displaced by climate change will be expensive and few low-income countries will be able alone to fund displacement and sustainable resettlement on the scale required for the displaced families to re-establish themselves elsewhere such that their level of livelihood is at least maintained. International funding involvement and support will, therefore, be critical to successfully resettle those people displaced by climate change. Individual nation-states have the primary role and responsibility for identifying communities that will be impacted by climate change and for which resettlement will eventually be necessary. However, the international community undoubtedly has a very important and increasing role for a number of reasons:

- Many of the countries, and communities within those countries likely to experience the most severe impacts of climate change, are poor and will not have the resources (financial, institutional, technological and human) to effectively respond to those impacts.
- More developed countries have been responsible for a disproportionately large share of the total greenhouse emissions which are the root cause of the change in the Earth's climatic conditions.

- The lack of an effective international regime to cope with climate change-induced migration could lead to significant negative humanitarian outcomes, development failure, and possibly conflict.

Moreover, according to the principle of 'common but differentiated responsibilities' the international community and especially higher-income nations will need to cooperate in order to ensure that appropriate technical, management, financial and political assistance is provided to nations and communities that are likely to experience climate change-induced forced displacement and resettlement.

There has been considerable discussion about the establishment of a global fund, or a number of regional funds, which will be contributed to by high-income nations and provide resources for low-income countries for adaptation, mitigation and responding to the impacts of climate change. Funding to support equitable and sustainable displacement and resettlement would be an appropriate use of such funding. However, the support would include a substantial element of capacity-building and assistance in setting up the appropriate structures and institutions to ensure that there is not only sufficient funding but also that it is allocated in a way that maximizes the benefits to the displaced populations.

Cernea (1995) has argued that in many resettlement schemes associated with large-scale infrastructure projects the costs of the resettlement are not internalized to the project. As a result the costs of resettlement are borne by the displaced population themselves. A basic principle of resettlement needs to be that the costs are fully borne by resettlement schemes. The World Bank (1994, p. 15) identified the following inadequacies of financial planning of resettlement programmes:

- The poor quality of financial reporting in project documents.
- Incomplete calculation of all the costs accruing to displaced people.
- Inadequate budgets for settlement activities.
- The failure to include the full costs of resettlement in the economic and financial assessment of the project.

2. Planning of the displacement-resettlement process

One of the assets which many of the low-income countries faced with the inevitability of eventually needing to resettle some communities due to the impact of climate change possess is time. While in some countries the displacement impacts of climate change are more imminent than others, in most cases the slow-onset effects mean that communities, nation-states and the international community have a significant period to plan for displacement and resettlement where it is considered to be eventually necessary. This is not to say that there is no urgency. Although the desired end point may be decades away, there is an urgency to begin the planning process. One of the clear findings from the resettlement literature is that time is required to put in place all of the institutions, structures and mechanisms to facilitate equitable and sustainable resettlement. Cernea (1995) identifies that a key barrier to the success of past resettlement projects has been weak governance with the institutions charged with the responsibility of resettlement lacking a political mandate and having poor institutional capacity. Too often, poor planning means that the displacement process is unnecessarily costly and distressing for those forced to move and there is inadequate preparation at the destinations to give them a reasonable opportunity to re-establish their livelihoods. Accordingly a significant effort needs to be put into the establishment of the institutions, structures and mechanisms to develop policy relating to displacement-resettlement and to operationalize that policy in a fair, efficient and effective way. This will involve considerable capacity-building based on appropriate training and development of a cadre of professionals who are well equipped to plan and operationalize each phase of the complex displacement-resettlement process.

The World Bank (1994, p. xiii) has identified a number of specific problems which recur in resettlement programmes and relate to planning:

- A failure to prepare satisfactory resettlement plans.
- Laxness in fulfilling responsibilities.
- Irregular or insufficient supervision.
- Insufficient follow-up when problems are identified.

Careful planning is required at all four of the stages in the resettlement process that were identified by Scudder (1981):

- Planning and design of new settlement and infrastructure provision.
- The transfer of displaced families to the new settlement during which the settlers are dependent on government support.
- Settlers begin to feel at home at the destination and take advantage of the opportunities that are available.
- Settlement is fully self-sufficient.

In particular insufficient planning for displaced persons to re-establish their livelihoods at the destination has been a consistent problem. Some of the issues are:

- Expecting agriculturalists to adjust to different soil, climatic, slope and other conditions without training.
- Not preparing sufficient infrastructure at the destination.
- Not giving settlers sufficient time to re-establish themselves at their destination with government supplementation of income and support.
- Selecting destination areas unsuitable for close settlement.
- Failing to replace off-farm work opportunities that were available at the origin in the destination areas.

As early as 1954, Lewis (1954, p. 3) identified the following planning requirements for resettlement to be successful:

- Selecting the right place for resettlement.
- Selecting the right settlers.
- Physically preparing the settlement site before the settlers arrive.
- Ensuring the settlers have sufficient economic and social capital.
- Facilitating group cohesion among settlers.
- Ensuring settlers have sufficient land to fully replace their livelihood.
- Ensuring settlers have secure tenure of that land.

All of these require sound planning, preparation and operationalization if they are to be achieved.

One issue of difference in climate change-related displacement to some land settlement schemes is that the latter are able to select settlers whose attributes are deemed to be most likely to facilitate successful adjustment at the destination. In much climate change-related displacement, as with resettlement associated with infrastructure projects, entire communities need to be re-established at a destination. This can be an advantage as it provides the potential at least for the social capital of a community to be transplanted. However, making special provision for disadvantaged groups – the poor, elderly, etc. – is an important element in planning resettlement.

In summary, one of the most important lessons emerging from the literature is that successful resettlement is possible only if a comprehensive strategy is developed to cover the entire process of preparation, displacement, arrival at the destination and adjustment there. Careful planning and operationalization by well-trained and properly resourced professionals is an important basic requirement.

3. Empowerment of the displaced people and communities

A clear finding of the literature on resettlement has been that too often the process has been a 'top-down' one in which the involvement of those being displaced has been limited. Displaced communities often perceive themselves as powerless and this erodes the resilience and social capital of resettled communities. Engagement of the communities to be affected from the earliest stages of planning and fully involving them at each stage in a way that gives them ownership of both the displacement and resettlement processes is one of the signal lessons that can be derived from the literature. Too often fully centrally conceived and operational-ized programmes fail to do this and as a result:

- Miss out on the insights of local informants about what strategies are most likely to be effective and those that are likely to fail.
- Miss the opportunity to gain the full cooperation of the displacees, especially their leaders.

Cernea argues that one of the major impoverishment hazards of reset-tlement which is most frequently overlooked is social disarticulation:

> Forced displacement tears apart the existing social fabric: it disperses and fragments communities, dismantles patterns of social organization and interpersonal ties; kinship groups become scattered as well. Life-sustaining informal networks of reciprocal help, local voluntary associ-ations, and self-organized mutual service arrangements are dismantled. The destabilization of community life is apt to generate a typical state of anomie, crisis-laden insecurity, and loss of sense of cultural identity . . . (Cernea, 1997, p. 1575).

The social and cultural dimensions of displacement and resettlement are often overlooked in the focus on re-establishing the economic live-lihood of the displaced persons. However, maintaining social capital of displaced communities is an essential part of them effectively re-establishing themselves at the destination. Powerlessness, dependency, vulnerability and lack of resilience can be major barriers to the success of displacement-resettlement schemes and involvement of those being displaced at each stage can reduce the chances that these negative developments will occur.

While it is not always possible, there should be an element of choice in the relocation process. It may be possible for some people to remain in the origin location. Clearly that is not always the case but sometimes displacement is perceived by policy-makers to be the only option to be made available to communities at risk, when in fact it is possible for a smaller number of people to have a sustainable livelihood in the origin. This may be the case, for example, on some Pacific island communities where the growing discourse on resettlement may be drawing attention away from other local adaptation alternatives. There also may be potential to offer displacees different alternative destinations or types of work to ensure that their particular skills and existing social networks are catered for and maximize their chances of successful settlement at the destination. For example, it may be more effective to facilitate some displacees moving to live close to relatives or friends at destinations rather than totally re-establish the whole community at the destination.

Part of the engagement with the community to be displaced must involve the development of a functional and effective relationship between the planners and relevant officials on the one hand and the communities on the other. Cernea (1997, p. 1577) drawing on several decades' experience of resettling displaced populations maintains that 'dysfunctional relationships between planners and groups affected by displacement are one of the roots of resettlement failure'.

Indeed, failure to develop such a relationship can lead to active opposition to resettlement within the community to be affected (Oliver-Smith, 1994). Cernea (1997, p. 1577) argues that relevant agencies frequently try to withhold key information from the communities, which results in distrust and a failure to harness the potential of the energy of the displacees which if effectively mobilized can contribute to reconstructing their livelihoods.

4.   Full engagement with destination communities

In most cases of resettlement there are communities already established at the destination and areas are rarely available where no local communities are to be impacted by the influx of displaced persons. A common problem with land settlement and infrastructure resettlement programmes is to neglect involving those pre-existing local communities in resettlement areas. Both in terms of taking into account their interests, as well as drawing on their experience and local knowledge to assist in resettlement, this is important. At the outset it is necessary for these communities to be fully engaged and consulted at every relevant stage of planning relocation and resettlement in the same way as for those being displaced. Not to do so will risk disaffection and resentment in that group, which can mobilize effective opposition to resettlement. A basic principle is that the origin community, like the displacees, should not experience a decline in their livelihood as a result of resettlement. Their rights need to be fully recognized and they should be properly compensated for any loss of property. It is important that resettlement not be attempted where the displacees and the pre-existing local populations have existing enmities or practices which may offend the other group. The experience of settling pig-raising Hindu Balinese transmigrants in Muslim Bugis communities in Sulawesi Indonesia is illustrative of the need to ensure there is comparability between natives and newcomers (Davis, 1976).

Another aspect relates to the resources and infrastructure made available to the displaced settlers. It is important that the native population do not feel that they are excluded or discriminated against by not having access to equivalent resources. An important distinction needs to be made between those infrastructure and services provided on a temporary basis to facilitate adjustment and those which are provided on a longer-term basis. The latter should be made available to both natives and newcomers. It could be, too, that involvement of the pre-existing community in the work required to prepare for settlement rather than to bring in workers from the outside can assist in the successful melding of natives and newcomers.

5. Making use of existing social networks

One of the consistent findings in migration research is the importance of the social networks established by migrants with their home communities in encouraging and facilitating further migration. However, social networks also have an important role in facilitating adjustment at the destination. Accordingly, where possible, resettlement of communities to locations where they have substantial social capital in the form of earlier generations of migrants should be encouraged. This is especially the case where resettlement involves rural to urban displacement – which will loom large in climate change-induced migration. Migrant communities are often instrumental in cushioning the newcomers' adjustment to the destination. This involves help in effectively entering the labour and housing markets but also in providing crucial social and cultural support and assistance.

While there is a substantial literature testifying to the importance of social networks in facilitating the adjustment of spontaneous internal and international migrants at the destination, there is little evidence of their use in planned migrations. One danger is that too much reliance can be placed upon that support and, as a result, weaken it. Clearly, there is a need for the development of policies and programmes which bolster and support the social capital embodied in social networks rather than seeing them as a substitute for government investment.

6.   Ensuring the reconstruction of the livelihoods of displaced persons

The bottom line of any resettlement programme must be that the level of livelihood of those displaced is, at least, re-established at the destination but preferably improved. However, the circumstances may be quite different to those at the origin so that the livelihood at the destination may necessarily be significantly different. The resettlers will often need retraining to equip them with the knowledge and skills required to earn a living at the destination, especially in situations where the economies are quite different to the origin. In the initial stages of establishment it is necessary to provide support through funding or access to work, such as in any work needed to establish the settlement – housing, infrastructure, land clearing and preparation, etc. This support must be available for a sufficient period to allow resettlers to re-establish themselves at the destination. There is some experience of this support being withdrawn prematurely so that settlers lapse into poverty before they get a chance to establish themselves.

In re-establishing livelihood at the destination, the experience has been that compensation for assets lost at the origin is rarely sufficient to allow settlers to establish an equivalent level of living at the destination. As Cernea points out:

> The cost of re-establishing a family and a community is generally bound to exceed the strict market value of the physical losses imposed on that family or community. Compensation alone, by definition, is therefore never sufficient for re-establishing a sustainable socioeconomic basis for resettlers (Cernea, 1997, p. 1579).

Re-establishment at the destination must involve the provision of appropriate infrastructure and services. Where the resettlement is based upon agriculture there are dangers that the newcomers will not be allocated sufficient land of sufficient quality to earn a livelihood. There is a tendency to make available land not wanted by pre-existing communities because of its more marginal quality. In addition, for many of those displaced, off-farm supplementary forms of income were critical to their pre-move livelihood but these are often overlooked by planners who focus totally on the establishment of the viable agricultural holding. Oberai (1986, pp. 141–61; 1988, pp. 8–19) found, in a review of land settlement schemes, that most such schemes have not been able to generate sufficient non-farm employment opportunities.

7. Recognizing differences in the displaced population

In resettlement there is no 'one size fits all' solution when it comes to facilitating the process. Cernea (1997, p. 1576) shows that some population subgroups are hurt more by displacement than others and the level and type of support and assistance they require also varies. In several programmes women are given less compensation than men and older people and children are neglected. One of the important issues in climate change-related displacement is that the poor are likely to be disproportionately affected as they have fewer resources available to make adaptations before resettlement becomes necessary. Vulnerable groups at the origin risk becoming even more vulnerable at the destination. Special attention to vulnerable groups is particularly necessary where entire communities are being resettled and there is no selectivity in who moves. Such groups will have the least resources, information and contacts at the destination to assist in the process of readjustment.

8. Re-establishing social and cultural capital at the destination

In the concentration on establishing the physical capital, natural capital and human capital lost by communities due to displacement, their loss of social capital is often neglected. As Cernea (1997, p. 1576) points out, strategies are required to assist displaced people to restore their capital in all its forms. The new settlers then need assistance to build up their social and cultural capital at the destination as part of the adjustment process. This will be facilitated where communities can re-establish themselves at the destination as a cohesive group but sensitivity and innovatory policies will be required if social capital is to be transplanted. In international migration some experience has been gained with multiculturalism policies which have focused upon new arrivals maintaining their language, culture and social networks while still embracing the main tenets of the host society. There would seem to be some transferability of these lessons to the internal migration resettlement context.

Conclusion

In the growing discourse on climate change and migration it is sometimes forgotten that the last half century has seen a massive redistribution of global population in response to a range of economic, social,
political and environmental processes. Most notably there has been a
massive shift of people from rural to urban areas. Since 1975 the urban
population of less-developed countries has increased by 3.3% while the
rural population has increased by 1.0% (UN, 2007). The scale of population redistribution anticipated to result from the future effects of
climate change is by no means unprecedented. Another dimension of
this neglect of existing knowledge and understanding of migration in the
climate change-migration debate discussed in this chapter relates to the
issue of forced displacement and resettlement. The considerable experience of low-income countries in this process needs to be taken into
account in planning for climate change-related displacement. It is especially important that the lessons drawn from this experience are heeded
because so few resettlement schemes in the past have been successful.
These lessons have been hard learned by previous generations of forced
migrants and to repeat their mistakes would not only be wasteful but
cause considerable suffering among those displaced.

In heeding the lessons of forced resettlement programmes of the past,
some key additional factors should be noted which are relevant in the
context of climate change-forced resettlement.

- There is a need for a comprehensive integration of climate change and
  disaster preparedness strategies.
- There is a crucial role for the international community to play in
  funding resettlement within poorer countries.
- The precautionary principle should apply with resettlement planning
  and execution occurring over an extended period and planning beginning long before the full impact of climate change is felt.
- Most resettlement will involve the poor and there are real dangers that
  resettlement could result in further impoverishment.

A final issue that should be mentioned in this context relates to the fact
that migration in response to the impact of climate change should be
seen as more than a survival mechanism. The growing international
discourse on the potential role of migration, both internal and international, as facilitating economic development would suggest that displacement and resettlement presents an opportunity for improving the

situation of people in poor countries. Hence planning that process effectively presents an opportunity not just to cope with the effects of climate change but actively to assist development in low-income countries and improve the standard of living of the people affected.

References

AidWatch, 2006. UN Office of the Special Envoy for Tsunami Recovery. *Reuters AlertNet*, 24 December. http://www.alertnet.org/printable.htm?URL=/db/crisisprofiles/SA_TID.htm

Arndt, H. W. 1983. Transmigration: achievements, problems and prospects. *Bulletin of Indonesian Economic Studies*, Vol. 19, No. 3, pp. 50–73.

Bahrin, T. S. 1981. Review and evaluation of attempts to direct migrants to frontier areas through land colonization schemes. In: *Population Distribution Policies in Development Planning*, Papers of the United Nations/UNFPA Workshop on Population Distribution Policies in Development Planning, Bangkok, 4–13 September 1979. New York, United Nations, pp. 131–68.

1988. Land settlement in Malaysia: a case study of the Federal Land Development Authority projects. In: A. S. Oberai (ed.), *Rural Migration Policies and Population Redistribution in Developing Countries, Achievements, Problems and Prospects*. New York, Praeger Publishers, pp. 89–128.

Basher, R. 2008. Disasters and what to do about them. *Forced Migration Review: Climate Change and Displacement*, No. 31, October, pp. 35–36.

Boano, C., Zetter, R. and Morris, T. 2008. *Environmentally Displaced People: Understanding the Linkages between Environmental Change, Livelihoods and Forced Migration*. University of Oxford, UK, Refugee Studies Centre. (Forced Migration Policy Briefing 1.)

Cernea, M. M. 1995. Understanding and preventing impoverishment from displacement: reflections on the state of knowledge. *Journal of Refugee Studies*, Vol. 8, No. 3, pp. 245–62.

1997. The risks and reconstruction model for resettling displaced populations. *World Development*, Vol. 25, No. 10, pp. 1569–87.

Cernea, M. and McDowell, C. 2000. *Risks and Reconstruction, Experiences of Resettlers and Refugees*, Oxford, UK, Berghahn Books.

Chambers, R. 1969. *Settlement Schemes in Tropical Africa: A Study of Organization and Development*. New York, Praeger Publishers.

Chinese Academy of Science. 2008. *Assessment of Carrying Capacity of Environment and Resources in the Worst Hit Earthquake Regions*. Beijing. (In Chinese.)

Davis, G. J. 1976. Parigi: a social history of the Balinese movement to central Sulawesi, 1907–1974. Unpublished doctoral dissertation. Palo Alto, Calif., Stanford University.

Deng, F. 1998. *Guiding Principles on Internal Displacement*, 11 February. (E/CN.4/1998/53/Add.2.)

Grier, P. 2005. The great Katrina migration. *Christian Science Monitor*, 12 September. http://www.csmonitor.com/2005/0912/p01s01-ussc.html

Hansen, N. 1981. A review and evaluation of attempts to direct migrants to smaller and intermediate-sized cities. In: *Population Distribution Policies in Development Planning*, papers of the United Nations/UNFPA Workshop on Population Distribution Policies in Development Planning, Bangkok, 4–13 September 1979. New York, United Nations, pp. 113–30.

Hardjono, J. M. 1977. *Transmigration in Indonesia*. Kuala Lumpur, Oxford University Press.

1986. Transmigration: looking to the future. *Bulletin of Indonesian Economic Studies*, Vol. 22, No. 2, pp. 28–53.

Heine, B. and Petersen, L. 2008. Adaptation and cooperation. *Forced Migration Review: Climate Change and Displacement*, No. 31, October, pp. 48–50.

Hugo, G. J. 2002. Pengungsi – Indonesia's internally displaced persons. *Asian and Pacific Migration Journal*, Vol. 11, No. 3, pp. 297–331.

2010. Climate change induced mobility and the existing migration regime in Asia and the Pacific. In: J. McAdam (ed.), *Climate Change and Displacement: Multidisciplinary Perspectives*. Oxford, UK, Hart Publishing, pp. 9–36.

Hugo, G. J., Bardsley, D. K., Tan, Y., Sharma, V., Williams, M. and Bedford, R. 2009. *Climate Change and Migration in the Asia-Pacific Region*. Final Report to Asian Development Bank, August.

Laczko, F. and Aghazarm, C. (eds). 2009. *Migration, Environment and Climate Change: Assessing the Evidence*. Geneva, International Organization for Migration.

Laczko, F. and Collett, E. 2005. Assessing the tsunami's effects on migration. *Migration Information Source*. Geneva, International Organization for Migration. http://www.migration information.org/Feature/print/cfm?ID=299

Lewis, W. A. 1954. Thoughts on land settlement. *Journal of Agricultural Economics*, Vol. 11, No. 1, pp. 3–11.

Lonergan, S. and Swain, A. 1999. Environmental degradation and population displacement. *Global Environmental Change and Human Security Project*. Research Report 2, Victoria, BC, May.

Lucardie, G. R. E. 1981. *The Geographical Mobility of the Makianese. Migratory Traditions and Resettlement Problems*, mimeo.

MacAndrews, C. 1979. The role and potential of land resettlement in development policies: lessons from past experience. *Sociologia Ruralis*, Vol. 19, Nos. 2–3, pp. 116–34.

Meze-Hausken, E. 2008. On the (im-)possibilities of defining human climate thresholds. *Climate Change*, Vol. 89, Nos. 3–4, pp. 299–324.

Nelson, M. 1973. *Development of Tropical Lands*. Baltimore, Md., Johns Hopkins University Press.

Oberai, A. S. 1986. Land settlement policies and population redistribution in developing countries: performance, problems and prospects. *International Labour Review*, Vol. 125, No. 2, pp. 141–61.

— 1988. *Land Settlement Policies and Population Redistribution in Developing Countries, Achievements, Problems and Prospects*. New York, Praeger Publishers.

OCHA. 2004. *Guiding Principles of Internal Displacement*. United Nations, Office for the Coordination of Humanitarian Affairs. http://www.reliefweb.int/idp/docs/GPs English.pdf

Oliver-Smith, A. 1994. Resistance to resettlement: the formation and evolution of movements. In: L. Kreisberg (ed.), *Research in Social Movements, Conflicts and Change*. Greenwich, Conn., JAI Press.

Paderanga, C. 1986. *A Review of Land Settlement Policies in the Philippines, 1900–1975*. Diliman, Manila, University of the Philippines. (School of Economics Discussion Paper No. 8613.)

Page, S. 2005. Many evacuees to stay away. *USA Today*, Fri/Sat/Sun, 14–16 October.

Pelzer, K. J. 1948. *Pioneer Settlement in the Asiatic Tropics*. New York, American Geographical Society.

Sastry, N. and Gregory, J. 2009. *Dislocation and Return of New Orleans Residents One Year after Hurricane Katrina*. Paper presented at XXVI IUSSP International Population Conference, Marrakesh, 27 September–2 October.

Scudder, T. 1981. *The Development Potential of New Land Settlement in the Tropics and Sub Tropics: A Global State of the Art Evaluation with Specific Emphasis on Policy Implication*. Washington DC, United States Agency for International Development (USAID).

Senaka-Arachchi, R. 1995. The problems of second generation settlers in land settlement schemes: the case of Sri Lanka. Unpublished Ph.D. thesis. Population and Human Resources Program, Department of Geography, University of Adelaide, South Australia.

Simkins, P. D. and Wernstedt, F. L. 1971. *Philippines Migration: The Settlement of the Digos-Padada Valley, Davao Province*. New Haven, Conn., Yale University Press. (Southeast Asia Studies, Monograph Series No. 16.)

Simmons, A. B. 1981. A review and evaluation of attempts to constrain migration to selected urban centres and regions. In: *Population Distribution Policies in Development Planning*, Papers of the United Nations/UNFPA Workshop on Population Distribution Policies in Development Planning, Bangkok, 4–13 September 1979. New York, United Nations, pp. 87–100.

Spitz, P. 1978. Silent violence: poverty and inequality. *International Social Science Journal*, Vol. 30, No. 4, pp. 867–92.

Tan, Y. 2008. *Resettlement in the Three Gorges Project*. Hong Kong, Hong Kong University Press.

Tirtosudarmo, R. 2001. Demography and security: transmigration policy in Indonesia. In: M. Weiner and S. Russell (eds), *Demography and National Security*, New York, Berghahn Books.

UN. 2007. *World Youth Report 2007: Young People's Transition to Adulthood: Progress and Challenges.* New York, United Nations.

UNHCR. 2006. *UNHCR Global Report 2006.* Geneva, United Nations High Commissioner for Refugees.

2009. *2008 Global Trends: Refugees, Asylum-Seekers, Returnees, Internally Displaced and Stateless Persons*, 16 June. Geneva, United Nations High Commissioner for Refugees.

UNICEF. 2009. *One Year after Cyclone Nargis: Myanmar on Hard Road to Recovery.* http://www.unicef.org/media/media_49541.html

Warner, K., Afifi, T., Dun, O., Stal, M., Schmidl, S. and Bogardi, J. 2008*a*. Environmentally induced migration. *The Bridge Magazine*, No. 10, pp. 32–43.

Warner, K., Afifi, T., Dun, O., Stal, M., Schmidl, S. and Bogardi, J. 2008*b*. Human security, climate change and environmentally induced migration. In: *Climate Change: Addressing the Impact on Human Security*, Hellenic Foundation for European and Foreign Policy (ELIAMEP) and Hellenic Ministry of Foreign Affairs, 2007–2008. Greek Chairmanship of the Human Security Network, Athens.

World Bank. 1988. *Indonesia: The Transmigration Program in Perspective.* Washington DC, World Bank.

1994. *Resettlement and Development*, 8 April. Washington DC, World Bank Environment Department.

11

Climate change and internal displacement: challenges to the normative framework

KHALID KOSER

Introduction

There is a general consensus that the majority of people likely to be displaced by the effects of climate change in the next century will be displaced inside their own countries, rather than across international borders. Such displacements are already taking place: according to the Internal Displacement Monitoring Centre (IDMC), in 2008 some 20 million people were displaced internally by climate-related, sudden-onset disasters (IDMC, 2009). The International Organization for Migration has concluded that '... environmental migration is likely to be mainly internal, with a smaller proportion taking place between neighbouring countries, and even smaller numbers migrating long distances' (IOM, 2009, p. 1). The United Nations Inter-Agency Standing Committee (IASC) on Migration and Displacement has also concluded that 'it is foreseeable that the majority of movements prompted by climate change and environmental degradation will occur within countries although increased cross-border movement of people is also likely' (IASC, 2008).

Yet the protection of those likely to be displaced internally has attracted far less academic or policy attention in recent years than those who will potentially cross borders. One reason is that international migration is usually viewed as posing more of a challenge to the international community than internal migrants, who are after all normally citizens of the country where they move. The rights of international migrants are curtailed as compared with those of citizens, and the international community plays an important role in bridging protection and assistance gaps where these occur, whether for labour migrants (for example through a range of International Labour Organization conventions) or refugees (especially as a result of the 1951 Convention Relating to the Status of Refugees and the subsequent 1967 Protocol). A second reason is that the lion's share of these internal movements will take place in developing countries, where the global political agenda is rarely set. In stark terms, the prospect of a relatively small proportion of those affected (although still potentially representing large numbers of people) crossing borders into the more developed countries is of more concern in these countries than the prospect of far larger displacements within countries in the developing world.

But a third reason is the assumption that there is a stronger legal and normative framework already in place to protect the rights of those displaced internally, than of those who will cross borders as a result of the effects of climate change. This latter group falls outside existing frameworks for protecting other international migrants or refugees, which do not specify environmental factors as a cause for migration. Although in a few cases ad hoc responses have developed to protect those who move across borders temporarily, the rights of those who are forced to migrate permanently across national borders – for example as a result of 'sinking' small island states – have yet to be addressed (Zetter, 2009). A number of options for filling these protection gaps for cross-border displacement are being debated, ranging from adapting or building on existing norms and instruments, to the development of guidelines on environmental migration in a 'soft law' approach, to the elaboration of a new binding instrument or convention (Martin, 2010).

Yet these debates risk distracting attention from the fact that gaps also exist in the current legal and normative framework for protecting those internally displaced by the effects of climate change. This chapter identifies five such gaps, relating to:

- the current definition of internally displaced persons;
- the inadequate translation of 'soft law' principles on internally displaced persons into national laws and policies developed to protect them;
- implementation challenges;
- shortcomings in institutional arrangements; and
- the unlikelihood of any realistic alternative framework emerging.

Before discussing these gaps in turn, I give a brief overview of the existing framework for the protection of internally displaced persons, based upon the *Guiding Principles on Internal Displacement*.

Guiding Principles on Internal Displacement

The *Guiding Principles on Internal Displacement* comprise the main 'international framework for the protection of internally displaced persons', as affirmed by the 2005 World Summit Outcome Document (UN, 2005). They were compiled under the mandate of the UN Representative of the Secretary-General on Internally Displaced Persons (Francis Deng) at the request of the UN Commission on Human Rights (now the UN Human Rights Council) to elaborate an 'appropriate normative framework'.

The reason for focusing on IDPs as a special category is that they often experience particular problems as a result of their displacement. The sorts of problems faced by IDPs but not usually faced by citizens who are not displaced from their homes include lack of shelter and problems related to camps; loss of property and access to livelihoods; discrimination because of being displaced; lack of identity cards and other formal documents that are left behind, confiscated, or destroyed; lack of access to services; lack of political rights; problems relating to the restitution of or compensation for lost property; and challenges relating to return and integration. As a result, IDPs often run a higher risk than those remaining at home of having their children forcibly recruited; becoming victims of gender-based violence; being separated from family members; being excluded from education; being unemployed; and suffering higher rates of morbidity and mortality.

There are thirty *Guiding Principles*, divided into four main parts. Principles 1–4 cover general principles, including significantly that the primary responsibility for protecting and assisting IDPs falls to national authorities. Principles 5–9 address the pre-displacement phase, and concern obligations under international law to prevent displacement unless it is unavoidable. In the context of climate change, for example, it may be possible to assist people to adapt to change and remain in their homes. Principles 10–23 focus on protection during displacement, regarding for example the rights to adequate living standards and to education. Principles 24–27 concern access to humanitarian assistance; and finally Principles 28–30 concern the right to return, resettle and reintegrate.

The conceptual underpinnings for the *Guiding Principles* are as follows. First, although IDPs have departed from their homes, unlike refugees they have not left the country where they are normally citizens. Second, they can therefore invoke all human rights and international humanitarian law guarantees available to the citizens of that country. Third, the applicability of refugee law is not possible and would be dangerous, in that it would limit the rights of citizens in their own country. Fourth, IDPs experience a very special factual situation and have specific needs. It is therefore necessary, finally, to restate in more detail those legal provisions that respond to their specific needs and to spell them out in order to facilitate application in situations of internal displacement.

The *Guiding Principles* are, therefore, in essence a compilation of existing human rights and international humanitarian law, and refugee law by analogy, as it applies to situations of internal displacement. The *Guiding Principles* do not comprise binding law – at best they are 'soft law'. They were drafted by experts and submitted to the Commission on Human Rights in 1998 as an expert text. They have not been negotiated or agreed by states. Instead the Representative of the Secretary-General on the Human Rights of Internally Displaced Persons (Walter Kälin) has promoted a 'bottom-up' approach to try to build consensus. His mandate has emphasized convincing states and regional organizations to incorporate the *Guiding Principles on Internal Displacement* into domestic law and to adapt their existing laws (Kälin, 2008).

Gaps in the definition of internally displaced persons

The definition of IDPs provided in the *Guiding Principles on Internal Displacement* is as follows:

> . . . persons or groups of persons who have been forced or obliged to flee or to leave their homes or places of habitual residence, in particular as a result of or in order to avoid the effects of armed conflict, situations of generalized violence, violations of human rights or natural or human-made disasters, and who have not crossed an internationally recognized state border (Deng, 1998, p. 2).

This definition is far wider than the refugee definition, incorporating a non-exclusive list of examples of causes of displacement that includes natural disasters, and thus covers many of the predicted effects of climate change which may precipitate movement.

At the same time there are at least three gaps, or weaknesses, in the definition. First, as it appears in a non-binding expert text, it is a descriptive definition rather than a legal definition. Unlike the refugee definition – narrow though it may be – that is legally binding upon more than 140 states that have ratified both the 1951 Convention and its accompanying 1967 Protocol, the IDP definition has no legal basis. Where states develop national laws and policies on internal displacement, they are under no obligation to accept the definition provided by the *Guiding Principles*, and thus may of course omit certain criteria in the definition.

Second, even the wide definition provided in the *Guiding Principles* excludes displacement on the basis broadly of economic motivations – for example to escape poverty or find work. The omission was deliberate on the part of the drafters of the text, for fear of rejection of the principles by states that were already concerned about their implications for the exercise of sovereignty, and in all likelihood would have baulked at even a non-binding text that pertained to the rights of millions of internal labour migrants. The problem is that a proportion of displacement resulting from the effects of climate change will be primarily economic in motivation, to escape a general deterioration of conditions of life and economic opportunities, for example as a consequence of the gradual encroachment on agricultural land by desertification or salinization, or the increasing frequency of floods and droughts. Even recognizing that in most cases the motivation to move is mixed, the omission from the *Guiding Principles* of economic motivations opens up the possibility of the exclusion from protection of at least some people displaced internally as a result of the effects of climate change, even if in theory they are protected by human rights law.

A third gap, or at least an issue open to interpretation, is the extent to which the definition extends beyond citizens of a country to international migrants who become displaced there. Given the tendency in many countries in the world for migrant workers to be in the most vulnerable position in the labour force, the first to be laid off in times of crisis, and often the most likely to face discrimination, for example in terms of housing rights, it is not unreasonable to assume that displacement as a result of the effects of climate change may affect migrant workers disproportionately. The experience of an estimated 80,000 migrants displaced in South Africa in 2008 – albeit in this case by a wave of anti-immigrant violence in South Africa – highlighted this particular protection gap (Koser, 2008). An argument that was made at the time in respect of the IDP definition was that for long-term migrants with legal status their homes or places of habitual residence were now in South Africa, although how long a migrant needs to have been present for this to be the case is unclear in law. Many of the migrants displaced in

South Africa, however, had been there for only a short period and did not have legal status. While international migration law defines the responsibilities that host states have towards migrants, for example as regards protecting their human rights and procedural guarantees in areas such as detention or expulsion, the rights of 'irregular' migrants are contested in international migration law. While the Office of the United Nations High Commissioner for Refugees (UNHCR) believed at the time that at least some of those migrants who originated in Zimbabwe may have been entitled to refugee status, none had formally applied for asylum. Finally, there were practical obstacles to protecting the rights even of those displaced migrants who were unequivocally entitled. The fact that many of the displaced did not have documentation, for example, made it almost impossible to discern legal from illegal migrants, and also difficult for them to prove their claims for asylum.

Challenges in translating 'soft law' to 'hard law'

To what extent has the effort to translate the 'soft law' represented by the *Guiding Principles on Internal Displacement* into 'hard law' that actually guarantees the legal rights of the internally displaced been successful?

On the one hand, it is a noteworthy success that some twenty countries worldwide have developed national laws or policies on internal displacement: In Africa these are Angola, Burundi, Liberia, Sierra Leone, the Sudan and Uganda; in Asia, India, Nepal, Sri Lanka and Tajikistan; in the Americas, Colombia, Guatemala, Peru and the United States; in Europe, Armenia, Azerbaijan, Bosnia and Herzegovina, Georgia, Serbia, the Russian Federation and Turkey; and in the Middle East, Iraq. In addition, Nigeria and the Philippines are reportedly close to developing laws or policies (Brookings-Bern Project on Internal Displacement, 2010). On the other hand, this total comprises less than half of the fifty-two countries currently listed by the Internal Displacement Monitoring Centre as hosting significant numbers of internally displaced persons; and omits some of the countries that are host to the largest IDP populations including Bangladesh (up to 500,000), Democratic Republic of the Congo (1.9 million), Ethiopia (up to 400,000), India (at least 500,000), Kenya (400,000), Lebanon (up to 390,000), Myanmar (at least 470,000), Pakistan (1.25 million), Syrian Arab Republic (433,000), and Zimbabwe (up to 1 million) (IDMC, 2009). A significant proportion of the world's documented IDPs, in other words, are not yet protected by national laws and policies.

There have been four main approaches to developing national law and policies (Wyndham, 2006):

- One is a brief instrument simply adopting the *Guiding Principles on Internal Displacement*, exemplified by Liberia's one-page Instrument of Adoption. The wholesale incorporation of the *Guiding Principles* may appear an effective way of ensuring the implementation of all provisions of the principles, suggesting absolute agreement with the principles and ensuring against the dilution of its provisions. Such an approach, however, denies national authorities, relevant governmental bodies, civil society, and IDPs themselves opportunities that the development of a more tailored law would present.
- A second approach has been to develop a law or policy to address a specific cause or stage of displacement. The Indian National Policy on Resettlement and Rehabilitation for Project Affected Families, for example, addresses displacement only as a result of development projects. The Angolan Norms on the Resettlement of the Internally Displaced Populations, as well as law and policies adopted in Azerbaijan, Bosnia and Herzegovina, Colombia, Nepal and Serbia, address only return and resettlement.
- A third approach is a law or policy developed to protect a specific right of the internally displaced; examples include the Turkish Law on the Compensation of Damages that Occurred due to Terror and the Fight Against Terrorism and the US Hurricane Education Recovery Act, which was developed following Hurricane Katrina and addresses, among other issues, the needs of displaced students and teachers.
- The final approach is a comprehensive law or policy addressing all causes and stages of internal displacement. The Colombian Law 387 and the Ugandan National Policy for Internally Displaced Persons most closely approximate a comprehensive law on internal displacement.

In other words, even in that minority of affected countries that have adopted national laws and policies on internal displacement, in many cases these laws and policies are incomplete.

At the same time a very important landmark in the emergence of 'hard law' on internally displaced persons has been the African Union (AU) Convention for the Protection and Assistance of Internally Displaced Persons in Africa (the so-called Kampala Convention), signed by African heads of state and government at a Special Summit of the AU in Kampala, Uganda, 22–23 October 2009. The convention is a legally binding instrument for Africa, that both references and reinforces existing international standards for protecting the human rights of IDPs established in the *Guiding Principles*, and has contributed to the universal authority of the Principles and their evolution from 'soft law' to 'hard law' (Solomon, 2010). The hope is that the convention will serve as a model instrument for other regions. Significant in the context of climate change, the convention stipulates the development at national level of early warning and disaster management systems; however by adopting the definition provided in the *Guiding Principles*, it does not overcome the gaps in definition outlined in the preceding section. At the same time a number of weaknesses in the convention have been noted, including for example a lack of effective enforcement mechanisms, insufficient guarantees for equality and non-discrimination, and the extent to which non-state actors are bound by its provisions.

Implementation gaps

In the particular case of the Kampala Convention, the next step is for states to ratify it. Thereafter the challenge, which is one that confronts many of the states that have already adopted national laws and policies on internally displaced persons, is implementation. It is striking that the three countries with the largest IDP populations worldwide – the Sudan (4.9 million), Colombia (up to 4.9 million), and Iraq (2.7 million) are also countries that have adopted national laws and policies. Shortcomings in these documents aside, clearly there is an implementation gap.

A recent Forced Migration Review assessment demonstrates how difficult it is to evaluate the impact of the *Guiding Principles* (FMR, 2008). In Myanmar they have been used to raise awareness about displacement and mobilize humanitarian assistance; on the other hand they have offered little diplomatic or political leverage with which to influence national authorities there. During elections in Bosnia and Herzegovina and Kosovo the *Guiding Principles* focused attention on the right to political participation for internally displaced persons; still worldwide IDP political participation remains inconsistent. They have inspired the peace agreement between the government and the Maoists in Nepal, but an effective IDP strategy is still not in place there. They informed the drafting of the African Union Convention for the Prevention of Internal Displacement and the Protection of and Assistance to Internally Displaced Persons in Africa; but its effectiveness will entirely depend on the level of compliance and degree of monitoring.

They were issued to all members of the Ministry of Refugees and Accommodation, designated by the Government of Georgia to provide assistance to IDPs resulting from the recent conflict there, although the response of the government to the crisis has been criticized. They form the basis for Uganda's National Policy for Internally Displaced Persons; but there is still a very significant implementation gap in that country.

Indeed Uganda is a good case study. It is a country that both Walter Kälin and Francis Deng have targeted in their advocacy, with the first visit by Deng in 2003 to discuss a draft of the national policy for internally displaced persons. The policy was adopted in August 2004, and has been praised by both Deng and Kälin as being comprehensive in its approach. Yet almost half a million people remain internally displaced in Uganda (there have been significant returns too), and a number of specific challenges to implementation have been identified (Koser, 2006).

One is the problem of ongoing insecurity in certain regions in Uganda, especially in the north. This has restricted access for the international community; it limits the extent to which the government can effectively deliver services, and by implication it means that national policy cannot be properly implemented in the affected areas. The reason this is of such great concern is that it is exactly in these areas that IDPs are most at risk and need most assistance.

Several commentators have also referred to a lack of political will on the part of the government of Uganda. This does not necessarily refer to a lack of will to implement the national policy. But it does refer to a lack of political will to create the conditions in which the policy can be effective. It has been suggested, for example, that there is a reluctance to address the root causes of the conflict that have created IDPs in the first place and keep them displaced. And as a result of not being able to implement the national policy, it has been suggested that the government is at times over-reliant on the international community and NGOs for protecting the rights of IDPs who are nevertheless Ugandan citizens.

Another challenge, certainly not unique to Uganda, is a lack of capacity effectively to deliver the commitments contained within national policy, especially at local government level. In northern Uganda there remain shortages of trained manpower, a lack of resources, and a poor communications and transport infrastructure. A particular aspect in the Ugandan context has been the inadequacy of the current structure for feeding funds from central to local government, relying as it does on conditional grants.

Another challenge identified in Uganda but that also recurs in a number of countries affected by internal displacement is a lack of coordination. Concerns have been expressed in Uganda that the distinction between short-term humanitarian aid and longer-term development and reconstruction is not always maintained. It has been suggested that there is a danger of overlap and competition as a result of the plethora of international agencies currently active in Uganda, especially in the north. A lack of coordination has also been reported both among NGOs and between NGOs, local authorities and government. A related issue has been found to be information gaps, and in particular a lack of awareness of the provisions of the national policy within all relevant government ministries, especially at local level. In many countries, and not just Uganda, IDPs themselves, and their representatives, are particularly poorly informed on the provisions of national policy and their rights.

Shortcomings in institutional arrangements

Institutional arrangements for operational assistance and protection for internally displaced persons fall to the international community, but only in support of national authorities. The *Guiding Principles* are clear that:

> National authorities have the primary duty and responsibility to provide protection and humanitarian assistance to internally displaced persons within their jurisdiction (Principle 3(1)).

At the same time:

> International humanitarian organizations and other appropriate actors have the right to offer their services in support of the internally displaced. Such an offer shall not be regarded as an unfriendly act or an interference in a state's internal affairs and shall be considered in good faith. Consent thereto shall not be arbitrarily withheld, particularly when authorities concerned are unable or unwilling to provide the required humanitarian assistance (Principle 25(2)).

The experience of international responses to humanitarian crises in the last few decades has demonstrated serious deficiencies, in part arising from 'institutional gaps' and a failure to coordinate to fill these gaps. While certain categories of people have benefited from the assistance of distinct international organizations mandated to serve their interests (e.g. UNHCR for refugees and UNICEF for children), others – in particular internally displaced persons – have fallen into 'protection gaps'.

No single agency has specific responsibility to protect IDPs, and the response to situations of internal displacement has tended to be ad hoc, with little predictability.

In order to try to rectify this, UN member states called in 2005 for more predictable, efficient and effective humanitarian action, and for greater accountability, when responding to humanitarian crises, especially in situations of mass internal displacement (OCHA, 2005). The context was the wider UN reform process and overall humanitarian reform agenda. The Principals of the Inter-Agency Standing Committee agreed as a result to a 'cluster leads' system where different UN agencies were appointed as leads in nine 'sectors' or areas of activity according to their areas of specialization. In December 2005 the 'cluster approach' was officially endorsed as a 'mechanism that can help to address identified gaps in response and enhance the quality of humanitarian action' (IASC, 2006). While not limited to IDPs, the new approach did have as one of its major aims to address the need for a more predictable, effective and accountable inter-agency response to the protection and assistance needs of IDPs.

The intention of the cluster approach is essentially to provide predictable action in analysing needs, addressing priorities, and identifying gaps in specific sectors in the field:

> It is about achieving more strategic responses and better prioritization of available resources by clarifying the division of labour among organizations and better defining the roles and responsibilities of humanitarian organizations within the sectors (IASC, 2006).

This approach attempts to raise standards in humanitarian response by ensuring that all sectors have clearly identified and accountable lead agencies that are mandated to be the 'first port of call' and 'provider of last resort' for their respective sectors. These responsibilities ultimately entail mobilizing relevant actors in a specific sector, developing response strategies based on needs, priorities and gaps in that sector, and implementing projects and activities in order to respond to those areas deemed important by the cluster membership. In addition, it is each cluster lead's responsibility to ensure that operational priorities shift with the context, such as from emergency relief to early recovery to development. A key element in the cluster approach's design is to strengthen strategic partnerships between NGOs, international organizations, UN agencies, and the International Red Cross and Red Crescent Movement in the field. It is incumbent on the lead agencies to find ways of involving all relevant sectoral actors in a collaborative and inclusive process so that they are given the opportunity fully to participate in setting and participating in the direction, strategies and activities of the cluster.

Within this system, the UNHCR agreed to be the cluster lead for IDPs during conflict-generated emergencies in three areas: protection, emergency shelter, and camp coordination/camp management. The UNHCR also joined as a member in several other clusters, such as water/sanitation/hygiene (led by UNICEF), logistics (led by WFP) and early recovery (led by UNDP).

The UNHCR has piloted the cluster approach among IDP populations in a number of African countries including Chad, Democratic Republic of the Congo, Liberia, Uganda and Somalia. Initial evaluations have been generally positive, although certain challenges have recurred. Some of these are operational, including identifying and deploying experienced staff; and the security of the environments in which they are working. Coordination between a significant number of partners in complex and dynamic situations has also proved difficult. More conceptual challenges have also emerged. One has been determining which people are of specific concern to the UNHCR as an agency, as opposed to defining the target population for the purposes of cluster members as a whole. The UNHCR has also been concerned that its activities with IDPs should not undermine its core mandate for refugees, either in terms of distracting resources, or by contradicting the right to asylum by 'anchoring' people in their own country. The transition from emergency relief to early recovery and development has proved hard to coordinate, in part because of a lack of targeted funding from the international community. Finally, there have been some criticisms that the cluster approach risks replacing the role of national authorities, rather than supporting or supplementing it.

Lack of alternatives

Without underestimating the significant achievements in the international legal and institutional response to the challenges of internal displacement, this chapter has also highlighted a number of shortcomings. A final gap in current responses is that there is very little prospect of a genuinely new approach emerging.

Ten years ago there were convincing reasons not to try to develop a binding UN Convention on the human rights of internally displaced persons. First, treaty-making in the area of human rights at the UN level had become very difficult and time-consuming. Francis Deng felt that something more immediate was required to respond to the needs of the growing numbers of IDPs worldwide, and wanted to avoid a long period of legal uncertainty resulting from drawn-out negotiations. This approach was, in particular, justified by the fact that the *Guiding Principles* were not creating new law but restated, to a very large extent, obligations that already existed under human rights and international humanitarian law binding upon states. In this context, there was also a concern that negotiating a text that draws as heavily from existing law as the *Guiding Principles*, might have provided some states the opportunity to renegotiate and weaken existing treaty and customary law. Furthermore, having a treaty approved would by no means have guaranteed its widespread ratification by states confronted with internal displacement. Finally, there were concerns that to draft a treaty that combines human rights and humanitarian law as do the *Guiding Principles* was probably premature. In legal, institutional and political terms, the distinction between human rights applicable mainly in peacetime and international humanitarian law for times of armed conflict still was so fundamental that strong opposition was likely from many states and organizations to any attempt to combine both areas of law in a single UN convention.

At the time of the presentation of the *Guiding Principles*, the governments of China, Egypt, the Sudan and India were particularly vocal in their opposition. It was not always easy to discern exactly what they were objecting to: political reactions to the *Guiding Principles* cannot be divorced from reactions to the wider issues of the appointment of a Representative, and from his advocacy for the idea that the international community has the right to intervene where states fail to protect their own displaced populations. The Government of China, for example, has regularly argued that no other state should interfere with the internal affairs of a sovereign state in the name of humanitarian assistance. Egypt, the Sudan and India have expressed concerns that humanitarian intervention may become the precursor for wider interference of powerful countries in the affairs of weaker states. These latter three countries have also questioned the purpose of and process for developing international standards on internally displaced persons.

Resistance also emerged among certain humanitarian groups. Some objected to the breadth of the IDP definition, preferring to limit it to those who would be defined as refugees had they crossed an international border. Proponents of the wider definition, however, argued that people displaced by natural or human disasters were often also in need of assistance and protection and ran the risk of neglect or discrimination by their governments. The International Committee of the Red Cross expressed reservations, which it still holds, that the *Guiding Principles* may risk privileging displacement over vulnerability, by focusing attention on internally displaced persons rather than other war-affected civilians in their country. The response has relied on the growing body of empirical evidence demonstrating the particular needs and vulnerabilities for the internally displaced, and cited above.

The Representative of the Secretary-General on the Human Rights of Internally Displaced Persons, Walter Kälin, has maintained that most of these reasons still stand today, and that it would still be too risky to try to formulate a binding UN Convention (Kälin, 2008). He has pointed out, for example, that negotiations on the 2005 World Summit Outcome document (UN, 2005) showed that while the *Guiding Principles* were welcomed by all governments, many among them were still not ready explicitly to recognize their binding character.

Underlying the reluctance on the part of many states to accept a binding convention is the question of sovereignty, which also probably limits the potential for more concerted intervention by international organizations and of the international community more broadly where states do not fulfil their obligations to protect their own citizens in situations of displacement. In many ways a precursor to the 'Responsibility to Protect' (R2P) concept, Deng developed the notion of 'sovereignty as responsibility', which stipulated that when governments are unwilling or unable to fulfil their responsibilities towards their own displaced citizens, they should be expected to request – and accept – outside offers of assistance. If they refuse, or deliberately obstruct access and put large numbers at risk, the international community has a right, even a responsibility, to intervene. Such intervention can range from diplomatic dialogue, to negotiation of access, to political pressure and sanctions; or in exceptional cases to military intervention.

This approach has been formalized through the R2P concept, which has the most significant normative development for responding to the needs of the internally displaced. One important aim, among others, is to provide a legal and ethical basis for 'humanitarian intervention': the intervention by external actors (preferably the international community through the UN) in a state that is unwilling or unable to prevent or stop genocide, massive killings and other significant human rights violations.

The R2P concept was elaborated by the International Commission for Intervention and State Sovereignty (ICISS), responding to former UN

Secretary-General Kofi Annan's challenge to the international community to chart a more consistent and predictable course of action when responding to humanitarian crises, particularly when international intervention on humanitarian grounds and the violation of state sovereignty are at odds. In essence, the Commission proposed a change in terminology and perspective from the notion of the international community's 'right to intervene', which was inherently flawed according to the principles of state sovereignty, to newly understanding it as their 'responsibility to protect', which substantively makes it incumbent on the international community to provide 'life-supporting protection and assistance to populations at risk' (ICISS, 2001, p. 17).

The R2P framework has gained widespread international accreditation. The UN Secretary-General's High Level Panel on Threats, Challenges, and Change in 2004 endorsed it as an 'emerging norm' to protect civilians from large-scale violence, and the Secretary-General also supported the concept in his *In Larger Freedom* report. In the 2005 World Summit Outcome Document heads of state also clearly and unambiguously accepted the collective international responsibility to protect populations from crimes against humanity. UN Security Council Resolution 1674, adopted on 28 April 2006,

> Reaffirm[ed] the provisions of paragraphs 138 and 139 of the 2005 World Summit Outcome Document regarding the responsibility to protect populations from genocide, war crimes, ethnic cleansing and crimes against humanity;

and commits the Security Council to action to protect civilians in armed conflict. One concrete output has been that UN Peacekeeping missions have increasingly been mandated with a 'Chapter 7' mandate[1] to aggressively protect civilians in conflict.

Supporters of R2P view it as a method of establishing a normative basis for humanitarian intervention and its consistent application. Detractors argue that R2P is a breach of the system of state sovereignty. Beyond the debate about its conceptual validity, the real challenge that confronts R2P is implementation. In two recent cases, genocide in Darfur and Cyclone Nargis in Myanmar, the responsible states were either culpable for or at least failed to respond to the assistance and protection needs of affected populations, including those internally

---

[1] United Nations Charter (1945), Chapter 7: Action with Respect to Threats to the Peace, Breaches of the Peace, and Acts of Aggression.

displaced, but at the same time refused international intervention. The R2P concept was invoked in both cases, but ultimately intervention was limited and ineffective.

Conclusions

It is a striking reversal that the normative framework for people displaced by the effects of climate change inside their own country is better developed than that for people displaced outside their country. Many of the former are internally displaced persons and their rights are variously protected by human rights law and international humanitarian law as articulated in the *Guiding Principles on Internal Displacement*. These principles, would, for example, apply to people displaced as a result of hydro-meteorological disasters (such as flooding, hurricanes, or mudslides); evacuated as the result of the designation of high-risk zones too dangerous for human habitation (e.g. coastal plains prone to flooding); and displaced by violent conflict that has been triggered by climate change. In contrast few of those displaced outside their country as a result of the effects of climate change would qualify for refugee status and international law does not currently protect their status in other countries.

While a priority is therefore to define the rights of people displaced outside their country by the effects of climate change, the prospect of increasing numbers of people displaced internally should also be a catalyst to address gaps and implementation challenges in the normative framework that applies to them. The rights of the majority of the 27 million people already internally displaced by conflict and many millions more by natural disasters and development projects are poorly protected now, and the effects of climate change will inevitably increase their number and further test protection in law and practice.

References

Brookings-Bern Project on Internal Displacement. 2010. National and Regional Laws on Internal Displacement. http://www.brookings.edu/projects/idp/Laws-and-Policies/idp_policies_index.aspx

Deng, F. 1998. *Guiding Principles on Internal Displacement*, 11 February. (E/CN.4/1998/53/Add.2.)

Forced Migration Review. 2008. *Ten Years of the Guiding Principles on Internal Displacement*. Oxford, UK, Refugee Studies Centre.

IASC. 2006. *Guidance Note on Using the Cluster Approach to Strengthen Humanitarian Response.* Geneva/New York, Inter-Agency Standing Committee, 24 November.

2008. *Climate Change, Migration and Displacement: Who Will Be Affected?* Working Paper submitted to UNFCCC Secretariat by the informal group on Migration/Displacement and Climate Change of the Inter-Agency Standing Committee, 31 October.

ICISS. 2001. *The Responsibility to Protect.* New York, International Commission for Intervention and State Sovereignty.

IDMC. 2009. *Internal Displacement: Global Overview of Trends and Developments in 2008.* Geneva, Internal Displacement Monitoring Centre.

IOM. 2009. *Migration, Climate Change, and the Environment: IOM's Thinking.* Geneva, International Organization for Migration. (Working Paper.)

Kälin, W. 2008. Hardening 'soft law': implementing the Guiding Principles on Internal Displacement. *Proceedings,* Washington DC, American Society for International Law.

Koser, K. 2006. Workshop on the Implementation of Uganda's National Policy on Internal Displacement. http://www.brookings.edu/fp/projects/idp/conferences/20060704_Uganda_Agenda.pdf

2008. *Protecting Displaced Migrants in South Africa.* http://www.brookings.edu/opinions/2008/0617_south_africa_koser.aspx

Martin, S. F. 2010. Climate change, migration, and governance. *Global Governance,* Vol. 16, No. 3, pp. 397–414.

OCHA. 2005. *Humanitarian Response Review.* New York/Geneva, United Nations Office for the Coordination of Humanitarian Affairs.

Solomon, A. 2010. *An African Solution to Internal Displacement.* http://www.brookings.edu/papers/2009/1023_african_union_solomon.aspx

UN. 2005. *World Summit Outcome 2005.* New York, United Nations.

Wyndham, J. 2006. *A Developing Trend: Laws and Policies on Internal Displacement.* http://www.brookings.edu/articles/2006/winter_humanrights_wyndham.aspx

Zetter, R. 2009. The role of legal and normative frameworks for the protection of environmentally displaced persons. In: F. Laczko and C. Aghazarm (eds), *Migration, Environment, and Climate Change: Assessing the Evidence,* Geneva, International Organization for Migration.

12

Displacement, climate change and gender

LORI M. HUNTER AND EMMANUEL DAVID

Introduction

Discussions within public, policy and academic realms regarding climate change and migration are often gender neutral (WEDO, 2008). As a result, important differences in the migration experiences of women and men are neglected. Yet migration is a social process – actually, migration is a social process embedded within a variety of other social processes. More specifically, gender-influenced cultural expectations, policies and institutions intersect to shape migration's causes and consequences. In this way, migration is inherently gendered and climate change will, therefore, yield different migratory experiences and impacts for women and for men. This chapter explores these potential gender dimensions.

The argument is informed and motivated by recent pioneering research presented in a special issue of *International Migration Review* (2006) focused on gender and migration studies, as well as a migration-focused special issue of *Gender, Technology and Development* (2008). Further, innovative scholarship linking gender and climate change has recently appeared in two issues of *Gender and Development* (2002; 2009). This important body of work adds nuance to earlier insight on the gendered aspects of migration – particularly as related to migration patterns, processes and policies.

In addition to these research advances, advocates continue to work tirelessly within international climate negotiations to bring gender more centrally into discussions of adaptation and mitigation. Years of effort reached a major turning point in 2007· at COP13 in Bali with the formation of 'Gendercc-women for climate justice' – a global network of women and gender activists working for gender and climate justice (Hemmati and Röhr, 2009). Also reflecting coalescence of concern, organizations such as the Women's Environment and Development Organization (WEDO) and the Global Gender and Climate Alliance (GGCA) – continue to work on engaging policy-makers, the public and researchers on gender aspects of development issues, including those related to climate.

Below we use a livelihoods framework to explore the ways in which climate change may differentially shape both migration's cause and consequence by gender. In addition, we offer discussion of two pathways through which climate change's gendered migration impacts may manifest:

- shifts in proximate natural resources and agricultural potential,
- increases in extreme weather events.

Finally, although we trust that this overview is useful, our effort in bringing together existing work has also made very clear that substantial gaps remain in both research and policy arenas. We close with a discussion of such gaps and related needs.

'Sustainable livelihoods' as a conceptual framework

Originating in work by the United Kingdom development organization, the Department for International Development (DFID), the 'Sustainable Livelihoods' framework has been successfully used in a wide variety of analytical endeavours including exploration of food security (Bank, 2005) and household diversification strategies. Central to the framework is the understanding that the relative availability of various 'capital assets' shapes household livelihood options (Figure 12.1). These assets include human capital (e.g. labour), financial capital (e.g. savings), physical capital (e.g. automobiles), social capital (e.g. kin networks), and natural capital (e.g. wild foods from communal lands). Of course, the relative availability of these assets is shaped by individual and household actions as well as broader socio-economic-political structures and processes. In addition, as suggested by the figure, livelihood strategies are further shaped by household vulnerability as impacted by shocks and stresses. As an example of gendered dimensions of livelihoods and livelihood decision-making, socio-culturally defined gender roles shape both perceived value of 'assets' (e.g. differential value placed on men's and women's human capital) as well as actual household and individual

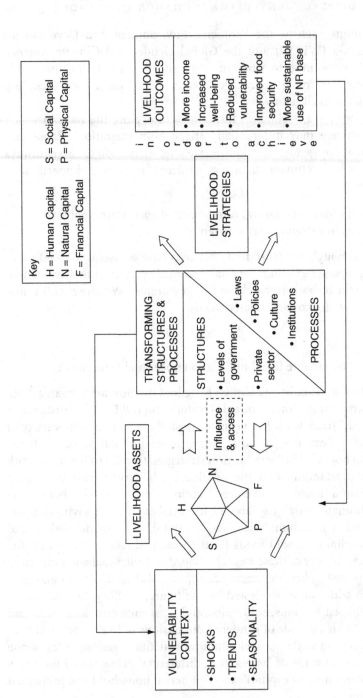

Figure 12.1 Sustainable livelihoods framework
Source: DFID (1999).

use of the assets themselves (e.g. natural resource collection perceived as 'women's' work).

Shifts in proximate natural resources and agricultural potential: a migration 'push' factor

The relative availability of particular assets ultimately shapes household livelihood strategies. Such strategies may include the use of human capital, such as migration in search of employment or engagement of local natural capital (e.g. making reed-based craft products for market sale) (Collinson et al., 2006; Pereira et al., 2006).

As 'natural capital' is often essential in meeting basic living requirements across rural regions of the world's less-developed nations, environmental change has immediate and direct impacts on livelihood options and, more directly, on the health and well-being of millions of households (Koziell and Saunders, 2001). Important natural resource-based activities include arable farming, livestock husbandry, and consumption and trade in natural resource products (e.g. firewood, wild herbs). Natural capital also acts as a 'buffer' against household shocks such as job loss and/or mortality (Hunter et al., 2009).

In the face of shifts in natural capital's availability, households may make adjustments in the use of other assets (Brown et al., 2006). Specifically, if lacking sustainable livelihood options due to cumulative processes of environmental degradation (Zweifler et al., 1994), households may strategically diversify with some members migrating in search of opportunity (Bilsborrow, 2002). In this way, a decline in natural-resource based livelihood options can act as a 'push' factor shaping patterns of out-migration (Bates and Rudel, 2004).

Of course, migration – residential relocation – is a complex phenomenon with many dimensions. For example, migration varies in its motivation (e.g. employment, amenities), level of permanence (e.g. relocation may be temporary), in addition to varying the number of household members involved (e.g. adult members may be sent towards employment opportunities). A continuum also exists as to the voluntary nature of relocation (e.g. natural disasters may force 'distress migration'). Other chapters in this volume offer more explicit detail on the potential pathways through which climate change may broadly shape future patterns and processes related to human migration (see e.g. Chapter 10). We focus here on the potential for gender differentiation in migration processes and patterns.

Gender and livelihood migration

> Any dramatic and unplanned change to the environment will present
> practical challenges to how people make their livelihoods, and this in
> turn will challenge or reaffirm women's and men's roles, and power, in
> their families, communities, and wider society (Sweetman, 2009).

Bringing the livelihood discussion back to gender, many contemporary
patterns suggest the 'environmental push' shaping livelihood labour
migration is not gender-neutral. Indeed, several powerful examples
illustrate the differential impacts and vulnerability across gender –
while also illustrating how various impacts are often culturally specific.
For example, Terry (2009, p. 7) illustrates how cultural forces create
gender-specific vulnerabilities through her description of migration
from West African nations. Here, she claims climate stress is ultimately
to blame for the recent deaths of young male migrants trying to reach
Europe by boat. Within this cultural setting, young men primarily
experience the 'push' of environmental decline and households send
young adult migrants in the hope of generating income through
remittances.

Environmental 'push' factors also impact on women. Recent work in
Nepal, one of the world's least-developed countries, provides evidence of
this gendered migration effect. In Nepal's Chitwan Valley, the vast
majority of households are dependent on proximate natural resources
through reliance on subsistence farming, animal husbandry, and locally
collected forest products. Using monthly panel data covering almost a
decade (1997–2006), Bohra-Mishra and Massey empirically model out-
migration as a function of environmental deterioration, controlling for
relevant social, economic and demographic variables. They find that,
for Chitwan Valley's women residents, increases in collection times for
fodder and firewood were both associated with increases in the probability of local migration. As argued by the researchers, 'Since fodder and
firewood are gathered from local forests, these results imply that deforestation is a significant cause of increased female mobility within the
Chitwan Valley' (Bohra-Mishra and Massey, Chapter 4).

Environmental 'push' factors also affect women even in settings where
they, themselves, less often migrate. As an illustration, in agriculturally
dependent rural Mexican communities, gendered migration and household divisions of labour often result in increased workloads for the
women left behind by male-dominated migration streams. In the
Sonoran state of north-west Mexico, women's livelihoods centrally

involve the conversion of locally produced fruits and vegetables into canned and candied products. These products are then used within the household, sold to market, and/or offered as gifts to strengthen social ties. Forecasted declines in water availability seriously compromise the viability of this local livelihood strategy with virtually no local alternatives for low-educated local women. Such declines in this natural capital-based livelihood option may push more male migration because distant alternative livelihood options tend to be more available to men. This is not to say, however, that such options are desirable – mining, for example, has proved particularly hazardous. Still, increased reliance on male-dominated migrant incomes has had negative implications for female empowerment – the empowerment which had recently been enhanced through the livelihood diversification offered by the local fruit industry (Buechler, 2009). Overall, in cases such as the Mexican Sonoran, we see livelihood strategies, including migration, shaped by environmental factors interacting with gendered labour processes and household division of labour.

In many cases, though, women do migrate – and often solo. Indeed, contemporary migration has been 'feminized' by the expansion of global markets and related socio-economic transformations. This dramatic shift has occurred in the post-1989 era of economic globalization (Piper, 2008), and the transformation is about more than simply changing proportions of migrant women and men. In fact, the female share of global migration streams is similar to that in the past – already in 1960, women represented 46.7% of international migrants, 49.6% by 2005 (Morrison et al., 2007). Beyond these aggregate numbers, however, what has more dramatically changed is the *type* of moves undertaken by women, as well as the potential impacts of those moves. Consideration of these changes is important as we reflect upon the potential gender dimensions of the migration–climate association.

As noted, compared with past decades, today's migrant women are far more likely to relocate on their own or with other migrants outside their family circle (UNFPA, 2006). A powerful example of the transformative effect of such migration is provided by the Philippines, an island nation with thousands of isolated, peripheral communities. Circular migration by island women for work both overseas and in the Philippines' urban areas has dramatically increased in recent years. Singapore, Hong Kong and the Arabian Gulf region are prominent destinations for Filipina overseas contract workers, often hired as live-in domestic workers (McKay, 2005). Migrant remittances sent back to origin communities are regularly invested in material goods such as agricultural tools, cars and motorcycles. Household renovations and development of home-based businesses (e.g. corner stores, tailor shops) are also fuelled by migrant income sent home. In general, female migration has become an important means of diversifying household livelihoods, funding additional diversification strategies, and therefore promoting overall household security (McKay, 2005).

More generally, migrant remittances are a critical source of income for vulnerable households across the globe. For 2007, recorded remittances flows worldwide were estimated at US$318 billion, of which US$240 billion went to developing countries. Importantly, informal channels of remittances are not included in these data – and hence these levels are no doubt significantly underestimated. In 2007, India received the largest volume of migrant remittances (US$27 billion), with China and Mexico ranking close behind (US$25.7 billion and US$25 billion respectively). International migrants from the Philippines sent back a total of US$17 billion, and migrants from France, US$12.5 billion.[1]

As suggested by the Philippines example, shifts in global labour demands underlie this dramatic change in the nature of male/female international migration. The sheer demand for 'reproductive' services (those associated with the domestic sphere such as cleaning and care-giving) within the more developed economies is, in part, fuelling female international migration. Examination of the professional profiles of male and female migration streams confirm this distinction. Specifically, the most highly skilled categories of migrants tend to be male (e.g. international recruitments), although women also comprise a substantial portion of this category. Still, household, entertainment and unskilled labour migrant streams are predominantly women. In addition, many skilled women undergo 'de-skilling' through the migration process as well-educated women undertake migration for domestic or other 'unskilled' employment opportunities.

As suggested by the gendered labour demands pulling international migrants, gendered distinctions within migration streams are embedded within, and the products of, broader cultural norms and processes. Indeed, these gendered migration patterns are clearly reflective of a broader social order – yet that social order is reciprocally influenced by migration processes.

---

[1] Lo (2008) offers a fascinating and nuanced account of the gendered and social impacts of remittances illustrated through West African migration.

Illustrating migration's influence on gender norms, early scholarship on gender and migration generally viewed migration as emancipating for women. Evidence in some settings suggests that migrant women may, indeed, experience greater independence and personal autonomy due to increased wage-earning potential and control over monetary resources (Foner, 2002). Young women (and men) from Kyrgyzstan, for example, migrate for income, education, as well as to escape cultural traditions such as early marriage (Thieme, 2008). Still, negative impacts have also been revealed as increases in responsibility do not always translate into increases in empowerment and/ or rights. As an example, given the highly feminized nature of Asian temporary migration streams, such livelihood decisions often result in a reversal of traditional gender roles whereby the wife becomes the family breadwinner (Piper, 2008). Marital conflict may result from the challenge brought to traditional masculinity as the migrant's husband then takes greater responsibilities for attending to the daily needs of the household and children (Piper, 2008). In some cultures, such visible role reversal is seen as a downgrading of the husband's social status (Pinnawala, 2008).

Of course, men's gender roles are not the only ones challenged by the new social arrangements. Solo migrant women must *themselves* renegotiate their sense of place within the origin household. Those women who have left children behind, for example, experience a phenomenon referred to as 'transnational motherhood' often within 'transnationally split' households (Piper, 2008). The experiences of women migrants from Sri Lanka, living in West Asia, illustrate these shifting social dynamics (Pinnawala, 2008). Indeed, over 70% of the total Sri Lankan contract workforce overseas are women, with the vast majority of these migrants engaged in domestic work classified as 'unskilled', particularly as housemaids (90% of female Sri Lankan overseas migrants since 1995). Research with this migrant population documents the emergence of new social positions and roles that often facilitate the management of the migrant's transnational families by proxy. In this case, a geographically absent mother remains engaged in the day-to-day life of the origin family through regular communication channels and the provision of instruction and advice to a local relative, often also female. In this way, a female migrant may use a proxy to manage her affairs by long distance – reporting back on confidential family issues such as misbehaving children or husbands (Pinnawala, 2008).

Yet there are no universals with regard to the impacts of migration on gendered roles and responsibilities. Such changes vary by cultural setting, household characteristics, and the migrant experience itself. Recent work focused on Mexican migrant households in the United States illustrates some of the nuances within the migration's impact on reconstruction of gender relations. Parrado and Flippen (2005) demonstrate how some gender relations found in Mexican origins are discarded upon migration while others are maintained and even reinforced. Specifically, while US residence expands employment opportunities for women, gains in this realm do not necessarily translate into more egalitarian household division of labour or more liberal gender attitudes within migrant households.

As suggested by the experience of US-based Mexican migrant households, in some cases, gendered migration streams reinforce as opposed to transforming traditional cultural norms. We can further consider the case of undocumented female Vietnamese migrants in Bangkok where, Yen and colleagues (2008) argue, gendered migrant labour results in the reproduction of patriarchal norms. In this case, women work invisibly behind the scenes, preparing food, cleaning and washing dishes, thereby dutifully undertaking the reproductive tasks 'assigned to women as something "natural"'. As women workers re-enact their ascribed roles within existent expectation and hierarchies, migration acts not as an empowering force, but rather as 'cultural imprisonment' (Yen et al., 2008, p. 378).

In all, early thoughts on the potential for gendered climate–migration connections can be informed by bringing together literature on the feminization of global migration streams and the various gender dimensions of livelihood migration. Overall, this intersection suggests the possibility of intensified pressure on livelihoods, particularly within impoverished households in regions with high dependence on local natural capital. As the local availability of natural resources becomes less predictable, and perhaps constrained, migration may be increasingly viewed as a livelihood option. In some cultural settings, men typically engage in such migrations (consider the West African example above), although global data suggest increasing numbers of solo female migrants. As evidenced by several examples provided here, there are gender dimensions to migration's determinants (e.g. gendered labour force 'pulls'), as well as gendered influences on the nature and experience of migration itself (e.g. 'skill' levels). Further, migration itself often results in a reshaping of gender norms and roles. All such gender-migration connections can logically be expected to shape the climate–migration process, and its impacts, as well.

Extreme weather events: learning from disaster research on migration and gender

In addition to livelihood research, disaster scholarship represents another body of existing literature ripe to inform thoughts on the climate–gender–migration nexus. Indeed, climate change research predicts increased occurrence and severity of extreme weather events, sudden-onset disasters as well as chronic disasters including drought and recurrent flooding (IPCC, 2007). Over the past few decades there has been a dramatic rise in the number of recorded disasters, although some researchers caution that increases in detection and recording reflect shifts in technology, classification, and media and communication (Eshghi and Larson, 2008). Since the reliable recording of disaster events in the 1960s, there has been a rise in the number of people affected by disasters (Hunter, 2005). In addition, a 2009 Oxfam report forecasts a 54% increase in the number of people affected by climate-related disasters such as floods and droughts. The report suggests that by 2015, an average of 375 million people will be affected by climate-related disasters annually, up from an average of 250 million per year.[2] In this context, the association between disaster risk reduction, development and climate change is increasingly discussed among researchers and policy-makers, especially around reducing socio-economic vulnerability to natural hazards (Helmer and Hilhorst, 2006; Thomalla et al., 2006; van Aalst, 2006). With mounting evidence of a link between climate change, migration and natural hazards, we must consider how social and economic vulnerabilities shape population movements before, during and after disaster.

Disasters and migration

There is a wide body of literature examining population mobility in disaster, and much of the research on the social dimensions of disaster is implicitly focused on the spatial flows of groups (Aguirre, 1983; Fowlkes and Miller, 1983). Like other migration in response to compromised livelihoods, one way to conceptualize disaster-related movements is in terms of level of permanence. Such temporal dimensions can range

---

[2] Oxfam International (2009) has classified drought, extreme temperature (e.g. heatwaves), wildfires, floods and meteorological storms as climate-related natural hazards. Geophysical natural hazards such as earthquakes, tsunami and volcanic eruptions were not included in the climate-related classification.

from short-term evacuation and temporary displacement to long-term, perhaps permanent, relocation. Oliver-Smith (2006) outlines several demographic movements relating to disaster, including flight (e.g. escape), evacuation (e.g. removal from harm's way), displacement (e.g. uprooting from home ground), resettlement (e.g. relocation to new homes), and forced migration (e.g. people must move to a new and usually distant place). Each form of movement also varies in relation to other factors in particular contexts. These compounding factors are presented in a series of multidimensional pairs, which overlap and intersect to shape the variable occurrences of movement in disaster: proactive–reactive; voluntary–forced; temporary–permanent; physical danger – economic danger; administered – non–administered. In other words, how and why people move varies in relation to a number of social forces, including time, risk and exposure, and level of initiation and control.

Embedded within these processes, there are both circular and trans-formative features of disaster migration as disaster-affected regions repopulate over time. Mileti and Passerini, for example, argue that relocation following disaster is not unidimensional and entails a variety of human activities, including (a) wholesale relocation of a population or community to an entirely new site; (b) intra-urban relocation within the pre-disaster location that moves human activities to areas with less risk; and (c) urban reconstruction on original sites, where 'relocation is no relocation at all' (Mileti and Passerini, 1996, p. 100).

While wholesale relocation of an entire group is relatively rare, several cases of mass migration after disaster are worth mentioning because of the scale and scope of the movements. Drawing on evidence from an instructive array of migration experiences, Reuveny (2008) examines the ways in which deteriorating environmental conditions shaped migratory flows in three cases: drought in the 1930s across the United States Great Plains; population movements from Bangladesh to India in the 1950s; and internal displacement from Hurricane Katrina which struck the southern United States in 2005. Hurricane Katrina, in particular, illustrates a variety of migratory impacts. When the hurri-cane made landfall on 29 August 2005, over 1.1 million people along the Gulf Coast were evacuated and 770,000 were temporarily displaced. By December 2005, over half a million evacuees had not returned to the affected area – with the event deemed 'the largest mass population displacement in U.S. since the "dust bowl" migrations of 1930s' (Falk et al., 2006, p. 116; White House Report, 2006).[3] But within this event, social structures linked to the intersection of race/ethnicity, class, gender, and age created conditions that put some groups more at risk than others. Oliver-Smith (2006) has argued that a single disaster can 'fragment into different and conflicting sets of circumstances', leading some groups to migrate and others to not migrate.

While there is an emerging literature on the ways in which disasters and environmental, natural and technological hazards function as 'push' factors in migration (Hunter, 2005), much less is known about the dynamics of slow-onset changes associated with climate change. The disaster social science literature provides a lens through which to frame potential climate-related movements. Furthermore, less is known about the gendered dimensions of movements driven by slow-onset environmental changes. The literature on the differential experiences of women and men in disaster provides yet another foundation to build empirical and theoretical knowledge about the gender–climate–migration nexus.

Gender and disasters

To better understand the association of gender and climate change, Terry (2009) argues that we need to draw on studies undertaken within cognate fields such as gender and disasters. Several instructive overviews of the gender-disaster literature provide an excellent starting point for exploration of this intersection (see e.g. Fothergill, 1996; Enarson et al., 2006). Much of this literature focuses less explicitly on the external agent (natural hazard, disaster, or for our purposes, climate change) and more on social vulnerabilities that expose different groups to different levels of risk (Cutter, 1996). These social vulnerabilities, it is argued, reflect the 'global distribution of power and human uses of our natural and built environment' (Enarson et al., 2006, p. 30) – and it is these vulnerabilities that are ultimately responsible for the gendered nature of disaster impacts.

---

[3] Interestingly, some research has also documented the non-occurrence of disaster-induced out-migration. Paul (2005), for example, argues that a 'constant flow of aid and its proper distribution by government and non-government organizations' led to the non-occurrence of out-migration following a 2004 tornado in Bangladesh. This claim is consistent with patterns found in other disaster research; permanent, rather than temporary, relocation following disaster is often driven by inadequate response by the state and government bodies (Oliver-Smith, 2006).

To put it more generally, although originating in ecological systems and natural hazards, disasters are inherently social phenomena. Their consequences are 'linked to who we are, how we live, and how we structure and maintain our society' (Fothergill, 2004, p. 27). Unequal distributions of hazard risks are thus linked to inequalities such as poverty, limited access to resources and mobility, as well as culturally constructed expectations that shape work patterns, household divisions of labour, and caretaking responsibilities, including those that are gendered.

Gendered differences in vulnerability can be broadly captured through the lens of poverty – and the global feminization of poverty has resulted in women and girls comprising over three-fifths of the world's 1 billion poorest people (UNFPA, 2008). As disasters disproportionately affect those already living in poverty (Fothergill and Peek, 2004), they have disproportionate effects on women across the globe and tend to leave poor women even more impoverished, especially when women's incomes are resource-dependent (Enarson, 2001). Thus gender and disaster researchers study differential experiences and losses among populations in particular social contexts, as well as the different vulnerabilities and capacities of women and men. Results suggest that at all phases of a disaster – before, during and after – women and men respond to risk differently, not because of biological sex differences, but because of location in social structures (Fothergill, 2004). Below we examine several ways in which women and men respond to disaster differently, with a specific focus on various forms of spatial processes. We do so with an eye towards informing understanding of the potential gender dimensions of climate impacts on disaster events and related migration.

Risk perception and evacuation patterns

Discussions of population movements and disasters often focus on displacement and forced migration after disaster. However, widespread population movements often occur before a disaster in the form of evacuation. Such movement is related to perception of risk, and research finds clear gender distinctions in hazard risk perception. In a qualitative study of gender, class and family in the 1997 Red River Valley flood in Grand Forks, North Dakota, Fothergill examined women's patterns of taking flood threats more seriously than men, especially if children were present, and engaging in preparation measures in the home. Several weeks after the flood, Fothergill travelled to the affected area, and, over a period of two years, interviewed forty middle-class women about their experiences of the disaster. She concluded that women perceive disaster and disaster threats as more serious than men because of social location: 'women have less control over their lives, have less power in the world, and therefore must take risks more seriously than men do' (Fothergill, 2004, p. 43). In a study of the same flood event, Enarson (1999) found that women evacuated earlier than men, experienced increased care-giving responsibilities when men in their families resisted evacuation, and were less likely to delay evacuation, in part due to concerns about the safety of youth. In both studies, social location was situated in relation to gendered division of labour, caretaking responsibilities and power.

We can turn to an August 1998 North Carolina hurricane for additional evidence of gendered risk perception shaping evacuation. Bateman and Edwards' (2002) cross-sectional survey of 1,050 coastal households administered in January 1999 found that socially constructed gender differences shape intention and capacity to evacuate. Drawn from data collected on household-level patterns of evacuation (rather than intra-household decision-making processes), the study revealed that women were more likely than men to evacuate because of differences in care-giving roles and family obligations, especially having a family member with special medical needs, their greater exposure to certain risks (e.g. women were more likely to live in mobile homes), heightened perception of risk, and engagement in preparedness activities such as developing an evacuation plan. Interestingly, however, researchers found that while men generally perceive less risk than women, once they do perceive heightened risk, they are more likely to evacuate than women with comparable risk exposure. As such, Bateman and Edwards outline the need to differentiate between intention to evacuate and the capacity to actually do so.

Similar patterns of the ways in which economic disadvantage and social vulnerability shaped mobility were found in other disaster contexts. Indeed, pre-existing vulnerabilities of women in New Orleans factored into the gendered experiences of Hurricane Katrina. Whereas 4.3% of working women in the United States report that they lack transportation, nearly 15% of women workers in New Orleans were without a vehicle in 2005 (Laska et al., 2008). According to Laska et al., lack of transportation serves as a barrier to women's successful evacuation, but it also constrains employment opportunities after disaster by limiting women's ability to return to work or to migrate in search of work. This point further emphasizes the need to examine the differences between intention and capacity to evacuate and the ways in which relative access to resources shape population mobility.

Risk perception differences between men and women also vary by disaster type, as well as potential severity. With respect to perceptions of climate risks, Terry (2009, p. 8) refers to a case study of small farmers in South Africa (conducted by Thomas et al., 2007) in which women farmers recognized heavy rainfalls as a distinct risk, whereas more men than women perceived drought as a distinct climate risk. Terry attributed these differences in perceptions of climate risk to broader livelihood patterns and relationships to livestock and agriculture. In other words, there is a connection between risk perception and livelihood migration, as explained above.

Much of our discussion thus far has focused on the cultural factors that have shaped mobility. However, cultural factors relating to gender also shape experiences of *immobility*, which in disasters can have deadly consequences. Again, we can turn to the gender and disaster literature to better understand how social norms, cultural expectations, and gendered processes put women at greater risk of disaster mortality and constrain ability to move out of harm's way.

To the extent that evacuation is related to early warning systems, we must bring attention to important gender differences in risk communication. For example, more women than men died in a 1991 cyclone in Bangladesh because warning signals did not reach many women in the home/household environment (Ikeda, 1995). Research has found that more women, on average, die in natural disasters than men (Neumayer and Plümper, 2007), although mortality rates vary by gender according to the type of disaster and the risks taken by human actors in particular contexts. Citing an Oxfam study of several countries affected by the 2004 Indian Ocean tsunami, Hyndman (2008, p. 109) points out that 'women were up to three times more likely to have died than men'.

Rather than attributing these disparities in mortality to biological sex differences in strength, size or physical capabilities, socially produced gender relations such as care-giving roles, childhood socialization and clothing norms affect women's ability to survive disaster. For example, many women's evacuation is constrained when they assume caretaking responsibilities for children and the elderly, and these responsibilities often put women more at risk than men of injury and death. Similarly, childhood socialization, which has encouraged boys rather than girls, for example, to learn to swim and climb trees, is cited as a factor in higher mortality rates among women during the 2004 tsunami (Oxfam International, 2005). Finally, gender norms relegating women's mobility in public and private spheres also contributed to higher death rates among women. For example, some traditional dress codes can restrict women's ability to quickly move from hazards. When debris from the 2004 Indian Ocean tsunami ripped away women's clothing, many women died indoors 'rather than allow themselves to be exposed to the shame of running outside naked to escape' (ActionAid International, 2008). In each of these examples risk perception and evacuation patterns, mobility and immobility were intimately tied to socially constructed notions of gender and differential location in social structures. These patterns must be examined in light of the urgent environmental shifts tied to climate change.

Temporary and permanent housing

When climate change pushes people to new residential locations, access to and experiences in temporary and permanent housing will probably vary by gender. The experiences of those displaced during disaster shed light on how housing concerns are not gender neutral and are at times even gender insensitive. Kinship networks frequently provide immediate temporary housing for those who evacuate from disaster, and it is often women's extended networks that pull members and resources to aid in successful evacuation (Litt, 2008). While some evacuees end up in the private homes of friends and family, many end up in government-provided temporary housing. Accommodations in shelters and temporary housing communities are not always designed around the needs of women and children (Enarson, 1999). In a comparative study of housing issues following Hurricane Andrew and the Red River Valley floods, Enarson (1999) found that women faced numerous challenges in temporary housing, including social isolation, lack of privacy, and insufficient or non-existent childcare, elder care or laundry facilities. Women were also fearful for their personal safety and concerned about overcrowding and lack of outdoor play spaces for children.

Bureaucratic procedures in government agencies, relief organizations and insurance companies can further disadvantage those whose pre-disaster living arrangements do not meet narrow definitions of 'family' or 'household'. Women-headed households often face obstacles to receiving aid from official relief programmes that draw upon models of single-headed households that privilege men (Morrow and Enarson, 1996). For example, in a study of African-American women displaced by Hurricane Katrina conducted one year after the storm, Murakami-Ramalho and Durodoye (2008) found difficulties receiving aid experienced by women and their adult female children living in the same home with separate incomes.

Just as scholars have begun to look at the intersection of race, class and gender, researchers of gender and disaster have begun to complicate categories of analysis by becoming more attuned to differences in bodies and sexualities (Enarson et al., 2006). Even basic definitions of male and female can create difficulties during displacement. Pincha (2008) conducted a qualitative study comprising over 150 focus-group interviews in more than forty-five areas affected by the 2004 Indian Ocean tsunami. Findings revealed that Aravanis, which do not fit within Western oppositional frameworks of female/women and male/men, were excluded from the relief process, temporary shelters and official death records, thus rendering this population invisible in many of the relief and reconstruction agendas.

These challenges are not limited to developing and non-Western contexts. After discussing the controversial case of a transperson who was jailed after showering in a women's restroom in a Texas shelter following Hurricane Katrina, despite being granted permission to do so by a shelter volunteer, D'Ooge (2008) concludes that housing challenges, both temporary and permanent, reveal not only institutionalized sexism, but also homophobia and transphobia. These examples reveal that notions of sex, gender and sexuality are always situated in specific socio-cultural contexts. As scholars interested in migration, we must consider the extent to which the specific meanings of these social categories travel across time and space. Similar challenges might arise for populations forced to migrate during more slow-onset environmental changes, especially if those movements are distant rather than local.

Differential out-migration and return-migration

Migration is not uniform with respect to women's and men's movements after disaster. Often there is frequent male out-migration, which can lead to an increase in female-headed households and women's care-giving responsibilities. There are also different patterns of return-migration. In the United States, for example, there are gender dimensions to Katrina-related migration which shed light on the broader association between migration, gender and disasters. There were several distinct migration flows as residents returned to New Orleans – and, in general, early analyses suggested that the return migration streams resulted in a city that was whiter, older and more affluent (Falk et al., 2006; Frey et al., 2007). Yet data later emerged revealing the greatest demographic change in post-Katrina New Orleans was the loss of roughly 60% of the city's female-headed households, especially those with the presence of children under 18 years (Newcomb College Center for Research on Women, 2008). Indeed, in the weeks and months following the storm, men returned in far greater numbers than women (Willinger and Gerson, 2008). One scholar observed that New Orleans had effectively become a militarized and masculinized 'city of men', with US soldiers, Homeland Security officers, contractors, migrant construction workers, Army Corps of engineer personnel, and volunteers, among others, converging on the disaster-affected area to engage in rescue and rebuilding efforts (Batlan, 2008). The rapid rise in the number of men in the city created conditions where many women were concerned about safety and the threat of gendered violence.

Despite some disagreement over the issue of disaster-related mass migration, the potential for migration remains real, and is documented in several case studies of disaster in a global context. Movement under these circumstances is political and linked to institutionalized arrangements of power that both enables and constrains mobility. As the cases above have demonstrated, these processes during disaster are linked to gendered power arrangements that shape how and where people move. In developing a 'politics of mobility and access', feminist geographer Doreen Massey (1994, p. 150) has argued that 'mobility, and control over mobility, both reflects and reinforces power'. This politics holds true in disaster to the extent that those who leave and those who are left behind are differentially situated within social structures that are linked to raced, classed and gendered hierarchies. But Massey notes that this is not just about 'unequal distribution', whereby some move more than others or wield more control than others. Instead, she argues that 'mobility and control of some groups can actively weaken other people. Differential mobility can weaken the leverage of the already weak' (Massey, 1994, p. 150). This was seen in the effects of men's reluctance to evacuate and of out-migration patterns on women's livelihood practices and increased domestic and community work following disaster.

Much of the existing literature on gender and disaster has aimed to reduce risk for women and girls. In addition to focusing on vulnerabilities, this literature has also aimed to highlight women's resilience to disaster. More studies are focusing on women's proactive work in disaster and women and women's groups are often at the centre of rebuilding livelihoods, households and communities. Similarly, while emergency managers and disaster officials tend to focus on prevention and preparedness activities, many in the climate change debates have framed similar activities as coping or adaptive strategies. While there may be differences in terms, climate change adaptation can readily draw upon the knowledge in the gender and disaster literature. For example, in 2008, the UN International Strategy for Disaster Reduction published *Gender Perspectives: Integrating Disaster Risk Reduction into Climate Change Adaptation; Good Practices and Lessons Learned*, which offers case-study data on women's leadership activities in managing natural and environmental resources, reducing disaster risks and adapting to climate change. It shows that gender-sensitive tools and practices used in disaster risk reduction (e.g. land and water use and management, alternative livelihoods) can be used to confront challenges, including those linked to livelihood migration, associated with climate change.

Bringing the literatures together: climate, migration and gender

In all, the examples provided above illustrate migration's many gender dimensions. In this way, they testify to the importance of embedding studies of migration within broader social and cultural contexts. We contend that gender considerations should be given central consideration as scholarship on climate change vulnerability and adaptation moves forward – indeed, gender-blind research very simply neglects the fundamental ways in which climate-shaped migratory experiences and impacts will differ for women and men.

As reflected within the example of young men's migration from West African nations, gender-influenced cultural expectations shape the 'push' of environmental degradation acting upon migration. Within disaster settings, gender further shapes evacuation migration – with women's social location as care-givers shaping risk perception and decisions for earlier evacuation. Further, gender distinctions in migration and return migration to/from climate-impacted regions differentially shape vulnerability and future livelihood options for men and women.

Of course, gendered social institutions also shape the 'pull' of migration – with dramatic recent increases in the migration of solo women reflecting cultural beliefs regarding their particular suitability for international domestic employment opportunities. In this way, women are increasingly carrying the burden of their households, which often remain in their origin communities (Piper, 2008).

Although much stands to be learned from existing livelihoods and disaster literature, progress in research on the migration–climate–gender nexus is challenged by the gender-blind nature of most data on international migration. International migration streams are, even in the aggregate, difficult to quantify – in part due to the large number of undocumented moves. Development of effective means of gathering and reporting gender-specific data adds yet another layer of challenge.

Even so, the challenge must be met as there is much work to be done in generating improved understanding of climate–migration–gender linkages. Some scholars argue that due to relatively higher levels of female poverty and broadly unequal power relations, climate change will disproportionately affect women (Buechler, 2009). In addition, an area not yet explored is the impact of women's out-migration on other women in the household and/or extended family (Piper, 2008). In many cases, however, it may be the men who bear the brunt of climate migration's trials (Terry, 2009). Others argue that the gender construct, itself, remains problematic (Pincha, 2008). In all cases, gender-informed research approaches are required to gain the nuanced understanding necessary to inform policy mitigating climate change impacts.

As to policy, international policy negotiations have often more centrally focused on the 'economic effects of climate change, efficiency, and technical issues' (Hemmati and Röhr, 2009), neglecting the social dimensions of climate change, including gender issues. According to Hemmati and Röhr (2009, p. 22), 'until today, no gender analyses have been conducted in relation to the instruments and articles of the UNFCCC and the Kyoto Protocol'. However, gender issues are increasingly considered in climate change negotiations, forums and conferences, in part because women and gender activists have been 'questioning the dominant perspective focusing mainly on technologies and markets' and putting 'caring and justice in the centre of measures and mechanisms' (Hemmati and Röhr, 2009, p. 26). A series of gender-focused conferences and forums have recently been held, each producing documents that aim to integrate gender into climate protection measures and instruments, including The Manila Declaration (22 October 2008), the Nairobi Action Plan for African Parliamentarians on Disaster Risk Reduction and Climate Change Adaptation (20 February 2009), and the Beijing Agenda for Global Action on Gender Sensitive Disaster Risk Reduction (22 April 2009).

To further inform such negotiations, scholars and activists should build upon the interesting and important body of scholarship reviewed above. The work spans multiple disciplines and cultural settings, some focused on gender and livelihoods, and some within the literature on gender and disasters. In general, cultural-specific gender norms will shape the specific pathways households choose in diversification of livelihoods faced with the implications of climate change. Migration, as a path to diversification in the face of climate change and climate-related disasters, is no exception. Explicit consideration of gender, among the many axes of gender differentiation within migration streams, adds an important nuance to our understanding of the ways in which the migration–climate nexus shapes individual lives, household decision-making and, in general, human well-being.

References

ActionAid International. 2008. Empowering women as community leaders in disaster risk reduction. In: *Gender Perspectives: Integrating Disaster Risk Reduction into Climate Change Adaptation. Good Practices and Lessons Learned*. Geneva, United Nations International Strategy for Disaster Reduction, pp. 34–38.

Aguirre, B. E. 1983. Evacuation as population mobility. *International Journal of Mass Emergencies and Disasters*, Vol. 1, No. 3, pp. 415–37.

Bank, L. 2005. On family farms and commodity groups: rural livelihoods, households and development policy in the Eastern Cape. *Social Dynamics*, Vol. 31, No. 1, pp. 157–81.

Bateman, J. and Edwards, B. 2002. Gender and evacuation: a closer look at why women are more likely to evacuate for hurricanes. *Natural Hazards Review*, Vol. 3, No. 2, pp. 107–17.

Bates, D. C. and Rudel, T. K. 2004. Climbing the 'agricultural ladder': social mobility and motivations for migration in an Ecuadorian colonist community. *Rural Sociology*, Vol. 69, No. 1, pp. 59–75.

Batlan, F. 2008. Weathering the storm together (torn apart by race, gender, and class). *National Women's Studies Association Journal*, Vol. 20, No. 3, pp. 163–84.

Bilsborrow, R. E. 2002. *Migration, Population Change and the Rural Environment*. Woodrow Wilson International Center for Scholars, Environmental Change and Security Project. pp. 69–94. (ECSP Report, Issue 8.)

Brown, M. E., Pinzon, J. E. and Prince, S. D. 2006. The sensitivity of millet prices to vegetation dynamics in the informal markets of Mali, Burkina Faso and Niger. *Climatic Change*, Vol. 78, No. 1, pp. 181–202.

Buechler, S. 2009. Gender, water, and climate change in Sonora, Mexico: implications for policies and programmes on agricultural income-generation. *Gender and Development*, Vol. 17, No. 1, pp. 51–66.

Collinson, M. A., Tollman, S. M., Kahn, K., Clark, S. J. and Garenne, M. 2006. Highly prevalent circular migration: households, mobility and economic status in rural South Africa. In: M. Tienda, S. E. Findley, S. M. Tollman and

E. Preston-Whyte (eds), *Africans on the Move: Migration in Comparative Perspective*. Johannesburg, Wits University Press.

Cutter, S. 1996. Vulnerability to environmental hazards. *Progress in Human Geography*, Vol. 20, No. 4, pp. 529–39.

DFID. 1999. *Sustainable Livelihoods Guidance Sheets: Framework*. London, UK Department for International Development. http://www.nssd.net/pdf/sectiont.pdf (Accessed September 2009.)

D'Ooge, C. 2008. Queer Katrina: gender and sexual orientation matters in the aftermath of the disaster. In: B. Willinger (ed.), *Katrina and the Women of New Orleans*. New Orleans, Newcomb College Center for Research on Women, pp. 22–25.

Enarson, E. 1999. Women and housing issues in two U.S. disasters: Hurricane Andrew and the Red River Valley flood. *International Journal of Mass Emergencies and Disasters*, Vol. 17, No. 1, pp. 39–63.

2001. What women do: gendered division of labor in the Red River Valley flood. *Environmental Hazards*, Vol. 3, No. 1, pp. 1–18.

Enarson, E., Fothergill, A. and Peek, L. 2006. Gender and disaster: foundations and directions. In: H. Rodríguez, E. Quarantelli and R. Dynes (eds), *Handbook of Disaster Research*. New York, Springer, pp. 130–67.

Eshghi, H. and Larson, R. 2008. Disasters: lessons from the past 105 years. *Disaster Prevention and Management*, Vol. 17, No. 1, pp. 62–82.

Falk, W., Hunt, M. O. and Larry, L. 2006. Hurricane Katrina and New Orleanians' sense of place: return and reconstitution or 'Gone with the Wind'? *DuBois Review*, Vol. 3, No. 1, pp. 115–28.

Foner, N. 2002. Migrant women and work in New York City, then and now. In: P. G. Min (ed.), *Mass Migration to the United States: Classical and Contemporary Trends*. Walnut Creek, Calif., Altamira Press.

Fothergill, A. 1996. Gender, risk and disaster. *International Journal of Mass Emergencies and Disasters*, Vol. 14, No. 1, pp. 33–56.

2004. *Heads Above Water: Gender, Class and Family in the Grand Forks Flood*. Albany, NY, State University of New York Press.

Fothergill, A. and Peek, L. 2004. Poverty and disasters in the United States: a review of recent sociological findings. *Natural Hazards*, Vol. 32, No. 1, pp. 89–110.

Fowlkes, M. and Miller, P. 1983. *Love Canal: The Social Construction of Disaster. Final Report*. Fort Belvoir, Va., U.S. Department of Defense, Technical Information Center.

Frey, W. H., Singer, A. and Park, D. 2007. *Resettling New Orleans: The First Full Picture from the Census*. Washington, DC, Brookings Institution. www.brookings.edu/reports/2007/07/katrinafreysinger.aspx

Gender and Development. 2002. Vol. 10, No. 2.

2009. Vol. 17, No. 1.

Gender, Technology and Development. 2008. Vol. 12, No. 3.

Helmer, M. and Hilhorst, D. 2006. Natural disasters and climate change. *Disasters,* Vol. 30, No. 1, pp. 1–4.

Hemmati, M. and Röhr, U. 2009. Engendering the climate-change negotiations: experiences, challenges, and steps forward. *Gender and Development,* Vol. 17, No. 1, pp. 19–32.

Hunter, L. M. 2005. Migration and environmental hazards. *Population and Environment,* Vol. 26, No. 4, pp. 273–302.

Hunter, L. M., Twine, W. and Johnson, A. 2009. Adult mortality and natural resource use in rural South Africa: evidence from the Agincourt Health and Demographic Surveillance Site. *Society and Natural Resources,* Vol. 24, No. 3, pp. 256–75.

Hyndman, J. 2008. Feminism, conflict and disasters in post-tsunami Sri Lanka. *Gender, Technology and Development,* Vol. 12, No. 1, pp. 101–21.

Ikeda, K. 1995. Gender differences in human loss and vulnerability in natural disasters: a case study from Bangladesh. *Indian Journal of Gender Studies,* Vol. 2, No. 2, pp. 171–93.

International Migration Review. 2006. Vol. 40, No. 1.

IPCC. 2007. Summary for Policymakers. In: *Climate Change 2007: The Physical Science Basis.* Contribution of Working Group I to the Fourth Assessment Report of the Intergovernmental Panel on Climate Change. Cambridge, UK/New York, Cambridge University Press. http://ipcc-wg1.ucar.edu/wg1/docs/WG1AR4_SPM_PlenaryApproved.pdf

Koziell, I. and Saunders, J. (eds). 2001. *Living off Biodiversity: Exploring Livelihoods and Biodiversity.* London, International Institute for Environment and Development.

Laska, S., Morrow, B. H., Willinger, B. and Mock, N. 2008. Gender and disasters: theoretical considerations. In: B. Willinger (ed.), *Katrina and the Women of New Orleans.* New Orleans, Newcomb College Center for Research on Women, pp. 11–21.

Litt, J. 2008. Getting out or staying put: an African American women's network in evacuation from Katrina. *National Women's Studies Association Journal,* Vol. 20, No. 3, pp. 32–48.

Lo, M. S. 2008. Beyond instrumentalism: interrogating the micro-dynamic and gendered and social impacts of remittances in Senegal. *Gender, Technology and Development,* Vol. 12, No. 3, pp. 413–37.

Massey, D. 1994. *Space, Place, and Gender.* Minneapolis, Minn., University of Minnesota Press.

McKay, D. 2005. Reading remittance landscapes: female migration and agricultural transition in the Philippines. *Geografisk Tidsskrift, Danish Journal of Geography,* Vol. 105, No. 1, pp. 89–99.

Mileti, D. and Passerini, E. 1996. A social explanation of urban relocation after earthquakes. *International Journal of Mass Emergencies and Disasters*, Vol. 14, No. 1, pp. 97–110.

Morrison, A. R., Schiff, M. and Sjöblom, M. (eds). 2007. *The International Migration of Women*. New York, Palgrave Macmillan/World Bank.

Morrow, B. H. and Enarson, E. 1996. Hurricane Andrew through women's eyes: issues and recommendations. *International Journal of Mass Emergencies and Disasters*, Vol. 14, No. 1, pp. 5–22.

Murakami-Ramalho, E. and Durodoye, B. A. 2008. Looking back to move forward: Katrina's black women survivors speak. *National Women's Studies Association Journal*, Vol. 20, No. 3, pp. 115–37.

Neumayer, E. and Plümper, T. 2007. The gendered nature of natural disasters: the impact of catastrophic events on the gender gap in life expectancy, 1981–2002. *Annals of the Association of American Geographers*, Vol. 97, No. 3, pp. 551–66.

Newcomb College Center for Research on Women. 2008. Summary of findings. In: B. Willinger (ed.), *Katrina and the Women of New Orleans*. New Orleans, Newcomb College Center for Research on Women, pp. 5–10.

Oliver-Smith, A. 2006. *Disasters and Forced Migration in the 21st Century. Understanding Katrina: Perspectives from the Social Sciences*. http://understandingkatrina.ssrc.org/Oliver-Smith/ (Accessed September 2009.)

Oxfam International. 2005. *The Tsunami's Impact on Women*. Oxfam Briefing Note, March. www.oxfam.org.uk/what_we_do/issues/conflict_disasters/ (Accessed September 2009.)

— 2009. *The Right to Survive: The Humanitarian Challenge for the Twenty-first Century*. Oxford, UK, Oxfam Publishing.

Parrado, E. A. and Flippen, C. A. 2005. Migration and gender among Mexican women. *American Sociological Review*, Vol. 70, No. 4, pp. 606–32.

Paul, B. K. 2005. Evidence against disaster-induced migration: the 2004 tornado in north-central Bangladesh. *Disasters*, Vol. 29, No. 4, pp. 370–85.

Pereira, T., Shackleton, C. M. and Shackleton, S. 2006. Trade in reed-based craft products in rural villages in the Eastern Cape, South Africa. *Development Southern Africa*, Vol. 23, No. 4, pp. 477–95.

Pincha, C. 2008. *Indian Ocean Tsunami through the Gender Lens: Insights from Tamil Nadu, India*. Mumbai, Oxfam America/NANBAN Trust.

Pinnawala, M. 2008. Engaging in trans-local management of households: aspects of livelihood and gender transformations among Sri Lankan women and migrant workers. *Gender, Technology and Development*, Vol. 12, No. 3, pp. 439–59.

Piper, N. 2004. Gender and migration policies in Southeast and East Asia: legal protection and socio-cultural empowerment of unskilled migrant women. *Singapore Journal of Tropical Geography*, Vol. 25, No. 2, pp. 216–31.

2008. Feminisation of migration and the social dimensions of development: the Asian case. *Third World Quarterly*, Vol. 29, No. 7, pp. 1287–303.

Reuveny, R. 2008. Ecomigration and violent conflict: case studies and public policy implications. *Human Ecology*, Vol. 36, No. 1, pp. 1–13.

Sweetman, C. 2009. Introduction. *Gender and Development*, Vol. 17, No. 1, pp. 1–3.

Terry, G. 2009. No climate justice without gender justice: an overview of the issues. *Gender and Development*, Vol. 17, No. 1, pp. 5–18.

Thieme, S. 2008. Living in transistion: how Kyrgyz women juggle their different roles in a multi-local setting. *Gender, Technology and Development*, Vol. 12, No. 3, pp. 325–45.

Thomalla, F., Downing, T., Spanger-Siegfried, E., Han, G. and Rockström, J. 2006. Reducing hazard vulnerability: towards a common approach between disaster risk reduction and climate adaption. *Disasters*, Vol. 30, No. 1, pp. 39–48.

Thomas, D. S. G., Twyman, C., Osbahr, H. and Hewitson, B. 2007. Adaptation to climate change and variability: farmer responses to intra-seasonal precipitation trends in South Africa. *Climate Change*, Vol. 83, No. 3, pp. 301–22.

UNFPA. 2006. *State of the World Population: A Passage to Hope, Women and International Migration*. New York, United Nations Population Fund.

2008. *State of the World Population: Reaching Common Ground, Culture, Gender and Human Rights*. New York, United Nations Population Fund.

UN/ISDR. 2008. *Gender Perspectives: Integrating Disaster Risk Reduction into Climate Change Adaptation. Good Practices and Lessons Learned*. Geneva, United Nations International Strategy for Disaster Reduction.

van Aalst, M. K. 2006. The impacts of climate change on the risk of natural disasters. *Disasters*, Vol. 30, No. 1, pp. 5–18.

White House Report. 2006. *The Federal Response to Hurricane Katrina: Lessons Learned*. Washington DC.

Willinger, B. and Gerson, J. 2008. Demographic and socioeconomic change in relation to gender and Katrina. In: B. Willinger (ed.), *Katrina and the Women of New Orleans*. New Orleans, Newcomb College Center for Research on Women, pp. 25–31.

Women's Environment and Development Organization (WEDO). 2008. Climate change and displacement: what it means for women. *Forced Migration Review: Climate Change and Displacement*, No. 31, October, p. 56.

Yen, N. T. H., Truong, T. D. and Resurreccion, B. P. 2008. Gender, Class and Nation in a Transnational Community. Practices of Identity among Undocumented Migrant Workers from Vietnam in Bangkok. *Gender Technology and Development*, Vol. 12, No. 3, pp. 365–88.

Zweifler, M. O., Gold, M. A. and Thomas, R. N. 1994. Land use evolution in hill regions of the Dominican Republic. *The Professional Geographer*, Vol. 46, No. 1, pp. 39–53.

13

Drought, desertification and migration: past experiences, predicted impacts and human rights issues

MICHELLE LEIGHTON

Introduction

Climate change is predicted to create hotter, drier climates, more variable rainfall and shorter growing seasons in the twenty-first century. The expected increase in duration and frequency of droughts is likely to lead to more widespread desertification, a loss in soil nutrients and fertility that diminishes agricultural production. These changes could cause developing countries in Africa, Asia and Latin America to lose one-third to one-half of their capacity for agricultural production by mid-century to end of the century (Cline, 2007).

The African continent may suffer most. With all but 4% of Africa's croplands being rain-fed (Heltberg et al., 2008, p. 5), drought has devastating impacts on the continent's rural people who subsist off the land and have little or no access to stored water supply for irrigation. The United Nations Development Programme (UNDP) estimates that up to 90 million ha of drylands in sub-Saharan Africa could experience drought (UNDP, 2009, p. 18; see also IPCC, 2007), and Cline suggests that Senegal and the Sudan could lose just over 50% of their agricultural capacity; Algeria, Ethiopia, Morocco and Mali 30% to 40% of their capacity; and the southern African region between 45% and 50% (Cline, 2007).

While drylands everywhere are routinely subject to moisture deficits, including droughts, and are thus susceptible to desertification processes, the concern today is that the intensity, incidence and severity of drought and desertification are accelerating with global warming. A growing number of experts believe that climate change is already occurring and leading to major socio-economic problems (IPCC, 2007).

The rural poor dependent on agriculture for their subsistence and employment are likely to suffer extreme hardship if there is a substantial loss of crops, livestock and demand for farm labour (UNDP, 2009; World Bank, 2008a; 2008b). Humanitarian agencies have long been concerned with the forced displacement and migration related to recurring droughts, most recently with those displaced among the 19 million people in the Horn of Africa affected, in part, from a drought in its fifth year in the region.[1] Although the relationship between drought and migration remains highly complex and context-specific, there is now a growing understanding that those in rural communities struggling with persistent drought and desertification may engage in migration as a coping strategy.

This chapter presents an overview of the case studies and local experiences that have documented the relationship between drought and migration. This relationship is complex and can be influenced by a number of variables. The chapter discuses the conclusions that recur most often from these studies and highlight the findings that are unique. Given the lack of study in this area, the gaps in scientific data and other constraints in methodological approaches that should be addressed with new research are considered. A number of human rights concerns in the context of drought-related migration are also raised, particularly for those forced to cross international borders. The lack of clear norms could leave many drought victims unprotected and subject to human rights abuse.

The chapter concludes with recommendations for the international community to adopt strategies to address the underlying causes of drought-related migration, to increase research on drought- and desertification-related migration patterns, and to clarify human rights and humanitarian standards to protect the victims of prolonged drought and food insecurity who are forced to migrate in order to survive.

Desertification and drought: past experience and predicted impacts

Desertification has been a major problem for dryland environments for centuries (Herrmann and Hutchinson, 2006). Fundamentally, it may be thought of as land degradation in arid, semi-arid and dry sub-humid

[1] The UN Office for the Coordination of Humanitarian Affairs (OCHA), Reliefweb database, has documented the crises, and its findings and related research can be found at http://www. reliefweb.int/rw/rwb.nsf/db900sid/HHOO-7YZUNM?OpenDocument&rc=1&emid= ACOS-635NZE.

areas, otherwise defined by international agreements as desertification resulting from climate and human activities which reduces soil fertility and the ability of vegetation to thrive (ibid.; see also UN, 1994). Contributing factors of human activity include the over-exploitation of lands and water supplies, deforestation, and agricultural land expansion causing severe soil degradation (Drigo and Marcoux, 1999). Low rainfall or extreme rainfall variability and high temperatures can create drought and exacerbate desertification. 'On the ground, droughts manifest themselves in vegetation stress and ultimately loss of green vegetation cover, decreases in stream flow, and the dying out and cracking of soil surfaces' (Herrmann and Hutchinson, 2006, p. 16). Drier soils are more susceptible to wind and water erosion. These biophysical changes result in lands with less fertile soils, less plant life and less agricultural production. The loss of vegetation means fewer areas in which livestock can graze, and thinner herds more susceptible to disease and less profitable.

Drought and desertification threaten rural household income sources and food security directly by affecting land assets or indirectly by contributing to the decline of agricultural income or employment. Many rural agricultural families are forced to diversify their income streams to survive. They can sometimes accomplish this by having one or more family members migrate (Bilsborrow, 1992; 2009).

Income decline and poverty are key determinants of migration from rural areas (Leighton, 2006; de Janvry et al., 1997), but they are not absolute. Those without any financial means are not likely to migrate. It is those with access to some financial resources who face the loss of livelihood that are *more likely* to migrate than those with either no resources or than those who are relatively well off (ibid., see also Massey et al., 1993; Bilsborrow, 1992). Moreover, a migrant's income or poverty relative to others in the same community can be more important than his or her absolute deprivation in determining migration: the increased likelihood or incidence of poverty can influence whether an individual or family (where migration is a household decision) consider migration as a coping strategy (de Janvry et al., 1997).

Researchers have examined this issue in a number of contexts and regions. While too little research exists to generate precise conclusions about what factors are the most significant in motivating migration in any given country, these studies suggest certain important recurring conclusions. Most have found a direct link or correlation between drought and/or desertification and migration from rural areas, though not all disaggregate drought factors (such as rainfall variability) from other variables such as income level, agricultural policy or local conflict. Certain anomalies among the studies suggest that the drought–migration relationship may be highly contextual.

Another recurring finding was that migration undertaken by those affected by drought and desertification is largely seasonal or temporary migration, rather than permanent, meaning that migrants return to their communities of origin at the end of the agricultural employment season or, at least, on a regular basis. In some cases, they may return to tend to their own household agricultural activity, using the migration income to support their investments.

The pattern of movement most documented is from one rural area to another, although some portion may move to urban destinations and, in a number of cases, over international borders in close proximity to the affected area, for example where seasonal farm labourers cross a nearby border of a neighbouring state for employment. This type of internal or closer-proximity migration relating to drought is better understood than longer-distance (overseas) migration relating to drought. It is generally assumed that international migration requires more planning, extended social networks, and is more costly than local migration (Bilsborrow, 1992; 2002). Several studies suggest that if there are sufficient networks, opportunities and motivation, migrants will undertake a more distant international move as a longer-term or even permanent survival strategy (Skeldon, 2008). Gender may also be an indicator of this type of migration (see e.g. studies of Nepal and Burkina Faso, discussed below). The key studies and their findings are highlighted below, divided between drought-related migration documented within countries or *within borders* and those documenting international migration.

Migration 'within' borders

Seasonal movements or circular migration in response to droughts, leading migrants home at the end of the growing season, appear to be fairly common. This type of migration has been documented in Burkina Faso (Henry et al., 2004), Ethiopia (Ezra, 2001), Mali (Findley, 1994) and Senegal (Seck, 1996) among other studies in sub-Saharan countries. Some researchers report that drought is a factor that has combined with high population growth to increase the incidence of poverty such that droughts in the Sahel that lasted years or decades, such as in 1968–1973 and 1982–1984, led to the use of migration as a systemic coping strategy (including in Burkina Faso, Chad, Djibouti, Ethiopia, Eritrea,

Mali, Mauritania, the Niger, Nigeria, Senegal, Somalia and the Sudan) (see Tamondong-Helin and Helin, 1991; Sahel Club, 1984).

In a study of rainfall and migration in Burkina Faso, Henry et al. found that men from regions with scarce or irregular rainfall are more likely to migrate from one rural area to another on a temporary basis (from two months to two years) to diversify their incomes than are those from wetter regions (Henry et al., 2004, pp. 446–53). Both average rainfall conditions and short-term rainfall scarcity will tend to push men to leave their village for other rural areas (ibid., p. 447). The inclination to migrate from one rural area to another locally is three times higher for men living in drought-afflicted areas (poor agro-climatic areas) than for those living in areas with higher rainfall averages (ibid., p. 446). The length of drought also appeared to play a role: the data suggested that if the drought had occurred in the prior three years, men in the drier areas had a 60% higher chance of migrating (ibid.).

The impetus to migrate before and during drought events is relayed in case studies of Ghana. Data gleaned from recent interviews with migrants suggest that both land failure and environmental degradation in the communities of origin, as well as the promise of better land in the south, prompted people to migrate (van der Geest, 2009; Black et al., 2008).

In Senegal, migration is a consistent coping response to long-term drought and desertification (Seck, 1996). Since the 1960s, more frequent drought and soil erosion have diminished crop yields, leading to a lack of farm employment and to large-scale emigration to Dakar and other urban centres. By the early 1990s, it was reported that 90% of the Tambacounda region's men between the ages of 30 and 60 had migrated at least once in their lifetime (ibid.). Since the 1980s, other communities have engaged in migration as a coping strategy in times of drought (see Knerr, 2004).

In Ethiopia, Ezra considered migration by individuals from poor households in forty ecologically fragile villages susceptible to drought, finding that they had a higher propensity to migrate compared with those from wealthier households in less ecologically vulnerable communities. Household surveys and sample data from selected drought-prone areas in Amhara and Tigray, the regional states with nearly one-third of Ethiopia's total population, demonstrate that rural out-migration is largely a consequence of environmental degradation and poverty that are structural and institutional in origin rather than from individual choice (Ezra, 2001).

Short-term, seasonal migration in response to drought is also documented in other parts of the world: Argentina (Adamo, 2003); India (Rogaly et al., 2002); Kazakhstan (Glazovsky and Shestakov, 1994; Shestakov and Streletsky, 1998); the Niger (Afifi, 2009); Turkey (Zeynep, 2008). Massey et al. (2007) found that in Nepal, for example, the impairment of natural resources led to migration to nearby communities. They reported a much weaker correlation with cross-border international migration, with migrants being members of lower castes and non-Hindus.

Gender may also be important in understanding migration undertaken in times of drought or other disaster in agricultural regions. Research suggests that men account for a greater share of seasonal migrants (Knabe and Nkoyok, 2006; Massey et al., 2007). In Nepal, for example, Massey et al. found that while environmental deterioration leads to short-distance moves within the immediate vicinity, this affected men more than women – more men than women typically migrate locally in a manner consistent with Nepal's gendered division of labour.

Migration across borders

A number of studies also suggest that rural communities affected by drought and desertification may respond by engaging in international migration, particularly where it is easy to cross a neighbouring border. The case studies, while fewer in number, reveal that this type of migration has become more important for rural communities, although overall it remains a much smaller portion of population mobility. Much of the documented international migration has been the movement of people to neighbouring states within the Sahel region, and from Mexico to the United States.

During the 1968–1973 Sahelian drought, an estimated 1 million people migrated from Burkina Faso, at least temporarily, to other countries of the Sahel (Sahel Club, 1984, cited in Tamondong-Helin and Helin, 1991, pp. 1, 86). Although it is believed that the majority of the populations moved within the Sahel region, it is uncertain how many of these migrants may have returned to their communities or settled permanently in other countries. Findley reports that Malians also moved to other areas in the Sahel roughly during this same period in response to drought (Findley, 1994, pp. 539–42).

In Senegal's Tambacounda region, drought-related migration is reported to have begun as a more local or seasonal form of migration, discussed above, but eventually grew and flowed to other African states and to Europe, in particular France where more extensive migrant networks exist (Seck, 1996).

Although many migrants eventually returned home, the absence of men for extended periods made it difficult to rehabilitate degraded lands and increased the economic burden on the remaining women and children. Remittances became critically important to the sending communities for survival, contributing 75% of family incomes in 1993 and helping to finance schools, post offices and social service centres (ibid.).

Not all studies confirm that as drought increases international migration increases, suggesting that drought-related international migration is highly contextual. For example, several studies document *declines* in longer-distance international migration during severe drought in particular communities. Rather, they report that more local and even proximate cross-border migration *increases*. Findley's study, for example, reports that local and Sahelian country migration increased while longer-distance international migration from Mali to France actually declined during the most significant drought years (Findley, 1994). Possibly, the financial capital needed to undertake such travel serves as a barrier after years of drought that reduces farm income. It has been postulated that community members wait for improved economic conditions before migrating overseas (see also Henry et al., 2004, p. 26; Kniveton et al., 2008; van der Geest, 2009), choosing to migrate locally instead.

Studies of desertification affecting communities in Asian countries such as Bangladesh, Kazakhstan, India, Islamic Republic of Iran, Syrian Arab Republic and Uzbekistan were similar to those reported in the African context above; that drought has played a role in both internal and international migration (India: Maloney, 1991; Kazakhstan: Glazovsky and Shestakov, 1994; Shestakov and Streletsky, 1998; Syrian Arab Republic: Escher, 1994; Goria, 1998). The level of influence of a drought in driving migration is uncertain as most studies report a variety of interrelated factors that can lead to migration. For example, Escher (1994) reports that a combination of drought, low agricultural production and population growth led to migration among the Syrian Druze. However, we do not know which of these factors was the *more* influential in driving this migration. Similarly, Shestakov and Streletsky (1998, p. 68) report that in Kazakhstan, migration was driven by several factors (none documented as more exceptional), including lower household income after the pollution of water resources and erosion of the Aral Sea which caused desertification of pastures and farmlands.

In Latin America, case studies in Mexico document a more prominent influence of drought and desertification in contributing to migration flows from rural-to-rural and rural-to-urban areas both inside the country and to the United States. Over 60% of the country's territory is considered drylands, with soil degradation, an indicator of desertification, reportedly affecting approximately 44% of these lands (SEMARNAT, 2008). Earlier studies of the country's desertification problems suggest that nearly two-thirds of the country's agricultural lands may have been affected by some form of soil degradation or other desertification (Leighton, 2006; 1997).[2] Recurrent drought, coupled with poor land-management practices, contributes to soil erosion in agricultural areas and reduces farm income. Many rural families find it necessary to undertake migration in order to cope with diminished incomes. In some cases, people may migrate directly from their rural town to a destination in the United States, particularly where they have a family member or strong social network already established (Leighton, 1997; Munshi, 2003). Some of the states undergoing rapid desertification are also those with rapidly accelerating rates of migration, particularly in Oaxaca, Tamaulipas and the Yucatan (Leighton, 1997).

In another Latin American region, the ability to obtain land for agricultural activity may determine migration. This is illustrated by Bilsborrow's study of migration in the Ecuadorian highlands, where between the late 1960s and early 1990s the population growth and agricultural activities led to the destruction of 70% of the forest stock, increasing soil erosion and diminishing arable land (Bilsborrow, 1992). From 1974 and 1982 there was a net out-migration of 7% of the population from the Sierra region, most of whom went to cities but a portion went to other rural areas of the Amazon. Many of these were the landless poor of the Ecuadorian highlands. Gaps in data made it difficult to determine how significant environmental factors were in decisions to migrate. However, Bilsborrow (1992, p. 21) reports that in a survey of 420 settlers in the Amazon from several communities in the highlands, two-thirds said that they migrated to gain access to and control their own land. He notes that the communities of origin, Loja, Manabi and Bolivar were not the poorest nor most intensely divided. A similar finding is reported by van der Geest (2009) in Ghana.

---

[2] In 1997, scientific studies and government agencies reported that approximately 70% of Mexico's agricultural lands were affected by some level of soil erosion and 1,000 square miles (260,000 ha) of potentially productive lands were taken out of production and 400 square miles abandoned each year (Leighton, 1997). The author undertook studies on desertification in Mexico with scientists from SEMARNAT and academic institutions from 1997 to 2000 as part of a Global Environment Facility funded project in three countries, and as a collaboration with the Natural Heritage Institute. The GEF reports are on file with the author.

Although few studies of international female migrants affected by drought have been undertaken, some observations are worth noting. In Burkina Faso, for example, Henry et al. found that women living in drier rural villages were more likely to migrate than those living in regions without drought and that stronger correlations existed between drought and migration to urban or international destinations than to other rural areas in the country (Henry et al., 2004, p. 446). This finding differs for males, as discussed above.

On a global level, women are increasingly engaging in international migration. In 2010, women are due to account for 49% of the expected 214 million international migrants. The growing interest of women in international migration was highlighted by Gila et al. in a study of Ecuador in 2008. They found that while men in Ecuador tend to migrate seasonally or temporarily within the country, women were more likely to undertake international migration, usually to Spain or Italy. While not directly linked to particular drought events, women were motivated to engage in international migration to urban areas because they believed that there would be more opportunity and equality in European Union countries than in Ecuador. Future studies of drought-related migration would benefit from deeper analysis of the gender differences in using domestic or international migration as a coping mechanism.

In sum, available research indicates that drought and desertification can influence migration when it affects livelihoods and income. The studies document better the relationship of droughts with rural-to-rural and rural-to-urban migration domestically, and in some cases with local, cross-border migration, as particularly demonstrated in countries of the Sahel where borders are more porous. Too little research exists to generate precise conclusions about what factors among rainfall, income level, or social networks in receiving communities, play the most significant role in motivating migration in a given country, or the extent to which agricultural or immigration policies play a role in internal migration in response to drought. The studies have generally been limited to findings in a particular community or group of communities and in a particular context. They have not been *scaled up* in most cases so as to apply on a national or regional scale. The studies of drought and migration in the Sahel region may come closest.

Other studies, much fewer in number, report the influence of drought on longer-distance or overseas migration. This relationship appears to be weaker but may in fact suffer more from a dearth of studies than from empirical findings as such. A few anomalies exist, too, suggesting that the severity of drought may not always instigate longer-distance migration. In these contexts, financing longer-distance migration may be less feasible after years of low agricultural returns so that more temporary or seasonal migration becomes a more feasible option. The role of 'climate' in overseas migration is thus quite context-specific. Finally, networks and financial means play a key role in all forms of migration but may be more important in determining international or overseas migration. Deeper research, as discussed below, is needed to close the gaps in present data and to better understand how these variables interact.

Gaps in data and methodology

Given that migration correlated to drought and desertification may be context- or site-specific, the dearth of case studies in this area leaves large gaps in the data available to analyse. Researchers have begun to evaluate traditional methodologies and suggest new approaches for better under-standing the environment–migration nexus (see e.g. Kniveton et al., 2009; Bilsborrow, 2009). These methods could improve efforts to document relationships between drought-related variables and socio-economic vari-ables (migration, poverty, etc.) and may help to determine more consistently in which contexts drought and desertification can stimulate migration.

The data deficits make it particularly difficult for scientists to build a more accurate picture of the potential for future drought-related migration, although some researchers have begun to do so by extrapolating existing data and using qualitative anlaysis. Black, Kniveton, Skeldon, Coppard, Murata and Schmidt-Verkerk, for example, used existing information rather than new fieldwork in considering the future of migration relating to climate change in Bangladesh, Ghana, Ethiopia and the Sudan.

A critical challenge in research on drought-related migration lies in capturing the decision-making process of migrants. Some household surveys are unable to do this as migrants or their family members, when interviewed, may not report all the factors influencing or leading to their decision to migrate, thereby potentially skewing the conclusions (Bilsborrow, 1992).[3] More generalized migration studies have found that

---

[3] For example, respondents in drought-prone areas may report that migration was due to work opportunities elsewhere, a lack of water, or the fact that 'the lands no longer give' (a factor the author frequently encountered in Mexico). These responses often belie a more complex interplay between biophysical phenomena, human activities and govern-ment policy: e.g. over-cultivated land, use of pesticides and lack of fallowing may combine with low rainfall, leading to soil infertility; government policy may also play a role by encouraging large agribusinesses to consolidate land, pushing smallholders out of the markets; and they may not have had access to farm credit, insurance or other benefits to reduce the risk of drought-induced losses.

how and why agricultural families engage in migration is often the result of a complex set of variables, including the viability of farm employment, land and water scarcity and stress, economic policies affecting the intensification of land use (that can lead to desertification or overdrafting of aquifers), family and community networks between sending and destination areas, and availability of financial resources to invest in migration. Further research and fieldwork that can disaggregate the complex variables influencing migration would be useful to better document how and when drought and desertification become a primary driver of migration in agricultural regions, and on the impacts of that migration on both the sending and receiving areas.

Moreover, relatively little is understood about the environmental 'tipping points' that can lead to migration decisions. Case studies suggest, for example, that while prolonged drought can instigate international migration in some cases, in others it influences shorter-distance, seasonal migration. These anomalies can exist among communities in the same country (see e.g. Munshi, 2003; Kniveton et al., 2008). In-depth and comparative case studies could help to better explain the drought–migration relationship as it affects different communities and cultures.

The complex interactions between environmental, agricultural, social and economic factors require analysis of data that transcend a single discipline or approach, and ideally that can capture both local and national patterns of migration (Heltberg et al., 2008). Even with such research, existing migration theory and methodologies pose other constraints in determining just how much of an influence environmental factors such as drought or desertification have or will have in population movements responding to climate change (see e.g. Kniveton et al., 2008).

The methods for analysing drought-related migration are constrained by a number of other factors. First, there is limited environmental time-series data and socio-economic data related to migration in developing countries. Environmental *time-series* data must go beyond simply considering rainfall patterns to consider land-use change and biodiversity loss comparable over time as these factors are also relevant to explaining the impacts of drought. Yet, this information is often not collected at community level, or at scales comparable to municipal-level demographic and socio-economic data collected at the same scale. In some of the earlier studies on drought- and desertification-related migration, researchers had to rely on incomplete data sets collected at earlier points in time and for different purposes, presenting limitations on the correlative relationships observed. These studies, although limited, provide insight into the use of demographic and bio-physical time-series data and GIS-based modelling that can be useful for future research in the field (see e.g. Leighton, 2002; de Janvry et al., 1997). They also reinforce the value of new fieldwork to collect household-level data with the goal of understanding the drought–desertification–migration nexus more concretely (see, e.g. a new study of Ecuador attempting to integrate various incomplete biophysical and demographic data sets, Bilsborrow, 2009).

Second, the available socio-economic data may not include information on migration in affected communities. Census data, for example, do not usually include information on the decision-making processes of migrants or their flows into more than one destination. Further, data that are available may not have been collected consistently in the communities sought to be studied. Local jurisdictions may have historically differed in their data collection processes over the last century. Information is often collected by different ministries at different points in time and at different scales (e.g. municipal versus statewide levels), depending on the goals for their end use, making it quite challenging to harmonize.

Third, once climate and socio-economic data are obtained, little consensus exists among scientists as to the most appropriate theories and methodologies to use in explaining the drought–migration relationship. Researchers can differ in their selection of the indicators they believe best represent this relationship. Among the biophysical variables the level of rainfall, vegetation loss, deforestation and or land-use change are all relevant but could yield different findings depending on the combination of variables chosen for analysis. Among socio-economic indicators, demographic data, levels of employment, agricultural policies, land development, potable water distribution, education levels, access to social services, and community infrastructure are among a myriad of possible variables that, depending on how they are combined, can also yield different conclusions.

Kniveton et al. suggest that while the use of accepted migration study methods, such as a *sustainable livelihoods* approach (considering the various strategies that people use to maintain their livelihoods from a holistic vantage) and the *new economics of labour migration* approach (considering that migration is used as a strategy to diversify rather than maximize individual income) could be appropriate for climate-related migration research, these methods may be too blunt to generate precise conclusions on the role of climate change (e.g. drought) in population movements (Kniveton et al., 2008, pp. 38–40). Rather, life history data-gathering techniques may be more appropriate, especially where time-series data are scarce, as well as other behavioural models (climate scenario-driven impact assessment, adaptation- and vulnerability-based approaches, and other integrated assessment techniques) (Kniveton et al., 2009). Migration histories were used, for example, in Ezra's study of drought-related migration in Ethiopia (Ezra, 2001) and in a new study begun in Ecuador (Bilsborrow, 2009).

In addition, better tools are needed to quantify the level of migrants from climate-affected areas. Estimating the number of migrants forced to move because of drought and desertification, or other extreme weather events relating to climate change, has been the subject of some controversy for the past decade (see e.g. Kniveton et al., 2008, pp. 29–30). Kniveton et al. argue that using various forms of regression analysis and agent-based modelling techniques can be a more precise means for measuring and understanding the level of population movement that might occur from drought or variable rainfall, conceding that these tools are still imperfect (Kniveton et al., 2008, pp. 41–53).

Hence, while a number of methodological hurdles exist in mapping drought-related migration, these hurdles are not insurmountable. If data can be collected and made available, new methods and models are evolving to help researchers integrate and correlate biophysical and socio-economic information which could be tested in environmentally induced migration contexts (see e.g. Kniveton et al., 2008; Bilsborrow, 2009; Leighton, 2002).[4]

Future studies and predicted impacts

Given the existing data gaps and methodological barriers, few researchers have considered the potential impacts of future droughts on migration flows. The increasing awareness of the impacts of climate change, however, has renewed interest in predicting the human migration relating to drought and desertification. The Environment and Human Security section of the United Nations University, for example, is working with researchers globally to consider how best to map existing trends and future scenarios, as reported by Koko Warner et al. (Chapter 8).

---

[4] The Monitor Systems Software model was developed to identify and monitor land degradation changes, biodiversity loss, and community socio-economic impacts. See e.g. Leighton (2002) regarding the findings and software developed by the multidisciplinary team assembled under the GEF Project No. GF/1040–00–10. See also Monitor Systems Software housed at the University of Chile, Santiago, with Prof. Fernando Santibanez.

In undertaking future study in this area, Black et al. (2008) and Hugo (1995) would caution that it is necessary to begin by varying the lens through which researchers have viewed these problems in the past, that is, migration should be viewed as a continuum from the most voluntary of movements to those forced by circumstance and climate change (Hugo, 1995). Thus, rather than consider first how the climate will change and second, how much migration may flow from those changes, Black, Kniveton, Skeldon, Coppard, Murata and Schmidt-Verkerk began their evaluation of four cases in Bangladesh, Ghana, Ethiopia and the Sudan by examining existing drivers of migration flows and then layering climate change predictions over those findings to determine how and to what extent the predicted changes might affect existing migration patterns as a means of considering future scenarios. (Black et al., 2008).

Their analysis led them to a series of predictions about the impact of climate change on drought, desertification and migration in the Sudan and Ghana. They concluded that in the Sudan, increased climate variability and drought is likely to intensify ongoing desertification of arable areas and to stimulate the movement of pastoral activity towards the south of the country. This would occur with the expected decline in production of millet and sorghum, the main staple crop of poor rain-fed farm systems. As employment opportunities declined, there would be an increase in rural out-migration from more marginal areas, 'exacerbating urbanization and long-term, more distant patterns of migrations for those who are able' (Black et al., 2008).

In Ghana, their analysis concluded that drought could affect commercial agriculture of cocoa in the south. Cocoa is very sensitive to drought. If the southern areas became drier this would probably reduce cocoa production and farm income. It could also alter the geographical distribution of cocoa pests and pathogens. The impairment of cocoa production due to drought impacts could result in migration from these areas (Black et al., 2008).

Beyond these areas, a number of countries are expected to be severely afflicted by droughts due to climate change in the next four to seven decades. The countries in which longer droughts are expected and in which migration from dryland areas is already increasing may be potential hot spots worthy of future, more intensive research and study. The cases of Ghana and the Sudan suggest that the focus of research on internal migration could be very fruitful in predicting impacts. However, the drought-prone countries due to experience higher rates of international migration are also key potential hot spots worthy of study.

For example, at least a dozen countries are predicted to suffer significant agricultural and food production declines due to prolonged droughts, e.g. between 30% and 60% declines, *and* are at the same time expected to have significantly higher *international* migration rates, irrespective of global warming (40%–150% increase over current levels) in the next three to seven decades. These countries include Ecuador, Equatorial Africa, Ethiopia, India, Iraq, Mali, Mexico, Morocco, Pakistan, Peru, Senegal, the Sudan, Syrian Arab Republic, Venezuela, Zambia and Zimbabwe.[5]

There is little question that as drought and desertification worsen over time, deeper community research would better equip decision-makers with knowledge to improve community adaptation strategies, and to identify communities that may need critical attention. The human rights issues highlighted below and discussed more thoroughly by other contributors to this volume suggest that future policy-relevant scientific research on migration related to drought and desertification is warranted.

Human rights implications of forced migration

The international community has become increasingly concerned about the human rights implications of people who may be displaced by drought and desertification as a result of climate change. A key issue is the significant normative gaps that leave these migrants unprotected if they move voluntarily or are forced to cross an international border in order to cope with drought disasters (IASC, 2008).

Victims of disaster who are forced to move within their country can avail themselves of human rights protection from their government, which has the primary responsibility to protect them in times of disaster. They may seek assistance in relocating, housing or employment from their government. Human rights principles governing internally displaced persons (IDPs), known as the *Guiding Principles on Internal Displacement*, have clarified the responsibility of governments to assist victims of natural disaster, including to protect their basic human needs, their free movement and to prevent discrimination (Deng, 1998; Kälin, 2009).

---

[5] These countries come from the author's analysis of agricultural models on production declines evaluated by Cline (2007), and from international migration projections by country from UNDESA (2004; 2006). A chart of these projections is on file with the author.

Governments unable to help victims can request assistance from the international community, which also has a responsibility to assist those unable to provide adequate protection (ibid.; UN, 2005). However, the *Guiding Principles* assume that displacement is forced, making it unclear as to whether governments have any responsibility to assist persons moving voluntarily because of drought or desertification events (Zetter, 2009).

African countries recently enshrined the *Guiding Principles* in the African Union Convention for the Protection and Assistance of Internally Displaced Persons in Africa ('Kampala Convention'), concluded in November 2009. The Kampala Convention further clarifies that climate change may cause internal displacement. It reaffirms government obligations regarding disaster victims and obligates other African countries to provide support when a state affected by disaster is unable to provide full assistance. As it has yet to come into force, it is uncertain whether the treaty is meant to protect all drought-related migrants. Further, it does not address the need for countries to accept drought disaster victims who are forced to cross an African border to find employment or other means of survival. There is no requirement to allow victims to take refuge in a country other than their own and the treaty does not apply beyond the African region. Thus, it leaves international migrants in the region vulnerable and unprotected.

In general, all persons, regardless of their status when entering another country, carry with them fundamental human rights guarantees (OHCHR, 2009, p. 20). These include civil, political, economic, social and cultural rights such as the right of freedom of movement, to choose their place of residence, to engage in religion or cultural practice, the right to life, privacy and health, the right to seek employment, and the right to be free from discrimination. However, these rights do not include the right to enter another country, to remain there, or to receive permanent legal status as may be accorded to refugees. This is largely because the definition of a 'refugee' in the main international treaty establishing refugee standards, the 1951 Convention relating to the Status of Refugees, was meant to include victims of conflict or widespread human rights abuse and not persons affected by environmental change, such as drought or desertification (UNHCR, 2009, p. 8).

Other refugee protections that have evolved over the years to encompass more victims of natural disaster also appear quite limited. For example, the 1969 Organization of African Unity (OAU) Convention on Specific Aspects of Refugee Problems in Africa provides protection to those fleeing 'events seriously disturbing public order in either part or the whole of his country of origin or nationality'. A similar provision is contained in the Latin American region under the Cartagena Declaration on Refugees (Section III(3), adopted 1984). These definitions presumably include climate disasters causing serious disruptions but it is questionable whether most droughts would qualify as disrupting national order. Where this occurs, receiving countries could grant temporary asylum to victims of potential harm. Temporary protected status (TPS), such as that allowed by legislation in Denmark, Finland, Sweden and the United States, could also be invoked to protect those already within these states if their countries could not protect them upon return. This special treatment however is highly discretionary and provides no long-term relief.[6]

Outwith refugee protection, each country determines on what basis to allow non-citizens to enter and to work in its territory. Drought-related international migrants are generally considered economic migrants. As such, they can be denied entry or, if they are in another country as an undocumented worker, can be arrested, detained, prosecuted criminally, and/or returned to their country of origin. They may also be denied other international assistance. An exception might be in the application of the humanitarian principle of *non-refoulement*, which would prevent a government's return of a person to their country regardless of legal status where the person's life or integrity is at risk, or where return would subject the person to the risk of cruel, unusual or degrading treatment (see e.g. UNHCR, 2009, p. 10). This principle could more easily be applied to persons forced across a border due to floods and hurricanes than in circumstances of prolonged drought. Persisting over time, droughts can impair livelihoods and food production capacity but may not pose an immediate threat of death. Drought victims would have a difficult burden to demonstrate an immediate threat to their life if forced to return home.

The complex nature of immigration regulations enforced at national borders means that ensuring the adequate protection of drought disaster migrants will require the international community to clarify standards and/or receiving states to adopt new policies. Without these, many would be left unprotected if both their home country and other countries fail to provide for basic needs such as food, shelter or employment, and they would have no status to appeal for

---

[6] See discussion of country TPS provisions in UNHCR (2009).

international assistance. As noted by the UN High Commissioner for
Refugees:

> [t]here are also cases in which displacement relates to a certain unwill-
> ingness to protect, or to prohibit discrimination. A normative gap could
> thus be considered to exist if both the country of origin and the host
> country obstruct or deny or are unable to ensure basic human rights
> (UNHCR, 2009, p. 10).

The concern is underscored by the fact that developed countries which
receive large numbers of immigrants annually have moved towards
policies of exclusion rather than entry over the past two decades
(Martin and Zucheher, 2008). Voluntary international assistance to
address the human impacts of drought and desertification has also
been provided only on an ad hoc basis. It has typically not come close
to the level of aid provided to governments affected by floods, landslides
or hurricanes.

Yet it is becoming better understood that the long-term investment in
efforts to help drought-affected communities either to adapt to climate
change or facilitate appropriate movements could ensure greater human
security and enhance the protection of human rights to life, food, health,
housing and culture for some of the world's most vulnerable groups
(OHCHR, 2009). The use of human rights standards as a framework for
protecting drought-related migrants could encourage the development
of more appropriate migration management strategies among sending
and receiving countries.

The need for governments to consider legal and normative standards in protecting victims of drought and desertification raises additional political issues, some of which are discussed in detail in other
chapters. For example, if climate change does lead to more frequent
and prolonged droughts, and significant agricultural capacity is lost
in developing countries, where will affected communities move to
survive if borders remain closed? Should human rights standards be
clarified to ensure protection of drought-related migrants, and what
standards are most appropriate to assist drought victims in migrating either internally or across borders? How will governments distinguish the protection of drought-related migrants from other
development-related or economic-type migrants who lack employment, food or clean water? The clarification of these issues would
help to ensure more appropriate policies on migration and human
security.

Conclusion

Existing research suggests that droughts and desertification as they affect livelihoods can contribute to poverty and play an important role in stimulating migration. Other variables, such as the availability of financial resources and social networks between sending and receiving communities, can facilitate migration from areas affected by drought and desertification. Generalizations in this area should be made with caution because the case studies are largely context-specific. Nonetheless, they can serve as a baseline for further research to better understand the linkages between migration decision-making processes and their relation to drought and land degradation.

In this regard, the cases studies documenting drought- and desertification-related migration suggest a number of recurrent conclusions with respect to migration. First, drought-related migration is most often internal, to other areas of the country, or across an international border if in close proximity to the drought-affected area. Longer-distance international migration related to drought is less documented but appears to be of growing significance.

Second, migration may be one of several adaptation strategies employed by households in response to drought and desertification. The slow-onset nature of these environmental problems may provide households with an opportunity to consider their options and thus migration may become a more voluntary undertaking. However, in some instances it may be the only viable solution for a household's survival. Finally, seasonal migration (returning to the rural area within two years) versus more permanent migration tends to be the predominant coping strategy among those affected by drought and desertification.

It is important to consider that notwithstanding the relevant foundations of existing case studies, whether and to what extent the increased frequency and duration of droughts that could result from climate change scenarios will alter these migration patterns remains highly uncertain. For example, if agricultural employment declines over an entire region, migrants may be forced to move to more distant destinations to find employment. The unsustainability of agricultural lands and increasing drought in regions around the world may already be contributing to a growing trend in this regard. Both regular and irregular migration is increasing to the western and northern African coasts, and from there into European jurisdictions (van Moppes, 2006).

Higher levels of international migration due to drought and desertification pose challenges for developing and developed countries alike. International migration from developing countries is now equally divided between movement to other developing countries and to developed countries. Developing countries could encounter more pressure to handle higher levels of both immigration and emigration flows if rural people affected by more frequent and severe droughts are forced to migrate longer distances or overseas to find adequate employment. The longer the duration of droughts, the greater the challenges for sending and receiving communities, particularly where local infrastructure, employment and social services are already stretched. Understanding how and where prolonged or *mega* droughts could *tip the balance* towards additional migration internally and internationally will be critical to addressing the human security impacts of climate change.

Recommendations for future research and policy reform

Today, wide gaps in the research on these issues exist in:

- content (how and when drought and desertification become a primary driver of migration, and which communities are most vulnerable);
- scale and methodology (breadth of studies, and methods for interdisciplinary analysis); and
- frameworks for appropriate migration management strategies of drought-related migrants.

The following recommendations could help to close these gaps and facilitate more policy-based scientific research.

As a relatively new topic among social scientists, little research capital has been invested in broad-scale drought-migration studies. Many of the studies undertaken have been community-level studies and largely, although not always, anecdotal in nature. While these help to clarify local socio-economic behaviours, their conclusions cannot readily be *scaled up* to assist policy at national level or readily used to create systems for *early warning*. The lack of statistically relevant data at national or regional levels constrains the design of policies that could build resilience and promote adaptation among vulnerable communities. Further research and fieldwork is needed on how and when drought and desertification become a *primary driver* of migration, and on the impacts of that migration for both sending and receiving areas. Investing in the development of both short- and long-term research, data collection on drought-affected communities and use of migration as a coping strategy could help to close these gaps.

In facilitating such studies, research and documentation on the most appropriate methodologies for drought-migration studies are also needed, including a better understanding of the theoretical models (as mentioned in this chapter), and greater emphasis on interagency and interdisciplinary data collection and data sharing. For example, environmental, water, agricultural and social welfare ministries have data important to understanding drought-related migration but often do not collect, share or analyse their information in a way that is meaningful to researchers in this field. Yet, as these ministries may all touch communities living in drought-affected regions, they could strengthen their cooperation to better observe and analyse migration patterns. They could open access to their data to other researchers from academic and non-governmental institutions (see e.g. Kniveton et al., 2009). Moreover, they could work together to support the design and integration of drought-related data in the national census process. If appropriately managed, census covering questions relating to drought and migration could yield data more easily comparable nationally and across jurisdictions, particularly as most countries undertake a national census.

A further critical need is a cogent analysis of options for countries to *manage* drought disaster migration. Anticipating that migration will continue to pose serious challenges, countries have already begun to discuss migration as an adaptation planning strategy within their National Adaptation Programmes for Action (NAPAs). As yet, however, there is little analysis on what standards, policies or programmes are most appropriate for managing this category of internal or international migration, particularly among receiving (or potential destination) countries. Research that undertakes a robust exploration of the potential frameworks on the ways and means for better managing drought-related migration flows *within* countries, and between sending and receiving countries, is critically needed. While global-level evaluation of best practices is clearly warranted, research on a regional, rather than global, scale may be of more immediate value to affected countries.

Finally, the forced movement of people due to prolonged drought and desertification poses serious human rights and humanitarian concerns, particularly for those forced to move across borders. The need for greater clarity of law and policy issues in this area, as well as best practice, is becoming more critical as climate change predictions worsen. Key issues that should be resolved include:

- How can drought-related migrants moving voluntarily within countries receive adequate human rights protection from governments in whose territories they move?
- How and in what way can norms be adopted to protect drought-related migrants crossing international borders and how would their status differ from other economic migrants?
- How can migration policies more constructively and appropriately treat some migration as an adaptation strategy within government responses to climate change?

Moreover, principles of international disaster cooperation should be further clarified or solidified into norms that can both facilitate government cooperation and provide greater accountability within the international community.

References

Adamo, S. 2003. Vulnerable people in fragile lands: migration and desertification in the drylands of Argentina – the case of the department of Jáchal. Ph.D. thesis. University of Texas, Austin.

Afifi, T. 2009. *Niger Case Study Report.* Bonn, United Nations University, Institute for Environment and Human Security. (UNU-EHS, EACH-FOR Project.)

Arnold, M., Chen, R. S., Deichmann, U., Dilley, M. and Lerner-Lam, A. L. (eds). 2006. *Natural Disaster Hot Spots: Case Studies.* Washington DC, World Bank. (Disaster Risk Management Series.)

Bilsborrow, R. E. 1992. *Rural Poverty, Migration, and Environment in Developing Countries: Three Case Studies.* Washington DC, World Bank. (Country Economics Department Paper 1017.)

2002. Migration, population change and the rural environment. *ECSP Report,* Vol. 8, Summer, pp. 69–94.

2009. Collecting data on the migration–environment nexus. In: F. Laczko and C. Aghazarm (eds), *Migration, Environment and Climate Change: Assessing the Evidence.* Geneva, International Organization for Migration.

Black, R., Kniveton, D. R., Skeldon, R., Coppard, D., Murata, A., and Schmidt-Verkerk, K. 2008. *Demographics and Climate Change: Future Trends and their Policy Implications for Migration.* Report prepared for the UK Department for International Development by the Sussex Centre for Migration Research.

Black, R., Natali, C. and Skinner, J. 2006. *Migration and Inequality.* Background paper prepared for *World Development Report 2006: Equity and Development.* http://siteresources.worldbank.org/INTWDR2006/Resources/477383-1118673432908/Migration_and_Inequality.pdf

Brown, O. 2008. *Migration and Climate Change*. Geneva, International Organization for Migration. (Migration Research Series No. 31.) http://publications.iom.int/bookstore/index.php?main_page=product_info&products_id=96

Castles, S. 2002. *Environmental Change and Forced Migration: Making Sense of the Debate*. Geneva, United Nations High Commissioner for Refugees. (*New Issues in Refugee Research*, No. 70.)

Clark, P. U. and Weaver, A. J. (co-lead authors). 2008. *Abrupt Climate Change: A Report by the US Climate Change Science Program & the Subcommittee on Global Change Research*. With E. Brook, E. R. Cook, T. L. Delworth and K. Steffen (chapter lead authors). Reston, Va., US Geological Survey.

Cline, W. 2007. *Global Warming and Agriculture: Impact Estimates by Country*. Washington DC, Peter G. Peterson Institute for International Economics.

Coleman, D. and Rowthorn, R. 2004. The economic effects of immigration into the United Kingdom. *Population and Development Review*, Vol. 30, No. 4, pp. 579–624.

de Haas, H. 1998. Socio-economic transformations and oasis agriculture in southern Morocco. In: L. de Haan and P. Blaikie (eds), *Looking at Maps in the Dark: Directions for Geographical Research in Land Management and Sustainable Development in Rural and Urban Environments of the Third World*. Utrecht/Amsterdam, Royal Dutch Geographical Society/Faculty of Environmental Sciences, University of Amsterdam.

2003. Migration and development in Southern Morocco: the disparate socio-economic impacts of out-migration on the Todgha Oasis Valley. Ph.D. thesis. University of Amsterdam.

2006. *Trans-Saharan Migration to North Africa and the EU: Historical Roots and Current Trends*. Oxford, UK, Migration Policy Institute. www.migrationinformation.org/Feature/display.cfm?id=484

de Janvry, A., Sadoulet, E., Davis, B., Seidel, A. and Winters, P. 1997. *Determinants of Mexico-U.S. Migration: The Role of Household Assets and Environmental Factors*. University of California at Berkley. (CUDARE Working Paper No. 853.)

Deng, F. 1998. *Guiding Principles on Internal Displacement*, 11 February. (E/CN.4/1998/53/add.2.)

Dia, I. 1992. Les migrations comme stratégie des unités de production rurale: Une étude de cas au Sénégal. In: A. Blokland and F. van der Staay (eds), *Sustainable Development in Semi-Arid Sub-Saharan Africa*. The Hague, Netherlands, Ministry of Foreign Affairs.

Drigo, R. and Marcoux, A. 1999. Population dynamics and the assessment of land use changes and deforestation. Rome, Food and Agriculture Organization. www.fao.org/sd/wpdirect/wpan0030.htm

El-Hinnawi, E. 1985. *Environmental Refugees*. Nairobi, United Nations Environment Programme.

Escher, A. 1994. *Migrant Network: An Answer to Contain Desertification. A Case Study of Southern Syria (Gabal al-Arab).* Paper presented at International Symposium on Desertification and Migrations, 9–11 February, Almeria, Spain.

Ezra, M. 2001. *Ecological Degradation, Rural Poverty, and Migration in Ethiopia: A Contextual Analysis.* New York, Policy Research Division Population Council. (Working Paper No. 149.) http://www.popcouncil.org/pdfs/wp/149.pdf

Findley, S. 1994. Does drought increase migration? A study of migration from Rural Mali during the 1983–1985 drought. *International Migration Review*, Vol. 28, No. 3, pp. 539–53.

Glazovsky, N. and Shestakov, A. 1994. *Environmental Migration Caused by Desertification in Central Asia and Russia.* Paper presented at International Symposium on Desertification and Migrations, 9–11 February, Almeria, Spain. www.geographytsu.freehomepage.com/CIS%20env%20migration.htm

Gila, O., et al., 2010. Migration and environment in Los Ríos, Ecuador (1997–2008), *Journal of Identity and Migration Studies*, Vol. 4, No. 1, 137–53.

Goria, A. 1998. *Desertification and Migration in the Mediterranean: An Analytical Framework.* Milan, Italy, Fondazione Eni Enrico Mattei.

Grinsted, A., Moore, J. C. and Jevrejeva, S. 2009. Reconstructing sea level from paleo and projected temperatures 200 to 2100 AD. *Climate Dynamics*, Vol. 34, No. 4, pp. 461–72.

Haug, S. 2008. Migration networks and migration decision making. *Journal of Ethnic and Migration Studies*, Vol. 34, No. 4, pp. 585–605.

Heltberg, R., Jorgensen, S. and Siegel, P. B. 2008. *Climate Change: Challenges for Social Protection in Africa.* Washington DC, World Bank. http://papers.ssrn.com/sol3/papers.cfm?abstract_id=1174774

Henry, S., Schoumaker, B. and Beauchemin, C. 2004. The impact of rainfall on the first out-migration: a multi-level event-history analysis in Burkino Faso. *Population and Environment*, Vol. 5, No. 5, pp. 423–60.

Herrmann, S. and Hutchinson, C. 2006. Links between land degradation, drought, and desertification. In: P. M. Johnson, K. Mayrand and M. Paquin (eds), *Governing Global Desertification.* London, Ashgate Press.

Hugo, G. 1995. Environmental concerns and international migration. *International Migration Review*, Vol. 30, No. 1, pp. 105–42.

IASC. 2008. *Climate Change, Migration and Displacement: Who Will Be Affected?* Working Paper submitted to UNFCCC Secretariat by the informal group on Migration/Displacement and Climate Change of the Inter-Agency Standing Committee, 31 October.

IPCC. 2007. *Climate Change 2007: Climate Change Impacts, Adaptation, Vulnerability.* Working Group II Contribution to the Intergovernmental Panel on Climate Change Fourth Assessment Report Summary for Policymakers, IPCC WGII Fourth Assessment Report. Geneva, Intergovernmental Panel on Climate Change.

Jäger, J., Frühmann, J. and Grünberger, S. 2009. *Environmental Change and Forced Migration Scenarios: Synthesis of Results.* Synthesis Report for the European Commission. http://www.each-for.eu/

Kälin, W. 2009. *Report of the Representative of the Secretary-General on the Human Rights of Internally Displaced Persons*, Walter Kälin, Addendum: Protection of Internally Displaced Persons in Situations of Natural Disasters. A/HRC/10/13/Add.1 (2009), p. 6.

Knabe, F. and Nkoyok, J. 2006. Overcoming barriers: promoting women's local knowledge. *KM4D Journal*, Vol. 2, No. 1, pp. 8–23.

Knerr, B. 2004. Desertification and human migration. In: D. Werner (ed.), *Biological Resources and Migration.* Berlin, Springer, pp. 317–38.

Kniveton, D., Schmidt-Verkerk, K., Smith, C. and Black, R. 2008. *Climate Change and Migration: Improving Methodologies to Estimate Flows.* Geneva, International Organization for Migration. (Migration Research Series No. 33.) www.iom.int/ jahia/webdav/site/myjahiasite/shared/shared/mainsite/published_docs/serial_ publications/MRS-33.pdf

——— 2009. Challenges and approaches to measuring the migration–environment nexus. In: F. Laczko and C. Aghazarm (eds), *Migration, Environment and Climate Change: Assessing the Evidence.* Geneva, International Organization for Migration.

Laczko, F. and Collett, E. 2005. *Assessing the Tsunami's Effects on Migration.* Oxford, UK, Migration Policy Institute.

Leighton, M. 1997. *Environmental Degradation and Migration: The U.S.-Mexico Case Study.* Washington DC, US Congressional Commission on Immigration Reform.

——— 1998. *The U.S.-Mexico Case Study, Environmental Change and Security Project Report.* Washington DC, Woodrow Wilson Center.

——— 2002. *An Indicator Model for Dryland Ecosystems in Latin America.* Final Terminal Report. Washington DC, Global Environment Facility. (GEF Project No. GF/1040–00–10.)

——— 2006. Desertification and migration. In: P. M. Johnson, K. Mayrand and M. Paquin (eds), *Governing Global Desertification.* London, Ashgate, pp. 43–58.

——— 2009. Migration and slow-onset disasters: desertification and drought. In: *Migration, Environment and Climate Change: Assessing the Evidence.* Geneva, International Organization for Migration.

Maloney, C. 1991. *Environmental Displacement of Population in India.* San Francisco, Calif., UFSI Inc./National Heritage Institute. (Field Staff Reports No. 14.)

Martin, P. and Zucheher, G. 2008. Managing migration, the global challenge. *Population Bulletin*, No. 3, pp. 1–3.

Massey, D. S., Arango, J., Hugo, G., Kouaouci, A., Pellegrino, A. and Taylor, J. E. 1993. Theories of international migration: a review and appraisal. *Population and Development Review*, Vol. 19, No. 3, pp. 431–66. www.jstor.org/stable/2938462

Massey, D. S., Axinn, W. and Ghimire, D. 2007. *Environmental Change and Out-migration: Evidence from Nepal.* Ann Arbor, Mich., University of Michigan. (Population Studies Center Research Report No. 07–615.)

Massey, D. S. and Singer, A. 1998. The social processes of undocumented border crossing among Mexican migrants. *International Migration Review,* Vol. 32, No. 3, p. 561.

Munshi, K. 2003. Networks in the modern economy: Mexican migrants in the U.S. labour market. *Quarterly Journal of Economics,* Vol. 118, No. 2, pp. 549–99.

OHCHR. 2009. *Report of the Office of the High Commissioner for Human Rights on the Relationship between Climate Change and Human Rights.* Geneva, United Nations High Commissioner for Human Rights. (A/HRC/10/61, 15 January.)

Papademetriou, D. 2005. *The Global Struggle with Illegal Migration: No End in Sight.* Oxford, UK, Migration Policy Institute. www.migrationinformation. org/Feature/display.cfm?id=336

Perch-Nielson, S. 2004. Understanding the effect of climate change on human migration: the contribution of mathematical and conceptual models. Diploma thesis. Zurich, Swiss Federal Institute of Technology, Department of Environmental Sciences.

Rogaly, B., Coppard, D., Ratique, A., Rana, K., Sengupta, A. and Biswas, J. 2002. Seasonal migration and welfare/illfare in Eastern India: a social analysis. *Journal of Development Studies,* Vol. 38, No. 5, pp. 89–114.

Sahel Club. 1984. *Environmental Change in the West African Sahel [Transformation de l'environnement dans le Sahel ouest africain].* Paris, Club de Sahel.

Seck, E. S. 1996. *Désertification: effets, lutte et convention.* Dakar, Environnement et Développement du Tiers Monde (ENDA-TM).

Selvaraju, R., Subbiah, A. R., Baas, S. and Juergens, I. 2006. *Livelihood Adaptation to Climate Variability and Change in Drought-Prone Areas of Bangladesh. Developing Institutions and Options.* Rome, FAO Rural Development Division.

SEMARNAT. 2008. *National State of the Environment Report for Mexico.* Secretaría de medio ambiente y recursos naturales/Mexican Secretariat for the Environment and Natural Resources. http://app1.semarnat.gob.mx/ dgeia/informe_2008_ing/index_informe_2008.html

Shestakov, A. and Streletsky, V. 1998. *Mapping of Risk Areas of Environmentally Induced Migration in the Commonwealth of Independent States.* Geneva, United Nations High Commissioner for Refugees/International Organization for Migration/Refugee Policy Group.

Skeldon, R. 2005. Migration and mobility: the critical population issue of our time. *Asia-Pacific Population Journal,* Vol. 20, No. 3, pp. 5–9.

_____ 2008. International migration as a tool in development policy: a passing phase? *Population and Development Review,* Vol. 34, No. 1, pp. 1–18.

Stern, N. 2007. *The Economics of Climate Change: The Stern Review.* Cambridge, UK, Cambridge University Press.

Tamondong-Helin, S. and Helin, W. 1991. *Migration and the Environment: Interrelationships in Sub-Saharan Africa. Environmental Change in the West African Sahel.* Washington DC, Natural Heritage Institute and National Academy Press. (Field Staff Reports No. 22.)

UN. 1994. United Nations Convention to Combat Desertification. www.unccd.int/convention/history/INCDresolution.php

—— 2005. Hyōgo Framework for Action 2005–2015: Building the Resilience of Nations and Communities to Disasters. In: *Final Report of the World Conference on Disaster Reduction.* (UN Doc. A/CONF.206/6.)

UNCCD. 2005a. *Reports submitted by Africa.* United Nations Convention to Combat Desertification. www.unccd.int/cop/reports/africa/africa.php

—— 2005b. *Combating Desertification in Africa.* United Nations Convention to Combat Desertification. www.unccd.int/publicinfo/factsheets/showFS.php?number=11

UNDESA. 1992. Agenda 21. New York, United Nations Department of Economic and Social Affairs, Division for Sustainable Development. www.un.org/esa/sustdev/documents/agenda21/index.htm

—— 2004. *World Population to 2300.* New York, United Nations Department of Economic and Social Affairs, Population Division. (ST/ESA/SER.A/236.) http://www.un.org/esa/population/publications/longrange2/WorldPop2300final.pdf

—— 2005. *Trends in Global Migration Stock: The 2005 Revision.* www.un.org/esa/population/publications/migration/UN_Migrant_Stock_Documentation_2005.pdf

—— 2006. *International Migration 2006.* Wallchart and table. (UN Publication, Sales No. E.06.XIII.6.)

—— 2008. *Trends in Global Migration Stock: The 2008 Revision.* www.un.org/esa/population/publications/migration/UN_Migrant_Stock_Documentation_2008.pdf

UNDP. 2009. *Human Development Report 2007/2008. Fighting Climate Change: Human Solidarity in a Divided World.* New York, United Nations Development Programme.

UNECA. 2006. *International Migration and Development: Implications for Africa.* Addis Ababa, United Nations Economic Commission for Africa.

UNFCCC. 2007. *Climate Change Impacts, Vulnerability and Adaptation in Developing Countries.* United Nations Framework Convention on Climate Change. http://unfccc.int/resource/docs/publications/impacts.pdf

UNHCR. 2009. *Forced Displacement in the Context of Climate Change: Challenges for States under International Law.* Geneva, Office of the United Nations High Commissioner for Refugees.

US Climate Change Science Program/Subcommittee on Global Change Research. 2009. *Final Report, Synthesis and Assessment Product 4.2*. Lead agency: US Geological Survey; contributing agencies: National Oceanic and Atmospheric Administration, National Science Foundation, US Department of Agriculture, Forest Service, US Department of Energy, Environmental Protection Agency, US Climate Change Science Program/USGS.

van der Geest, K. 2009. *Migration and Natural Resources Scarcity in Ghana*. Bonn, United Nations University, Institute for Environment and Human Security. (UNU-EHS, EACH-FOR Project.)

van Moppes, D. 2006. *The African Migration Movement: Routes to Europe*. Working Paper. Radboud University, Nijmegen, Netherlands, Research Group Migration and Development. (Migration and Development Series, Report No. 5.)

Vargas-Lundias, R., Lanly, M., Viallareal, M. and Osorio, M. 2007. *International Migration, Remittances and Rural Development*. Rome, International Fund for Agricultural Development/Food and Agriculture Organization. www.ifad.org/pub/remittances/migration.pdf

World Bank. 2008a. *Migration and Remittances Fact Book*. http://go.worldbank.org/QGUCPJTOR0

2008b. *World Development Report 2008: Agriculture for Development*. Washington DC. http://go.worldbank.org/ZJIAOSUFU0

Yang, L. 2004. *Unequal Provinces but Equal Families? An Analysis of Inequality and Migration in Thailand*. Essays on the Determinants and Consequences of Internal Migration. University of Chicago, Department of Economics.

Zetter, R. 2009. The role of legal and normative frameworks for the protection of environmentally displaced people. In: F. Laczko and C. Aghazarm (eds), *Migration, Environment and Climate Change: Assessing the Evidence*. Geneva, International Organization for Migration.

Zeynep, K. 2008. *Turkey Case Study Report*. Bonn, United Nations University, Institute for Environment and Human Security. (UNU-EHS, EACH-FOR Project.)

14

The protection of 'environmental refugees' in international law[1]

CHRISTEL COURNIL

Introduction

In early 2009, in its report on the relationship between climate change and human rights, the Office of the United Nations High Commissioner for Human Rights (OHCHR, 2009, p. 22) encouraged the international community to find political solutions for population displacement linked to climate change. These population displacements are still little known and gave rise, at Bonn in October 2008, to a major international seminar of experts and researchers entrusted with drawing up the first studies on migratory flows resulting from environmental degradation.

Although environmental displacement has existed since the dawn of humanity, the expression 'environmental refugees' emerged[2] in the mid-1980s. Other terms are sometimes preferred: ecological refugees, climate refugees, environmental migrants, climate evacuees, eco-refugees, persons displaced due to a natural disaster, environmentally displaced persons, etc. The description of these population movements varies depending on the persons and the bodies expressing themselves: United Nations agencies and international organizations (UNDP, UNHRC, IOM, etc.), political institutions and actors (European Parliament, Council of Europe, Belgian Senate, Norwegian Refugee Council, political leaders and parties, etc.), civil society and NGOs (OXFAM, Red Cross, Christian Aid, etc.), experts and academics. These actors use terms based on their aims regarding expertise, 'academic theorization', activism or political actions, and depending on the forms of protection they wish to see emerging. Thus,

---

[1] This chapter is based on a contribution to the French EXCLIM project (Managing displacement of populations as a result of extreme climate phenomena), which is funded by the Management and Impacts of Climate Change programme (GICC).
[2] For the emergence of the concept and its use, see Chapter 9.

the emphasis on 'environmental refugees' by NGOs is above all an appeal to the international community and to those in power to raise awareness and take action on the climate change risks hanging over some communities. The International Organization for Migration is more prudent about using the term 'environmental migrants' in its publications (particularly edifying are: IOM, 2007a; 2007b; 2008a; 2008b; 2008c; 2009), owing to the difficulty in identifying and ranking the factors which have led to the departure of the displaced populations (voluntary or forced migration). Academics[3] (often jurists and political scientists) and experts[4] formulate theoretical constructions around the terms 'environmentally displaced persons' and 'climate refugees', frequently with the aim of reflecting on the amendment of existing legislation or the creation of new forms of legal protection systematized in our contribution and summarized in Table 14.1. The UNHCR (2008) has clearly positioned itself on the erroneous use of the expressions 'climate refugees' or 'environmental refugees' which, in its opinion, leads to serious confusion with existing international law.

Indeed in the field of law, the expression 'environmental or climate refugees' is not established. No legal text uses such terms. The use of this expression leads rather to ambiguity (IASC, 2008) in relation to the definition of refugee established by international law. The Geneva Convention relating to the status of refugees gives a definition in international terms in Art. 1, Section A, as:

> A person who owing to a well-founded fear of being persecuted for reasons of race, religion, nationality, membership of a particular social group or political opinion, is outside the country of his nationality and is unable or, owing to such fear, is unwilling to avail himself of the protection of that country; or who, not having a nationality and being outside

---

[3] See e.g. Williams (2008, p. 502), call for a regional effort in the United Nations Framework Convention on Climate Change (UNFCCC); Docherty and Giannini (2009), call for a new convention; Moberg (2009), call for an Environmentally Based Immigration Visa (EBIV) programme; Cooper (1998, p. 480), call for an environmental refugee definition in the Geneva Convention; King (2006), call for the establishment of an International Coordinating Mechanism for Environmental Displacement (ICMED); McCue (1993–1994, p. 151), call for the creation of a new convention and fund; and Zartner Falstrom (2001–2002, p. 15), call for a new convention. See also articles by Kolmannskog (2009), Lopez (2007), Mercure (2006) and Marcs (2008); also Westra (2009) and Ammer and Stadlmayr (2009).

[4] See e.g. Biermann and Boas (2007), call to draw up a protocol to the United Nations Framework Convention on Climate Change (UNFCCC); Hodgkinson et al. (2008), call for the negotiation of a Convention for Persons Displaced by Climate Change; Byravan and Chella (2006), response based on greenhouse gas emissions.

Table 14.1 *Summary of the main protection options for 'environmental refugees'*

Various forms of protection	Authors	Actors	Geographical application	Actions	Type of law	Type of proposal
Reinterpret or amend Art. 1 A Addendum to the Geneva Convention	Cooper (1998) Experts (Maldives)	Academic Experts	International	Reinterpret or amend existing law	Hard law	Protection
Amendment or reinterpretation of international law on statelessness	UNHCR approach Ammer and Stadlmayr (2009)	UN Agency Experts	International	Amend or reinterpret existing law	Hard law	Protection
Extend subsidiary protection (Swedish model)	UNHCR and NRC Kolmannskog and Myrstad (2008)	UN Agency and national institution Experts	National or regional (Europe)	Supplement existing law	Hard law	Protection
Reinterpret, establish or amend temporary protection (European and US model)	–	–	National or regional	Reinterpret, supplement existing law	Hard law	Protection

Proposal	Author	Actor	Level	Approach	Type	Function
Use, extend the *Guiding Principles* on IDPs	Kälin	Special Representative for IDPs	International	Reinterpret or supplement existing law	Soft law	Protection
Add an addendum to the *Guiding Principles* on IDPs	–	–				
Define the law on intervention in the event of disasters (ILC work and the idea of a Framework Convention on protection of persons in the event of disasters)	–	–	International	Create a new right	Hard law	Protection Responsibility
International Convention on Refugees	McCue (1994)	Academic	International	Create a new right	Hard law	Protection Distribution
International convention on environmentally displaced persons	Magniny (1999) Limoges Project (2008)	Academic Academics				Responsibility
Institutional mechanism for the coordination of environmentally displaced persons	King (2006)	Academic	International	Supplement existing law Create a new right	Political and Institutional	Protection Responsibility
International convention on climate refugees	Docherty and Giannini (2009)	Academics	International	Create a new right	Hard law	Protection Distribution Responsibility

backed up by a multidisciplinary regime

International convention on climate displacement	Hodgkinson (2009)	Experts	International	Create a new right	Hard law	Protection	Distribution	Responsibility
Regional agreements for climate refugees	Williams (2008)	Academic	Regional or continental	Create a new right	Hard law	Protection	Distribution	Responsibility
Extend the Pacific Access Category to climate refugees	–	–	National	Supplement existing law	Hard law	Protection		Responsibility
Visa Programme for climate refugees	Nettle amendment (2007); EBIV project Moberg (2009)	Politician; Academic	National	Create a new right	Hard law	Protection	Distribution	Responsibility
Reception mechanism based on states' greenhouse gas emissions	Byravan and Chella (2006)	Scientists	International	Create a new law	Hard law	Protection	Distribution	Responsibility
Protocol on climate refugees, appended to the Framework Convention on Climate Change	Biermann and Boas (2007)	Experts	International	Amend existing law and create	Hard law	Protection	Distribution	Responsibility

> the country of his former habitual residence as a result of such events, is
> unable or, owing to such fear, is unwilling to return to it . . . .

This definition does not then explicitly refer to environmental degradation. The sudden emergence of the concept of an 'environmental refugee' disrupts the traditional categories of migration law. It is quite legitimate to ask whether the emergence, even the establishment, of this concept will undermine the subtle edifice under the Convention and, as a result, the whole logic of post-war international refugee law; and what about the concept of environmental migrants? The difficulty of isolating the environmental reasons among the circumstances of a migrant's (for example, voluntary) departure is a real impediment to formalizing this concept and developing a possible status.

Although the conceptual difficulties are debated, the related question of legal protection is also crucial. Today, no legal instrument defines or offers *direct*, clear and relevant protection to all 'environmental refugees': it remains to be constructed. 'Specialized' texts pertaining to foreigners and refugees are generally disappointing (Cournil, 2006; Cournil and Mazzega, 2007): the Geneva Convention is unsuitable, the OAU Convention falls short and there are weaknesses in European Union instruments relating to asylum and immigration (see above the directives on giving temporary protection in the event of a mass influx of displaced persons and on subsidiary protection),[5] etc. The same applies to texts on international human rights law although they do contain some openings.[6] Yet prevention, assistance, protection and the resettlement of displaced persons mobilize several aspects of international law: international environmental law, international humanitarian law, international refugee law, international human rights law and international disaster response law (IDRL). This chapter aims to show not only the few potentialities of these laws but, above all, to set out the necessary legal expectations in order to deal with this question more fully. Thus, in addition to the existing legal instruments, we explore the prospective directions currently being debated in academic, political, associative and expert circles. A critical typology of this 'legal fiction' is proposed. The degree of realism, relevance and feasibility of these legal avenues are examined, while recognizing that the latter are but one aspect of the political mitigation and adaptation measures required.

Three approaches are explored successively with emphasis on the inadequate and difficult legal prospects for changing international law

---

[5] For an analysis of these two directives in relation to environmental migration see Kolmannskog and Myrstad (2008).

[6] Annual report on the relationships between climate change and human rights, op. cit.

on refugees and stateless persons in the following section; on the emergence of overall protection for displaced persons or victims of disasters and the changing of protection alternatives to the Geneva Convention in the next section. Finally, we discuss the possibility for the international community to develop new tools in order to offer a status specific to environmental or climate refugees.

Amending international law on refugees and stateless persons: an inadequate and difficult process

The first conceivable approach to defining and protecting environmental refugees would be to amend international refugee law, therefore the Geneva Convention, and reconsider the law on stateless persons for individuals whose country is in danger of disappearing because of climate change. Both of these perspectives seem to be inadequate and difficult.

The inadequate perspective of extending the definition of a refugee under a convention

The inclusion of a protocol in the Geneva Convention or the extension of its Art. 1 A were raised at symposia for academics in Limoges (France) in 2005 and in Maldives[7] in 2006. The American jurist Jessie Cooper (1998, pp. 480–88) analysed the problem of environmental refugees and their protection under the Geneva Convention by reinterpreting Art. 25[8] of the Universal Declaration of Human Rights (UDHR). She considered that the definition of a refugee could be extended by adding to Art. I A of the Geneva Convention the degraded environmental conditions that endanger life, health, livelihoods and the use of natural resources. Although she stated that the definition would incorporate the values and principles stemming from human rights (UDHR, the International Covenant on Civil and Political Rights (ICCPR) and the International Covenant on Economic, Social and Cultural Rights (ICESCR)) into the definition of a refugee, she nevertheless said that this apparently easy task was difficult. First for legal

---

[7] First meeting to discuss a Protocol on environmental refugees: recognition of environmental refugees in the 1951 Convention and the 1967 Protocol relating to the status of refugees, 14 and 15 August 2006, Maldives.

[8] Art. 25(1): 'Everyone has the right to a standard of living adequate for the health and well-being of himself and of his family, including food, clothing, housing and medical care and necessary social services, and the right to security in the event of unemployment, sickness, disability, widowhood, old age or other lack of livelihood in circumstances beyond his control.'

technical reasons, as human rights vocabulary did not yet sufficiently include environmental dynamics. Second, for political reasons, it was unlikely that such an extension of the text would be envisaged at the global level as the initiative would meet considerable resistance.

The main advantage of an amendment to the Geneva Convention lies in its application, which would be easily implemented because the States Parties to the Geneva Convention have set up an already operational system of recognition (for example, OFPRA in France). On the other hand, that kind of mechanism would rule out including persons internally displaced for environmental reasons who are now by far the largest group. Moreover, it is by no means sure that many 'environmental refugees' could cross borders and successfully invoke new protection under the convention. At the same time, the application of the Geneva Convention by the recognition bodies of the countries of the North has for several years been very restrictive. It is therefore unlikely that this solution could offer 'mass' protection, especially in a context of identity-based isolationism and the closing of borders. Let us not forget this although the Geneva Convention is based on an international definition, it is implemented through the determination of status by the relevant national authorities.

In addition, would this new definition under a convention not undermine the subtle treaty edifice and, as a result, the whole logic of post-war international refugee law? Would there not be a risk of disrupting the right of asylum with a review of the definition of refugee? Lastly, would the introduction of a protocol annexed to the Geneva Convention, or a new definition relating to the international protection of environmental refugees, not hide the thorny issue of economic, ecological and political responsibility at the root of such displacement? It would then be necessary concomitantly to devise an accountability mechanism.[9]

Difficult revision of international law relating to stateless persons

The law relating to stateless persons could be revised in terms of populations whose state is in danger of disappearing because of climate change. The predicted disappearance of some island nation-states with the rise in sea levels will provoke the departure of tens of thousands[10]

---

[9] The question of accountability is crucial but is not addressed here. Reference should be made to Chapter 15 by Astrid Epiney.

[10] It should be made clear, however, that there will be considerably more departures of populations, hundreds of millions of people from flood-prone areas and without the disappearance of the state, e.g. Bangladesh.

of people (for example the 11,000 inhabitants of Tuvalu). It is in this sense that, during the climate negotiations, the UNHCR submitted a note[11] on the need for rapid action to prevent statelessness linked to the effects of climate change. Indeed it is noteworthy that, in 1996, the United Nations General Assembly[12] entrusted to the UNHCR a global mandate to prevent and reduce statelessness. The latter has therefore just begun, as part of the Bali Action Plan,[13] a reflection process on the links between climate change and 'stateless persons'.

The UNHCR organized expert roundtables on displacements related to climate change in Bellagio in February 2011, as part of the 60th anniversary of the 1951 Geneva Convention and of the 40th anniversary of the 1961 Convention on the Status of Stateless Persons. These roundtables brought together politicians, academics and experts to identify existing gaps in the protection of displaced populations due to climate change, evaluate the legal options available to address these shortcomings, explore practical arrangements raised by 'sinking islands', and finally consider strategies and common perceptions regarding the answers to these migratory movements.

Under Art. 1, para. 1, of the New York Convention of 28 September 1954, a stateless person is 'a person who is not considered as a national by any state under the operation of its law'. In practice, in international law, statelessness can result from: a legal dispute between several countries that prohibits the obtainment of a nationality either by *jus soli* (a territorial nationality right) or by *jus sanguinis* (a blood right); a loss of nationality; or an administrative error. The Convention on the Reduction of Statelessness of 30 August 1961 supplements the definition of a stateless person by compelling the States Parties to reduce cases of statelessness and facilitate the naturalization of stateless persons. At the regional level, the European Convention on Nationality stresses that situations of statelessness must be avoided and that states must endeavour to find solutions.

In the light of these considerations, the 'physical' disappearance of a state (disappearance of state territory) is not provided for in the texts or in their present-day interpretations (McAdam, 2008, p. 28). Indeed, the

---

[11] See document submitted by UNHCR, supported by IOM and the Norwegian Refugee Council (NRC), to the 6th session of the Ad Hoc Working Group on Long-term Cooperative Action under the Convention (AWG-LCA 6), 1–12 June, Bonn (UNHCR, 2009).

[12] General Assembly Resolution 50/152, 9 February 1996.

[13] The reference is to the conclusions of the 13th Conference of the Parties (COP) to the United Nations Framework Convention on Climate Change (UNFCCC) adopted in Bali. This plan adopts a roadmap to reach a long-term international agreement in order to replace the Kyoto Protocol when it expires in 2012.

special character of island nation-states in danger of disappearing for ever seems, in our view, to depart from the 'traditional' concept of statelessness in international law in that there is no denial of nationality. How then can this international law be amended without changing the original meaning? How can the legal category of a stateless person be reinvented or updated so as to foster the emergence of 'climate statelessness'?

This theory will probably be difficult to put into practice,[14] especially as the issue raised by climate displacement is precisely to protect, even extend, state links in spite of the physical disappearance of a state because of climate change. Moreover, the rights of a population and not just of a few individuals should be recognized collectively. The fact remains that these populations will always be 'de facto stateless persons' for whom political and legal solutions will need to be devised.

In its study, the UNHCR proposes that consideration be given to multilateral agreements that would ideally allow populations at risk to settle elsewhere with a legal status (dual citizenship, respect for culture, right of residence, social benefits, etc.). Thus, the UNHCR sets out a number of conditions for concluding these agreements, such as the participation in the negotiations of the populations concerned and the governments of the island states affected. Above all, the UNHCR's note encourages the recognition needed for this form of climate stateless-ness[15] in international law, especially in the United Nations Framework Convention on Climate Change into which population displacement and statelessness should be incorporated.

Extending protection alternatives and protection of internally displaced persons or victims of disasters: plausible medium-term options

Amending protection alternatives to the Geneva Convention, extending the protection of internally displaced persons (IDPs) and the creation of a protection regime for victims of disasters are three plausible medium-term options, provided that states take on board the issue of environmental refugees and integrate it into international regulations relating to human-itarian or human rights in their national legislation.

---

[14] It is possible that small island states, even uninhabitable, will at least conserve their territorial waters and therefore their inhabitants will not lose their passport.

[15] On this point, see the ideas developed by Ammer and Stadlmayr (2009, pp. 66–70) on national minority.

Developing protection alternatives to the Geneva Convention

Protection alternatives are intended to protect certain categories of asylum-seekers who do not meet the criteria of the Geneva Convention. They were developed in particular in the 1990s and have even been harmonized in Europe (EU subsidiary protection). For example, a comparative study conducted in 2002 by Daphné Bouteillet-Paquet on subsidiary protection revealed that only four countries (Greece, Sweden, Finland and Italy) had a protective legislative mechanism for people fleeing environmental disasters (Bouteillet-Paquet, 2002, p. 181).

Consequently, rather than renegotiating the Geneva Convention, current thinking (UNHCR, 2009) at the UNHCR and Norwegian Refugee Council (NRC) focuses on whether it would be 'less risky', easier and faster to develop national or regional (European) protection alternatives for relatively wide-ranging humanitarian or environmental reasons in order to be able to anticipate cross-border or inter-state climate migration. The American, European and Nordic models are illustrative of this approach.

In 1990, the United States established a temporary protected status (TPS) for those who do not meet the legal definition of refugee, but who still need protection because of the risks they would incur if they returned to their own country. The specific feature of this temporary protection is that it may be granted if there were a serious environmental disaster resulting in disruption to living conditions. The affected state must be unable to cope with the return of its own nationals, and must therefore be designated a disaster area. Temporary protection may last for six to eighteen months and may be extended if living conditions do not improve in the affected country. In order to avoid the 'pull factor', TPS only applies to persons already resident in the United States at the time of the incident and only when a formal request for protection has been made by the State of Origin (cf. Moberg, 2009, p. 1127). Applications have included those made for Nicaraguans and Hondurans after Hurricane Mitch in 1998, but also for migrants in the wake of volcanic eruptions (see McAdam, 2009; USCIS, 2004). The possibility of granting TPS to Haitians is currently being discussed. As it stands, this protection regime is still very ad hoc and totally inadequate to cope with the expected migration scenarios.

The EU Directive on temporary protection in the event of a mass influx of displaced persons, adopted in 2001, could be updated or amended to take account of the situation of environmental refugees. As will be remembered, this protection is for situations of mass influx of displaced persons 'who come from a specific country or geographical area, whether their arrival in the Community was spontaneous or aided, for example through an evacuation programme'.[16] This definition could include environmental migration and enable refugee reception to be shared out across Europe. To this end, Member States have the option either to directly include the various hypothetical environmental migrations by substantially amending the directive to create a new 'temporary protection' category, or to interpret the directive broadly and thus update the scope of the directive. However, both extension and amendment appear unlikely at present. Member States are particularly 'cautious' when it comes to hosting refugees and the directive was not intended to be extended to 'environmental refugees' as it was designed to deal with the influx of people fleeing in or near Europe from conflicts similar to the war in the former Yugoslavia.

New legislation in the Nordic countries should also be looked at with a view to harmonizing regional cooperation – not least European – on environmental refugees. For example, Swedish legislation provides a protection alternative to refugee status for persons who need protection as a result of an environmental disaster in their country.[17] Vikram Kolmannskog and Finn Myrstad (2008) have shown that a recent Finnish Aliens Act proposal could clarify humanitarian protection by guaranteeing that people will not be returned to their countries of origin in the event of environmental disasters. Denmark has granted protection to unmarried women and families with young children from areas where living conditions are regarded as extremely difficult owing to famine or drought, for example. Finally, in the draft Aliens Act in Norway, the Ministry of Immigration has recognized the need to grant residence permits to applicants who come from a region affected by a humanitarian catastrophe, including a natural disaster. We would point out, however, that while the idea of expanding and developing subsidiary protection is a possible short-term solution, experience has shown that diversification of asylum protection is often synonymous with insecurity

---

[16] Council Directive 2001/55/EC of 20 July 2001 on Minimum Standards for Giving Temporary Protection in the Event of a Mass Influx of Displaced Persons and on Measures Promoting a Balance of Efforts between Member States in Receiving Such Persons and Bearing the Consequences thereof [2001] OJ L212/12, Art. 2d.

[17] http://www.migrationsverket.se/english.jsp

(Cournil, 2005; Bouteillet-Paquet, 2002), compared with the rights available under the particularly protective Convention status.

Strengthening the rights of IDPs by including environmentally displaced persons: towards a global definition of IDPs?

Even though they are migrating and in need of protection, displaced persons within their own countries are not called 'refugees' because they do not enjoy international protection under the Geneva Convention. They have been known as internally displaced persons[18] (IDPs) since the adoption of the *Guiding Principles on Internal Displacement* in February 1998 as a result of the campaign by Francis Deng, former Representative of the Secretary-General on the Human Rights of Internally Displaced Persons. This non-legally binding instrument (Deng, 1998) was an attempt to bring together rights and obligations in a single document; but, more importantly, it provided an international definition of IDPs by clarifying existing ambiguities and filling the gaps in international texts on internal migration. Several countries have already incorporated the *Guiding Principles* into their national legislation. Approximately twenty governments have adopted policies on IDPs (including Turkey, Angola, Burundi, Colombia, Liberia, Peru, Philippines and Sri Lanka). In June 2008 the Organization of American States (OAS) adopted Resolution 2417 (XXXVIII-O/08) on internally displaced persons that referred to the *Guiding Principles*. These principles have been recognized as a useful element of *soft law*, but they are not binding and, notwithstanding the few examples mentioned above, have only been applied to a limited extent in most of the countries concerned. In 2009, the African Union adopted a Convention on the Protection and Assistance of Internally Displaced Persons,[19] the text of which draws heavily on the *Guiding Principles*. This regional treaty is the only binding instrument on the issue. It remains to be seen whether this future hard law treaty will be effective on the African continent and whether it will apply to environmental migration.

In the *Guiding Principles*, IDPs are defined as

---

[18] On the origin and prospects of IDPs, Dubernet (2007); Orchard (2010).

[19] The African Union Convention on the Protection and Assistance of Internally Displaced Persons in Africa (Kampala Convention), adopted on 22 October 2009. See also Jaksa and Smith (2008). See, in particular, Article 5(4), according to which 'States Parties shall take measures to protect and assist persons who have been internally displaced due to natural or human made disasters, including climate change.'

persons or groups of persons who have been forced or obliged to flee or to
leave their homes or places of habitual residence, in particular as a result
of or in order to avoid the effects of armed conflict, situations of generalized violence, violations of human rights or natural or human-made
disasters, and who have not crossed an internationally recognized state
border (Deng, 1998).

While people displaced as a result of natural disasters are already
mentioned, one day we might also explicitly include other reasons
such as people displaced owing to the effects of climate change, for
example. A redefinition of the *Guiding Principles* would provide a
comprehensive definition and protection of IDPs regardless of their
reason for leaving and would overcome the difficulties involved in a
non-consensual definition of climate or environmental refugees. The
inclusion of a 'rider' or a specific article on environmental displacement might thus be envisaged in the *Guiding Principles*. As far as the
UNHCR (2008) is concerned, the *Guiding Principles* already provide a
normative framework for dealing with the protection of certain environmentally displaced persons.

However, expanding the definition of IDPs would have the disadvantage of 'diluting' the protection of climate refugees into that covering
internally displaced persons and of not enabling the perpetrators to be
identified. Nevertheless, a 'climate justice' element could still be developed in tandem with a common but differentiated accountability mechanism and a compensation fund. That still leaves the problem of the
effectiveness of the law and the fact that it would be difficult to implement in poor countries particularly exposed and vulnerable to climate
change and disasters. The fact is that the application of these principles
suffers from a lack of institutional recognition: no organization or
agency is currently solely responsible for the protection of IDPs. It is
true that the 'cluster approach' was adopted in 2005 (Diagne and
Entwisle, 2008) and aid delivery is now shared between several United
Nations agencies (e.g. UNHCR deals with the legal aspects, camp management and accommodation for IDPs; UNDP, WHO, UNEF, UNICEF,
etc. are also active); nevertheless, this cluster or modular approach
presents serious difficulties in terms of coordinating aid (Dubernet,
2007; Charny, 2005; Eschenbächer, 2005). The academic Tracey King
(2006) suggests setting up an International Coordinating Mechanism for
Environmental Displacement (ICMED), which would allow agencies
and organizations already working on some of the different aspects of
the issue of environmental migration to coordinate their efforts.

Enhance the 'protection of persons in the event of disasters': a relevant but relatively unexplored option

Although not all disasters are a result of climate change, more and more of them are to be expected due to the secondary effects of climate change, which will lead to population displacement.

At its 60th session on 4 November 2008, the International Law Commission (ILC) reflected broadly on the notion of 'protection of persons in the event of disaster'. The Commission's remit is to consider improvements and the progressive development and codification of international law. In early 2008, a preliminary report by Mr Valencia-Ospina[20] was published,

> tracing the evolution of the protection of persons in the event of disasters, identifying the sources of the law on the topic, as well as previous efforts towards codification and development of the law in the area, presenting in broad outline the various aspects of the general scope with a view to identifying the main legal questions to be covered and advancing tentative conclusions (ILC 2008, Chap. 9, p. 332).

An earlier ILC report (ILC, 2007) adopts a broad definition of disasters[21] (both natural and man-made); disasters relating to climate change can be included here. The various phases of a disaster, i.e. the pre-, in- and post-disaster phases (prevention, mitigation, response and rehabilitation) are discussed. The concept of protection is given a broad definition as encompassing response, relief and assistance. Consequently, the issue of environmentally displaced persons is not alien to this work; on the contrary, it could enrich it and give it a new dimension. The report shows that the right of disaster victims to assistance is not straightforward or easy to establish. The Special Rapporteur proposed a two-pronged approach towards establishing bona fide *international law of intervention in the event of disasters*. On the one hand, he suggests considering consolidating the law in the form of a possible Framework Convention on the protection of persons in the event of disasters. This would involve identifying and clarifying the principles of common law regarding the provision of aid or assistance from states and competent international organizations and the conditions under which an affected state would accept such an offer. On the other hand, the report proposes considering the case for consolidating the existing rules to provide

[20] Preliminary Report of the Special Rapporteur (A/CN.4/598) and the Secretariat Memorandum, focusing primarily on natural disasters (A/CN.4/590 and Add.1–3) 5 May 2008.

[21] There still remains the issue of gradual environmental degradation, which has received little attention.

for genuine international cooperation in the interests of disaster victims and in compliance with the *principles of humanity, neutrality, impartiality and sovereignty*. In sum, this recent doctrinal work on disaster victims could provide support for the issue of environmental or climate displacement. Even if the reports do not mention it explicitly, sooner or later the issue will be addressed as one of the possible consequences of disasters.[22] Would it not be appropriate, then, to consider creating a proper 'disaster victim' status in international law (with a section devoted to the displacement ensuing from disasters)? This would provide an opportunity to respond globally to the issue of disasters and thus indirectly to 'environmental displacement'.

Building *sui generis* protection: a logical but tricky option

Three novel options are currently being discussed: a new international treaty on environmental refugees and environmentally displaced persons; specific protection for climate refugees; and regional agreements. Attractive as they are, these solutions cannot be realized overnight. They require real political awareness and strong commitment from states and the international community; this does not yet seem to be the case.[23] Moreover, they will inevitably give rise to new legal consequences.

International treaty on environmental refugees and environmentally displaced persons: custom-made protection

In her 1999 thesis, the French jurist Véronique Magniny discussed in some depth the case for the creation of a new international convention (Magniny, 1999). She posits that international solidarity is the only appropriate response to the protection of environmental refugees. The complex impact of disasters is such that we can no longer think in terms of state borders and national protection; we must think in terms of global protection. Véronique Magniny suggests opting for a multilateral international treaty ('hard law') so that states will be bound by their international commitments. The

---

[22] The UNHCR believes that for some environmental displacements, especially in connection with natural disasters, the *Operational Guidelines* drafted by the Inter-Agency Standing Committee (IASC, XXXX) on the protection of persons affected by natural disasters and the accompanying handbook are the frames of reference.

[23] In this connection, see the difficulties in implementing ILO conventions on migrant workers and the current limited scope of the United Nations International Convention on Migrant Workers and Members of Their Families.

academic Gregory S. McCue (1993–1994) has proposed a new international convention that would include the principles of international refugee law and environmental law such as an obligation to prevent, assist, alert and provide information about the environmental situation. He has suggested setting up a compensation fund for the resettlement of refugees; the fund could be managed by the secretariat of the new convention. In the same vein, the Swedish MP Tina Acketoft, rapporteur of a report on environmental refugees for the Council of Europe's Committee on Migration, Refugees and Population, recently came out in support of a European convention. Her report of 23 December 2008 encourages Europe to be a pioneer by adopting an original regional legal text (see below).

Those in favour[24] of a new text call for the new legal instrument to be autonomous and specific. Indeed, it would seem desirable that international protection should be achieved through the creation of a specific instrument rather than by simply amending or extending the Geneva Convention. The December 2008 draft convention on the international status of environmentally displaced persons (published in *Revue européenne du droit de l'environnement*, 2008, pp. 381–93) drafted by environmental law specialists at the University of Limoges (OMIJ/CRIDEAU) is currently the most complete protection proposal package. While this is not the only initiative of the kind, as a group of Australian experts are also calling for the creation of such a convention,[25] it is the most comprehensive effort to date (drafting of the convention, explanatory report with reference to additional protocols). The draft reads like a genuine international convention (with preamble, chapters and articles, etc.). The consensual term 'environmentally displaced persons' was chosen and these are defined in Art. 2 as:

> ... individuals, families and populations confronted with a sudden or gradual environmental disaster that inexorably impacts their living conditions and results in their forced displacement, at the outset or throughout, from their habitual residence and requires their relocation and resettlement.

The draft combines protection, assistance and responsibility by incorporating the principles of environmental assistance, proximity, proportionality,

---

[24] See footnote 4 and UNHCR Policy Paper (2008).

[25] Hodgkinson et al. (2009, p. 2). See also http://www.ias.uwa.edu.au/new-critic/eight/?a=87815. An initial presentation of the draft convention was made at the University of Copenhagen during the conference on Climate Change: Global Risks, Challenges and Decisions, March 2009.

effectiveness and non-discrimination. The most important of these is the principle of common but differentiated responsibilities 'with the aim of prevention and reparation' (Art. 4, covered by a specific protocol). Ten fundamental rights[26] which are common to temporarily and permanently displaced persons are set out in Art. 5: the right to information and participation; to be assisted; to water and food aid; to shelter; to health; to legal personality; the right for any natural person displaced to another state to retain their own civil and political rights in their home state; the right to respect for family unity; the right to education and training; and the right to earn a living. The draft convention aims to establish a mechanism for granting the status of environmentally displaced person and develops cooperation with several international and regional institutions. It proposes establishing a World Agency for Environmentally-displaced Persons (WAEP) comprising an administrative council, a high authority (which would have a monopoly on the interpretation of the convention and decide appeals against any national commission's decision to grant or refuse the status of environmentally displaced person), a scientific council, a secretariat and a World Fund for the Environmentally Displaced (WFED). As with the system for granting refugee status under the Geneva Convention, the draft convention provides that each State Party must create a national commission for granting the status of environmentally displaced person. While this convention can be criticized in terms of the options selected, it does at least provide a practical starting point for cross-sectoral discussion so that a substantive debate can be launched on the principles and definitions that need to be added to a new international convention. A 'hard law' solution nevertheless remains a long-term option; states do not (yet) appear ready for this type of solution. However, the countries of the Southern Hemisphere will have more incentive to support this type of project as a large majority of them will be harshly affected by future disasters and the impact of climate change, particularly the island states (see in this regard the expectations of the Alliance of Small Island States). They thus have every interest in defending a convention that focuses on international solidarity.

This 'new convention' approach is by no means far-fetched; indeed, in January 2009 the Committee on Migration, Refugees and Population and the Committee on Environment, Agriculture and Regional Affairs of the

---

[26] We feel that the principle of non-refoulement should be included here.

Parliamentary Assembly of the Council of Europe adopted a resolution[27] and a recommendation[28] calling for:

> ... a further investigation of existing gaps in law and protection mechanisms with a view to an eventual elaboration of a specific framework for the protection of environmental migrants, either in a separate international convention or as parts of relevant multilateral treaties.

They also encourage progress to be made institutionally by calling for:

> ... an effective coordination structure to be established that would pull together the various international agencies and stakeholders focusing on risk reduction, humanitarian response, adaptation and development.

This regional institution approach is highly ambitious and unique; its reception remains to be seen.

Protocol or specific convention on climate refugees: innovative protection arrangements that are difficult to put in place

A group of researchers and experts (Biermann and Boas, 2007) has suggested granting specific protection to 'climate refugees' only by enshrining the term in international law. They want a *sui generis* protection to be established through a climate refugee protocol to be appended to the 1992 United Nations Framework Convention on Climate Change. The text would be drafted around five principles:

- principle of relocation or resettlement;
- principle of resettlement rather than temporary protection;
- principle of collective rights granted to local people;
- principle of international assistance within states;
- principle of sharing the burden of hosting refugees internationally.

An executive committee for the recognition and resettlement of climate refugees would ensure the effective implementation of this protocol through the establishment of a specifically created fund (Climate Protection and Refugee Resettlement Fund – CPRRF). The main advantage of this solution is that it ties the protection of climate refugees in with state responsibility for climate change. However, a disadvantage is

---

[27] Resolution 1655 (2009), Environmentally Induced Migration and Displacement: A 21st Century Challenge.

[28] Recommendation 1862 (2009), Environmentally Induced Migration and Displacement: A 21st Century Challenge.

that the plan limits the category of refugees to be protected by offering international protection to climate refugees alone. For the latter, the definition has still to be finalized and is likely to give rise to heated debate.

In 2009, an interesting, reasoned proposal from two American lawyers to establish an 'international convention for climate refugees' provided further support for a specific, narrow definition.[29] The authors, Bonnie Docherty and Tyler Giannini (2009, pp. 368–72), base their definition of climate refugees on six factors:

- the existence of forced migration (non-voluntary, i.e. linked to the survival of the person);
- temporary or permanent resettlement of the climate refugee;
- cross-border migration only (i.e. the definition is restricted to those who cross an international border);
- an environmental disruption related to and consistent with the effects of climate change;
- a sudden *or* gradual environmental upheaval; and finally
- a flexible link with human action in the environmental upheaval.

The authors advocate a binding instrument, rather than a general policy instrument, thus circumscribing the definition of climate refugees in order to be able – in their view – to provide a better response to humanitarian needs and highlight more effectively the unique nature of climate migration. Furthermore, their idea that the protection of climate refugees must be accompanied by a genuine multidisciplinary regime is one that we can endorse. The authors suggest three key principles: assistance, shared responsibility and administration of an international regime for climate refugees, each of which refers to specific actions, nine in total (ibid., pp. 373–91). 'Assistance' would involve consolidating standards for determining climate refugee status, protection of human rights and humanitarian assistance. 'Shared responsibility' would clarify the role of the host state and that of the affected state; cooperation and assistance measures would need to be established here. The administrative aspects of the regime would include the creation of a

---

[29] 'It defines a climate change refugee as an individual who is forced to flee his or her home and to relocate temporarily or permanently across a national boundary as the result of sudden or gradual environmental disruption that is consistent with climate change and to which humans more likely than not contributed' (Docherty and Giannini, 2009, p. 361).

global fund, a coordinating body and, finally, a body of scientific experts.

Hodgkinson and his co-authors (Burton and Hodgkinson, 2009) also suggest a new convention for 'persons displaced by climate change' to enable the collective and regional recognition of populations at risk and with a flexible definition based on six categories of displaced person.[30] The authors envisage a number of requirements such as:

- long-term resettlement measures;
- assistance based on the common responsibilities of states but differentiated with regard to greenhouse gas emissions;
- adaptation and mitigation measures implemented by the host states through international financial assistance;
- the creation of an assistance fund and regular scientific studies of people exposed to climate change risks.

As can be seen, all these proposals attempt to address the lack of direct legal protection for people displaced by climate change. Noting this legal vacuum and the inadequacy of the existing instruments, the lawyers and other experts mentioned suggest unprecedented forms of protection in an attempt to go beyond the current legal limitations. They believe that the protection of these people will necessarily require a legal recognition of the phenomenon. Most of these approaches include an 'integrated approach' combining issues of prevention, protection, relief, responsibility and new institutions (administrative and financial). These various proposals show the need for a holistic approach in order to grasp the complexity of this global issue. The proposals are extremely interesting and deserve to be discussed and enhanced by the competent international bodies; they could be the start of an international *sui generis* protection for climate refugees and of global governance of climate migration.

The international community is increasingly alive to the issue of climate refugees. The talks held in Bonn in June 2009 steered for the first time, albeit in a marginal way, international discussions towards consideration of climate-induced migration. Several proposals for

---

[30] Burton and Hodgkinson (2009, pp. 13–14): temporary displacement, permanent local displacement, permanent internal displacement, permanent regional displacement, permanent intercontinental displacement, temporary regional and international displacement.

action[31] – which are still to be debated – have been submitted to the climate change negotiations. The 16th Conference of the Parties held in December 2010 in Cancun made the first explicit reference to migration and displacement in relation to climate change in a document of the States Parties to the UNFCCC. Admittedly, however, this constitutes an acknowledgement of the emergence of the issue in the debates, but is quite imprecise and does not set up any legally binding obligations for States. The climate refugee issue could be a double-edged sword in the climate negotiations: it could either encourage states to act quickly and establish the legitimacy of climate refugee protection in international law; or, conversely, it could be counter-productive and encourage some reluctant countries to withdraw, thus slowing down climate progress. In any event, it is still premature to say that we are moving towards a specific protection for climate refugees; nevertheless, the idea is slowly gaining ground.

Special or bilateral agreements to protect the populations of threatened island states: a regional solution

As already mentioned, the disappearance of a state and the migration of its population lead us to contemplate a new form of statelessness and the survival or political and legal continuity of a state when its territory disappears. In this context, there is a need to envisage reception and genuine regional integration of these people by means of a bilateral agreement between a host state and the state threatened with disappearance; this must be done before the territory is submerged.

New Zealand has established a quota migration system (Pacific Access Category programme)[32] for Tuvaluans, based on economic criteria (for companies, employment and study). The programme is currently based on economic rather than climate migration. This type of partnership could be extended with one or several states undertaking to accommodate climate refugees over time. However, in practice, the Pacific Access Category programme has not been a great success (see also the field study by Shen and Gemenne, 2008), as those at risk do not usually wish to leave

---

[31] 'Activities related to national and international migration/planned relocation of climate [refugees] [migrants] [displaced persons by extreme climate events . . .]' in Framework Convention on Climate Change, Ad Hoc Working Group on long-term cooperative action under the Convention, Sixth Session, Bonn, 1–12 June 2009, FCCC/AWGLCA/2009/INF.22 June 2009, p. 45.

[32] http://www.immigration.govt.nz/migrant/stream/live/pacificaccess/

their island, preferring adaptation and mitigation measures to be planned instead.

A new form of regional protection would be an example of an early show of solidarity by a neighbouring state towards a state threatened with disappearance. Taking the form of a bilateral or multilateral agreement, it would establish a predetermined reception policy (number of people admitted, rights granted, reception areas, right to work, respect for local culture, language, recognition of traditions, etc.). The establishment of such an agreement would undoubtedly be the most pragmatic solution; it does, however, have certain limitations. Indeed, why should the 'burden' of hosting the refugees fall only on the neighbouring or volunteer country? What about neighbouring countries which do not have the economic resources to host these people in dignified conditions? The burden of reception could weigh heavily on the host country if there is no regional or international compensation fund, thus leading to political stumbling blocks in the negotiations unless the international community decides to share the financial burden of hosting these populations. Furthermore, if the host destination is decided unilaterally, the refugee is deprived of the choice of future place of migration. The fact remains that voluntary distribution is feasible when applied to the hundreds of thousands of people from small island countries but will be difficult to replicate in other contexts, such as in Africa, for example.

There could be a 'softer' short-term solution but it has not been adopted thus far. The Australian Senator Kerry Nettle has unsuccessfully proposed amending the Immigration Act by calling for a new visa category[33] for climate refugees, which would enable 300 people per year from Tuvalu, Kiribati and other Pacific islands to be admitted. The establishment of such a mechanism would undoubtedly be the swiftest solution but it is left to the goodwill of the host state. The academic Kara K. Moberg (2009, pp. 1135–36), has suggested creating an Environmentally Based Immigration Visa (EBIV) programme and believes that international discussions on greenhouse gas emission reduction requirements should consider adopting such a programme. The number of immigrant visas issued by each country would depend on the percentage of their greenhouse gas emissions, with the biggest producer issuing the most visas. However, this solution will be very difficult to implement in international negotiations, given the current tensions on the amendment of the Kyoto Protocol.

[33] http://www.kerrynettle.org.au/files/campaigns/extras/Climate20Refugee%20Bill%20brief%2020.6.07.pdf

The proposal from the academic Angela Williams (2008, p. 30) calls for a regional system (bilateral or regional agreements) because it seems unlikely that states will accept a global binding package on the recognition and protection of persons displaced by climate change. The main advantage of regional cooperation lies in the fact that states can frame appropriate policies in a relatively short time and according to their abilities. Regional cooperation on environmental protection provides a good example. That author believes that although the problem of climate displacement is an international one, the immediate impact of climate change will be felt at regional level first. She therefore suggests that partners such as the African Union, the Organization of American States, the European Union and the Association of Southeast Asian Nations should launch a regional debate on their migration risk. Displaced persons will of course tend to seek resettlement or temporary accommodation solutions that are close to where they live. A regional cooperation framework would enable states to sign up to different levels of commitment and development, depending on the individual capacity of the countries affected and the severity of the problem. Williams is optimistic about the chances of a regional scheme such as this being adopted in the South Pacific.

Conclusion

The complexity of climate change and its impact on the way ecosystems operate, the diversity of interdependencies between societies and their environments and the random dimension of risk lead us to contemplate, now more than ever, the measures that must be designed and implemented in order to address the human impacts of climate change. Population displacement will undoubtedly be one of the most difficult human impacts to manage. We have shown throughout that this is one of the international community's most difficult and urgent challenges. Will the local, regional and international politicians of tomorrow be up to these challenges? We have shown that, while the legal instruments are sometimes inadequate, they can be improved to meet the new challenge. Some proposals for the protection of displaced populations also show that a legislative rethink is required if we are to be able in the future to understand and regulate the complexity of these intractable human situations. The legal expert, Jane McAdam, posits the idea of a genuine Environmental Migration Governance (McAdam, 2009). Indeed, the complexity arising from the vast number of displacement hypotheses, the uneven geographical impact of climate change and the different resilience levels of countries will no doubt lead the international community to design and create new legal instruments, including in international law.

All the approaches described here will need to be discussed in terms of their degree of relevance and effectiveness and, in particular, in the light of empirical research into the regions that give rise to environmental displacement. Nevertheless, while reflecting on the legal protections that need amending or framing, we must not lose sight of the fact that legal solutions only materialize as a result of strong political choices; this is hardly the case today. While the sheer numbers predicted would appear to make the protection of environmental refugees a matter of considerable urgency, the international community and states still need to move up a gear with regard to climate change adaptation and mitigation measures.

Three paths must be pursued simultaneously: mitigation of climate change; reducing the risks caused by climate change; and management and protection of persons displaced by climate change. The adoption of an ambitious new climate protocol, the establishment of specific adaptation, integration and human displacement management measures under the disaster prevention and risk reduction action plans of the Hyōgo Framework for Action[34] and legal protection for environmental refugees will be the key measures needed to meet the global challenges of climate change.

References

Ammer, M. and Stadlmayr, L. 2009. *Climate Change and Human Rights: the Status of Climate Refugees in Europe*. Research project on climate change. Vienna, Ludwig Boltzmann Institute of Human Rights (BIM).

Basher, R. 2008. Disasters and what to do about them. *Forced Migration Review: Climate Change and Displacement*, No. 31, October, pp. 35–36.

Biermann, F. and Boas, I. 2007. *Preparing for a Warmer World: Towards a Global Governance System to Protect Climate Refugees*. Global Governance Project. (Global Governance Working Paper 33.)

---

[34] This action framework is the main instrument adopted under the auspices of the United Nations to reduce disaster risks. Its goal is to build up states' resilience to disasters by 2015. It follows on from the World Conference on Disaster Reduction (Kobe, Hyōgo, Japan, 2005) which was held after the International Decade for Natural Disaster Reduction. In 2000 the United Nations launched the International Strategy for Disaster Reduction (ISDR). See Basher (2008, p. 36).

Bouteillet-Paquet, D. 2002. La protection subsidiaire: progrès ou recul du droit d'asile en Europe? Une analyse critique de la législation des États membres de l'Union européenne. In: D. Bouteillet-Paquet (ed.), *La protection subsidiaire des réfugiés dans l'Union européenne: un complément à la Convention de Genève?* Brussels, Bruylant.

Burton, T. and Hodgkinson, D. 2009. *Climate Change Migrants and Unicorns: A discussion note on conceptualizing climate change displaced people.* http://www.hodgkinsongroup.com/documents/PeopleDisplacedByClimateChange.pdf

Byravan, S. and Chella, R. S. 2006. Providing new homes for climate change exiles. *Climate Policy*, No. 6, pp. 246–52.

Charny, J. R. 2005. New approach needed to internal displacement. *Forced Migration Review*, Supplement, October, pp. 20–21.

Cooper, J. B. 1998. Environmental refugees: meeting the requirements of the refugee definition. *New York University Environmental Law Journal*, Vol. 6.

Cournil, C. (ed.). 2005. *Le statut interne de l'étranger et les normes supranationales.* Paris, L'Harmattan, pp. 317–51. (Logiques juridiques.)

——— 2006. Vers une reconnaissance des 'réfugiés écologiques'? Quelle(s) protection(s), quel(s) statut(s)? *Revue du droit public*, No. 4, pp. 1035–66.

Cournil, C. and Mazzega, P. 2007. Réflexions prospectives sur une protection juridique des réfugiés écologiques. *Revue européenne des migrations internationales*, Vol. 23, No. 1, pp. 7–34.

Deng, F. 1998. *Guiding Principles on Internal Displacement*, 11 February. (E/CN.4/1998/53/add.2.)

Diagne, K. and Entwisle, H. 2008. UNHCR and the Guiding Principles. *Ten Years of the Guiding Principles on Internal Displacement. Forced Migration Review*, GP10, December, pp. 33–34.

Docherty, B. and Giannini, T. 2009. Confronting a rising tide: a proposal for a convention on climate change refugees. *Harvard Environmental Law Review*, Vol. 33, No. 2, pp. 349–403.

Dubernet, C. 2007. Du terrain au droit, du droit sur le terrain? Origines et trajectoires du label 'déplacé interne'. TERRA-Ed. (Coll. Esquisses.) http://terra.rezo.net/article644.html

Eschenbächer, J.-H. 2005. Making the collaborative response system work. *Forced Migration Review*, Supplement, October, pp. 15–16.

Hodgkinson, D., Burton, T., Dawkins, S., Young, L. and Coram, A. 2008. Towards a convention for persons displaced by climate change: key issues and preliminary responses. *The New Critic*, No. 8, September. http://www.ias.uwa.edu.au/new-critic/eight/hodgkinson

——— 2009. Towards a convention for persons displaced by climate change: key issues and preliminary responses. *Climate Change: Global Risks,*

Challenges and Decisions. Bristol, UK, IOP Publishing. (Earth and Environmental Science No. 6, 562014.)

IASC. 2006. *Protecting Persons Affected by Natural Disasters.* IASC Operational Guidelines on Human Rights and Natural Disasters. Washington DC/ Geneva, Brookings-Bern Project on Internal Displacement/Inter-Agency Standing Committee.

2008. *Change, Migration and Displacement: Who Will Be Affected?* Working Paper submitted to UNFCCC Secretariat by the informal group on Migration/Displacement and Climate Change of the Inter-Agency Standing Committee, 31 October.

ILC. 2007. *Protection of Persons in the Event of Disasters.* International Law Commission Secretariat Memo, 11 December.

2008. *International Law Commission Report, 60th session* (5 May to 6 June and 7 July to 8 August 2008), General Assembly, Supplement No. 10 (A/63/10).

IOM. 2007a. *Development and Natural Disasters: Insights from the Indian Ocean Tsunami.* Geneva, International Organization for Migration. (Migration Research Series No. 30.)

2007b. *Migration and Climate Change.* (Migration Research Series No. 31.)

2008a. *Climate Change and Migration: Improving Methodologies to Estimate Flows.* (Migration Research Series No. 33.)

2008b. *Migration Development and Environment.* (Migration Research Series No. 35.)

2008c. *Expert Seminar: Migration and the Environment.* (International Dialogue on Migration, No. 10.)

2009. *Migration, Climate Change and the Environment.* (Policy brief, May.)

Jaksa, B. and Smith, J. 2008. Africa: From voluntary principles to binding standards. *Ten Years of the Guiding Principles on Internal Displacement. Forced Migration Review*, GP10, December, pp. 18–19.

King, T. 2006. Environmental displacement: coordinating efforts to find solutions. *Georgetown International Environmental Law Review*, Vol. 18, No. 3, pp. 543–66.

Kolmannskog, V. 2009. *Climates of Displacement – Investigating Protection Possibilities in Climate Change-Related Forced Migration.* Copenhagen/ Oslo, Danish Institute for International Studies/Norwegian Refugee Council.

Kolmannskog, V. and Myrstad, F. 2008. Förvaltningsutskottets betänkande 26/2008 rd, ogranskad version 1.0 FvUB 26/2008 rd – RP 166/2007 rd, Regeringens proposition med förslag till lagar om ändring av utlänningslagen och av vissa lagar som har samband med den. http://www.eduskunta.fi/faktatmp/utatmp/ akxtmp/fvub_26_2008_p.shtml (Finnish Aliens Act, 301/2004, amendments up to 973/2007 included.)

2009. Environmental displacement in European asylum law. *European Journal of Migration and Law*, Vol. 11, No. 4, pp. 313–26.

Lopez, A. 2007. The protection of environmentally displaced persons in international law. *Lewis & Clark Law School's Environmental Law Online*, Vol. 37.

Magniny, V. 1999. Les réfugiés de l'environnement. Hypothèse juridique à propos d'une menace écologique. Doctoral dissertation. Paris, University Paris I Panthéon-Sorbonne, Department of Law.

Marcs, C. 2008. Spoiling Movi's River: towards recognition of persecutory environmental harm within the meaning of the refugee convention. *American University International Law Review*, Vol. 32, pp. 31–71.

McAdam, J. 2008. Climate change 'refugees' and international law. *Judicial Review Today, Climate change and environmental planning law, Journal of the NSW Bar Association*, Winter.

____ 2009. *Environmental Migration Governance*. Paper 1, p. 31, fn 143. Sydney, University of New South Wales.

McCue, G. S. 1993–1994. Environmental refugees: applying international environmental law to involuntary migration. *Georgetown International Environmental Law Review*, Vol. 6.

Mercure, P.-F. 2006. The quest for a legal status for transboundary environmental migrants: the issues involved in the concept of refugee. *Revue de droit de l'Université de Sherbrooke*, Vol. 37, pp. 1–39.

Moberg, K. K. 2009. Extending refugee definition to cover environmentally displaced persons displaces necessary protection. *Iowa Law Review*, Vol. 94, pp. 1107–37.

OHCHR. 2009. *Report of the Office of the High Commissioner for Human Rights on the Relationship between Climate Change and Human Rights*. Submitted for the tenth session of the Human Rights Council. Geneva, United Nations High Commissioner for Human Rights. (A/HRC/10/61, 15 January.)

Orchard, P. 2010. Protection of internally displaced persons: soft law as a norm generating mechanism. *Review of International Studies*, Vol. 36, No. 2.

Revue européenne du droit de l'environnement. 2008. No. 4, pp. 381–93.

Shen, S. and Gemenne, F. 2008. Tuvalu's Environmental Migration to New Zealand. Environment, Forced Migration & Social Vulnerability, International Conference 9–11 October, Bonn, Germany.

UNHCR. 2008. Climate Change, Natural Disasters and Human Displacement: a UNHCR Perspective. Geneva, United Nations High Commissioner for Refugees. (Policy Paper, 23 October.)

____ 2009a. Forced Displacement in the Context of Climate Change: Challenges for States under International Law. Submission to the UNFCCC Ad Hoc Working Group on Long-term Cooperative Action under the Convention by the UNHCR in cooperation with the Norwegian Refugee Council, Representative of the Secretary-General on the Human Rights of Internally Displaced Persons and the United Nations University, 20 May.

2009*b*. Climate Change and Statelessness: An Overview. UNHCR, supported by IOM and the Norwegian Refugee Council (NRC), Note to the 6th session of the Ad Hoc Working Group on Long-term Cooperative Action under the Convention (AWG-LCA 6), 1–12 June, Bonn. http://www.unhcr.org/refworld/docid/4a2d189d3.html

USCIS. 2004. *DHS Concludes Temporary Protected Status for Nationals of Montserrat*, 6 July. Washington DC, Department of Homeland Security, US Citizenship and Immigration Services. http://www.uscis.gov/files/pressrelease/MontserratTPS_7_6_04.pdf

Westra, L. 2009. *Environmental Justice and the Rights of Ecological Refugees.* London, Earthscan.

Williams, A. 2008. Turning the tide: recognizing climate change refugees in international law. *Law & Policy*, Vol. 30, p. 502.

Zartner Falstrom, D. 2001–2002. Stemming the flow of environmental displacement: creating a convention to protect persons and preserve the environment. *Colorado Journal of International Environmental Law and Policy*, Vol. 1, No. 13.

15

'Environmental refugees': aspects of international state responsibility

ASTRID EPINEY

Introduction

It seems indisputable today that environmental degradation, in general, and global warming, in particular, are a major cause of migration worldwide, which will increase in scope considerably in the next few years.[1] Global warming may even lead to the disappearance of some island states, as sea levels rise (WBGU, 2006, p. 62; Voigt, 2008, pp. 1ff.). However, estimates regarding the number of people affected vary significantly,[2] which in no way diminishes the seriousness of the situation quantitatively (in a few years, the number of environmental refugees could be much higher than the number of people falling within the scope of the Geneva Convention (Biermann and Boas, 2008, p. 10)) or qualitatively – if it is taken into account that environmental degradation may affect vulnerable population groups (women, children, the sick and minorities) in particular.

This observation raises the question of the extent to which current international law admits (or has admitted) obligations relating to the prevention of situations giving rise to 'environmental refugees' and/or the reaction thereto. The problem involves a whole range of issues (see Epiney, 2010). This chapter therefore focuses on aspects relating to the law of international responsibility and, more precisely, on the extent to which and conditions under which the recognized rules of state responsibility for

---

[1] See e.g. WBGU (2007, pp. 125ff.); Zerger (2009); Biermann (2001, pp. 24ff.); Nuscheler (2004); IPPC (2007); Biermann and Boas (2008, pp. 10ff.); Lassailly-Jacob (2006, pp. 374ff.); Chemillier-Gendreau (2006, pp. 447ff., 450); Cournil (2006, p. 1036); Greminger and Jakob (2008, p. 7).

[2] 50 million in 2010 and 150 million in 2050 by some estimates. See Myers (2002, pp. 609ff.) and the summary of the current state of this discussion in Biermann and Boas (2007, pp. 9ff.); Zerger (2009, pp. 85f.).

wrongful acts are applicable in 'environmental refugee' situations. However, as these 'secondary' norms are closely linked to 'primary' norms – which contain obligations with which states must comply – it is equally necessary to refer to primary norms to a certain extent. Therefore, we address the problem – first clarifying the concept of 'environmental refugee' as understood here (Section I), followed by a short reminder of the principles of international responsibility (Section II). In Section III these principles are considered in relation to (1) the potentially violated primary norms, and (2) two specific aspects of state responsibility in this context. Section IV concludes with a few remarks outlining the prospects for the development of international law on the subject.

The analysis focuses, however, on issues regarding international responsibility for wrongful acts (therefore, conduct violating international law) and not on issues regarding the scope of the various state obligations; in other words, we concentrate on the principles of international responsibility rather than the detailed interpretation of individual obligations. Furthermore, we do not address international responsibility for lawful acts (state behaviour in accordance with international obligations), the role of international organizations, in particular the United Nations, and the potential implications of the application of the 'responsibility to protect' concept.[3] Therefore, we do not discuss the institutional aspects (institutions likely to contribute to solving expected problems and currently active in refugee, environmental and funding matters – for this last point, see Biermann and Boas, 2007, pp. 16ff.), without denying their crucial importance.

I The concept of 'environmental refugee' – definitions

There is no official definition of 'environmental refugee' in international law. On the contrary, quite different concepts exist (see summary and other references in Biermann and Boas, 2007, pp. 2ff.; Lassailly-Jacob, 2006, pp. 376f.), some of them even focusing on 'climate refugees', specifically excluding, in certain cases, people who migrate because of environmental degradation for reasons other than global warming (see e.g. Biermann and Boas, 2007, p. 4).

---

[3] According to this concept, in cases of serious and apparent violations of human rights within a state, the international community, the United Nations Security Council and/or third states should (be able to) intervene to protect people affected, by resorting to the use of force if the situation so requires. For the importance of the 'responsibility to protect' concept and the role of the United Nations within the context of 'environmental refugees', see Epiney (2010).

This is neither the place nor the time to review the entire discussion on this subject, particularly because the definition also depends on the objective and the context in which it is used. Therefore, it is merely specified here that, drawing on UNEP's 1985 definition, we understand 'environmental refugees' to mean people forced to leave their residence temporarily or permanently on account of severe environmental degradation that seriously affects their living conditions or makes living in that region (almost) impossible or, at least, unbearable.[4] Such being the case, this definition includes only migration caused directly and primarily by environmental deterioration (due directly or indirectly to human behaviour or natural disaster). For example, in our opinion, the term 'environmental refugees' applies in the following situations to:

- persons fleeing because of a rise in sea level;
- persons fleeing an extreme natural disaster (e.g. a hurricane);
- persons fleeing following a major accident (e.g. of industrial origin) that makes living in the particular region almost impossible or unbearable;
- persons fleeing because environmental degradation – for example, owing to water scarcity or inability to cultivate fields – leads to living conditions that make survival difficult or impossible in the region of origin.

However, it should be noted that migration is generally the result of many factors. The matter cannot be examined in detail here; our reflection is thus limited to individual cases in which a decisive link can be established between environmental degradation and migration, the establishment of this link being problematic and not the subject of discussion below.

Although 'environmental refugees' are obviously not refugees within the meaning of the Geneva Convention,[5] we believe that it is justifiable to use the term 'refugee' in the environmental context. Indeed, the threats

---

[4] See El-Hinnawi (1985, p. 4): 'Environmental refugees are defined as those people who have been forced to leave their traditional habitat, temporarily or permanently, because of a marked environmental disruption (natural and/or triggered by people) that jeopardized their existence and/or seriously affected the quality of life.' For elements for a definition, see further Zerger (2009, p. 88), who, for the most part, agrees with the definition proposed in this chapter.

[5] The Convention of 28 July 1951 relating to the Status of Refugees, RS 0.142.30, and the Protocol of 31 January 1967 relating to the Status of Refugees, RS 0.142.301. Under Art. 1A, the term 'refugee' may apply only to persons who fear being persecuted for reasons of race, religion, nationality, membership of a particular social group or political opinion. As such the 'classic' legal regime in international law on refugees cannot be applied to environmental refugees. For details, see Cournil (2006, pp. 1041ff.). A similar view is given, e.g., in Lassailly-Jacob (2006, p. 378); Cournil and Mazzega (2006, p. 419).

faced by affected persons – based on the elements of the definition provided – are in fact comparable to those faced by persons considered 'refugees' under the Geneva Convention. That said, use of the term 'environmental refugees', 'environmental migrants', 'environmentally displaced persons' or any other is in no way prejudicial to these persons' legal status. Ultimately, it is simply a matter of definitions.

II International responsibility of states for wrongful acts – principles

The rules governing the responsibility of states for internationally wrongful acts have developed over the years[6] and today, the principles set out in the articles formulated in 2001 by the International Law Commission (ILC) – and noted by the United Nations General Assembly (UN doc. A/Res./56/83, 1983) – seem for the most part to be recognized by states as falling under customary international law, although questions on some aspects of these articles are still controversial. The difficulty regarding international responsibility lies, in this context, not in the question of whether or not a particular principle of international responsibility has been recognized as being part of customary international law, but rather in being able to apply the principles of international responsibility within the framework of the (legal) problems raised by the issue of environmental refugees.

This chapter therefore merely outlines the main thrusts of the principles governing international responsibility – that is, the rules that determine the conditions, content and consequences of state responsibility for wrongful acts ('secondary norms') which must be distinguished from 'primary norms' which contain state obligations under international law (for this distinction, see Epiney, 1992, pp. 25ff.). In this regard, the following principles, based on the ILC Articles, should be mentioned (see references in Felder, 2007; Ziegler, 2002):

- The principle of responsibility: all states are responsible for the breach of an international law obligation. In other words, every internationally wrongful act of a state entails the international responsibility of that state (Art. 1 of the ILC Draft Articles).[7]

---

[6] For principles of international responsibility, see e.g. (in addition to the relevant chapters in the various international law books) Felder (2007, pp. 46ff.); Ziegler (2002, pp. 85ff.).

[7] See Permanent International Court of Justice (PCJI), the Chorzow Factory case (1927), PCJI, Series A; No. 8, p. 21.

- The international responsibility of a state is subject to two conditions (Art. 2 of the ILC Draft Articles):
  - first, if the act or omission must be attributable to the state, that is, it must be considered state conduct. Art. 4ff. of the ILC Articles specify the conditions of attribution of conduct to a state. In principle, only the conduct of state organs (Art. 4) or of persons or entities empowered by the state to exercise elements of governmental authority (Art. 5 and 6) may be attributed to that particular state. Art. 8ff. addresses specific questions, in particular state responsibility for 'de facto organs' (Art. 8);
  - second, if the act or omission[8] thus attributable to the state constitutes a breach of an international obligation of that state; this means that there must be a 'wrongful act'. All international law obligations incumbent upon states at the time of the act must be taken into account, regardless of their source – customary law, an international treaty or any other source of international law (see Art. 12ff. of the ILC Articles for details).
- 'Injury' beyond that caused 'automatically' by each breach of international law is not required for finding that a case of international responsibility exists.
- Similarly, there is no requirement to show any 'fault' whatsoever in order to establish international responsibility.
- Art. 20ff. of the ILC Articles identify circumstances in which state responsibility is precluded even though the conditions under Art. 2 are met.
- Art. 28ff. of the ILC Articles codify the content of states' international responsibility. In the present context, it is particularly important to highlight that reparation (see e.g. the wording of Art. 31 and 34ff. of the ILC Articles) will be made only for the injury caused by a wrongful act, in other words, the consequences of the wrongful act, which means that a causal link is required.

These general principles of state responsibility are applicable in the context considered here. Indeed, Art. 55 of the ILC Articles provides that specific obligations under special rules shall be applicable only to the extent that they govern the questions of responsibility under discussion. However, it can also be inferred from this article that the existence of special rules do not exclude, in principle, the use of the general principles

---

[8] An omission is always particularly important if international law admits obligations to act under certain conditions. Some aspects of this issue are discussed in Section III.

unless the specific rules are genuinely considered a 'self-contained regime', which is, all things considered, an exception. Be that as it may, the provisions regarding the consequences of non-compliance with state obligations contained in the texts of international environmental law and international human rights law cannot be considered exhaustive, because the mechanisms provided are generally too weak for it to be considered that they preclude the application of the general principles of state responsibility.[9]

III   State responsibility for 'environmental refugees'

Currently, there is no specific legal instrument regulating the legal situation of environmental refugees or involving states' obligations in their regard or concerning prevention obligations (Cournil, 2006, p. 1040). For this reason, we have directed our reflection to the scope which can be given to a number of non-specific international obligations (with regard to state responsibility for causing situations promoting the emergence of environmental refugees). We first demonstrate that international obligations do exist,[10] which may be particularly important in the present context, because they require states to adopt conduct that is (also) supposed to prevent the emergence of environmental refugees. These are the primary norms that are potentially relevant in this context. We then focus on questions common to a certain number of these primary norms which arise in the context of (potential) state responsibility for causing migratory flows. Therefore, as mentioned in the introduction to this chapter, we highlight aspects of international responsibility, the interpretation of primary rules not being the subject of our reflection, apart from some specific aspects relating to the environmental refugee problem.

[9] In the same vein with regard to international environmental law, see Voigt (2008, pp. 3f.). With regard to human rights, see Simma (2007, p. 365).

[10] Specific community law issues are not addressed in this paper. However, it should be underlined that community law currently has no specific instrument on refugees, although Council Directive D 2001/55 on minimum standards for giving temporary protection in the event of a mass influx of displaced persons and on measures promoting a balance of efforts between Member States in receiving such persons and bearing the consequences thereof (OJ 2001 L212/7), may also apply to environmental refugees. See Cournil (2006, pp. 1052ff.).

1   Relevant international obligations (primary norms)

In our view, a breach of the following international obligations may have a 'causal' link to environmental refugees: (a) human rights obligations, (b) international treaty obligations, (c) customary law obligations, and (d) state responsibility for environmental refugee flows.

However, in our opinion, a general obligation (independently of a violation of international law) on states to assist other states that cannot cope with major difficulties, that is, in this context, states that cannot take preventive, precautionary or adaptive measures to deal with the environmental refugee problem, is not admissible. Currently in international law, there is no general obligation to assist other states.

(a)   Human rights

Binding international texts on human rights – only the 1966 Universal Covenants are considered here[11] – do not provide for any rights relating to a 'healthy environment' or any specific rights regarding the protection of human beings in the event of serious damage to the environment. However, a number of rights guaranteed by the Covenants may be of some significance where there is environmental damage. Mention may be made, in particular, of the following rights guaranteed by one of the 1966 Covenants:

- Art. 12, para. 4, of the International Covenant on Civil and Political Rights (Covenant II) provides that no one shall be arbitrarily deprived of the right to enter his own country. When environmental degradation is such that persons affected must leave their region of origin and go abroad, this right may come into play. Similarly, this may prove important should a state disappear on account of a rise in sea level.
- Art. 6, para. 1, of Covenant II enshrines a right to life. This right is protected under Art. 2 of the European Convention on Human Rights and the European Court of Human Rights has developed in respect of Art. 2 and Art. 8 of the Convention (protection of private life) case law enabling persons to enforce these rights in connection with

---

[11] However, it should be recalled that regional human rights texts may also be very relevant in the context of this paper. The European Court of Human Rights rendered several opinions on the issue of the conditions under which some rights under the European Convention on Human Rights could be important in regard to environmental issues. See references in note 12.

environmental degradation on a certain scale.[12] Environmental degradation can effectively threaten people's lives and may even cause their death.

- Art. 12, para. 1, of Covenant II provides that all persons are free to choose their residence, and that right could be curtailed on account of environmental degradation.
- Art. 11, para. 1, of the International Covenant on Economic, Social and Cultural Rights (Covenant I) recognizes everyone's right to adequate food. As a result of environmental degradation, it may become harder, or even impossible, for some people to feed themselves properly.

These rights – as other human rights – put states under (at least) two 'categories of obligations':[13] (i) first, the obligation for the state to refrain from certain courses of conduct; (ii) second, the obligation to take protective measures under certain conditions; and lastly, (iii) the question arises of whether an obligation not to deny entry can be inferred from these rights.

(i) **Obligation to refrain from certain courses of conduct** In the present context, the obligation to refrain from conduct that violates or may violate human rights may be important in cases in which the state itself – through acts attributable to it under the principles of state responsibility[14] – causes environmental degradation that triggers migration. For example, certain state conduct leading to environmental degradation, thus infringing one or more of the above-mentioned rights, may include continuation by a state of a major industrial project that seriously damages a region's environment and consequently endangers the life or health of a number of people. This obligation therefore serves to prevent environmental degradation that may trigger migratory flows (within a country or towards other countries). Consequently, it is

---

[12] See the Court's (Grand Chamber) judgment of 30 November 2004, Öneryildic/Turkey, No. 48939/99; the Court's judgment of 20 March 2008, Budayeva/Russia, No. 15339/02. For details on the scope of the ECHR as regards prejudice to the environment, see Meyer-Ladewig (2007, pp. 25ff.); Heselhaus and Marauhn (2005, pp. 549ff.); Loucaides (2004, pp. 249ff.); Merino (2006, pp. 55ff.).

[13] State obligations only are considered here, as it is generally recognized that human rights obligations are addressed primarily to states.

[14] See Art. 4ff. of the ILC Draft Articles on state responsibility for internationally wrongful acts, United Nations Document. A/Res./56/83. Regarding doctrine, see e.g. Felder (2007, pp. 46ff.); Ziegler (2002, pp. 85ff.).

incumbent upon the state to refrain from such conduct, although the exact scope of human rights in regard to state actions detrimental to the quality of the environment has not yet been determined.

All the same, the scope of this obligation is limited in the present context:

- First, it only concerns acts by the state itself. However, environmental degradation is caused most often by private individuals.
- Second, there are often several causes of environmental degradation in a given region. It is therefore often impossible to pinpoint a specific and 'isolated' state cut as the sole or primary reason for the environmental degradation, and by extension, for migratory flows of environmental refugees (see details on the chain of causation in III.2(b) below).

(ii) States' obligation to take protective measures Under human rights laws, states are under an obligation not only to refrain from actions that violate these rights, but also to take adequate measures to prevent other people (in particular private individuals) and events (natural) from infringing these rights.[15] As a result, this aspect of the scope of human rights requires states to intervene if it is feared with a measure of probability that a human right may be breached.[16] However, the obligation here is merely an 'obligation of due diligence'. Therefore, the state must be aware of the risk of infringement of a human right. This condition is fulfilled if it is considered that the state could have been made aware by using the precautionary criteria adapted to the situation. Similarly, the measures to be taken are those warranted by the circumstances to prevent a breach of human rights. This entails an obligation of conduct and not of result (for more on these obligations, see III.2(a) below).

This obligation may involve states in which environmental deterioration occurs (states of origin) and third states.

---

[15] See e.g. Kälin and Künzli (2008, pp. 118ff.); Künzli (2001, pp. 215ff.); Nowak (2005, Art. 2, No. 3ff.); Klein (2000, pp. 298ff.); Rensmann (2008, pp. 116ff.); Leckie (1998, pp. 81ff.); Epiney (2003, p. 389).

[16] For the right to life, see the judgment of the European Court of Human Rights in Strasbourg of 30 November 2004 (Grand Chamber), Öneryildic/Turkey, No. 48939/99. In this judgment, the Court recognized that state responsibility may be engaged because adequate protective measures were not taken and it specified the conditions for such a responsibility.

States of origin   In all situations in which environmental degradation is such that affected persons risk losing their lives or being deprived of food necessary for their survival, and are consequently forced to leave their homes, the state in which these vulnerable populations live must, in principle (if the above-mentioned conditions are fulfilled), take appropriate measures (for example, warn of natural disasters, minimize environmental degradation or enable affected populations to leave the region in question in order to settle elsewhere within the state), even in cases where the state itself has not caused the environmental degradation.

However, the scope of this principle is somewhat limited regarding migration due to environmental degradation, because of strict conditions regulating states' obligation to take protective measures.

- First, it should be noted that in many situations, preventive or adaptive measures cannot (adequately) prevent environmental degradation that triggers migration and are simply not at all apt to prevent environmental deterioration. This is especially the case for a whole range of consequences linked to global warming such as the rise in sea levels, extreme natural events and water scarcity in certain regions owing to desertification. Furthermore, even preventive and adaptive measures demanded by the circumstances that usually should prevent environmental degradation often do not suffice. In other words, they simply cannot prevent environmental degradation in all cases. For example, dykes are built to prevent flooding, but they may not hold if the water rises higher than expected (for example because of unusual winds).
- Second, certain events that cause environmental degradation cannot be foreseen.
- Lastly, it must above all be noted that states are only obliged to take measures as required under the due diligence standard. Therefore, states should generally have a legislative and administrative system that enables them to take appropriate measures should there be a threat of infringement of human rights. However, obligations which they are unable to fulfil may not be formulated (see further III.2(a) below). If a state lacks the financial and technological resources necessary to take the preventive or adaptive measures required, in principle, it may not objectively be held responsible under international law for not taking such measures (for details on these principles, see Epiney, 1992, pp. 217ff.). Furthermore, if it is taken into account that most situations involving environmental refugees arise in countries

of the South, in particular Africa and Asia (see Gouguet, 2006, pp. 382ff.), the de jure and de facto limitations of states' obligation to take sufficient protective and adaptive measures become evident.

Third states Nonetheless, the question remains as to the extent to which other states or some of them (that is, third states, not the state on whose territory the environmental degradation and 'environmental migration' take place) are obliged under international law to take some protective measures. Thus third states that themselves contribute to environmental degradation in other countries or fail to take appropriate action to prevent such behaviour by private parties might be obliged to desist from such actions or to take appropriate steps to prevent private parties from committing such acts. One such example might be the major part that the industrialized countries play, through greenhouse gas emissions, in climate warming and thereby in environmental degradation in some other countries (especially those of the South). Another might be where state or private activities in one state jeopardize some of the aforesaid human rights.

However, such an obligation only takes effect for a third state if it can be established clearly that acts committed on the third state's territory are indeed the chief cause of the environmental degradation and, thereby, of a violation of one or more of the aforesaid human rights and the cause of migratory flows: it is the causal relationship which is central in this context (see III.2(b) below). If the third state in question merely contributes 'slightly' to environmental degradation in another state, the current state of international human rights law does not allow such an obligation to be upheld as, in principle, the prime and sole responsibility rests with the state of origin. In this instance, any human rights aspect does not materially add to the general obligation not to cause significant ecological damage on the territory of other states. Nonetheless, where there has been a violation both of this general rule of public international law (see also the comments in III.1(d) below) and of a human right obligation, that aspect might take on a measure of importance in view of the relatively strong enforcement machinery in the field of human rights.

Be that as it may, we believe that a third state's responsibility cannot be repudiated on the sole grounds that it has no effective control on the territory of the state of origin. True, as a general rule the state of origin bears the chief responsibility for ensuring respect for human rights on its territory. However, to the extent that human rights are jeopardized following an action or omission by a third state, we find that there is no argument that allows the applicability of human rights to be repudiated, regardless of extraterritoriality. States are bound by human rights in respect of all state behaviour whether or not some effects of that behaviour are felt in other states (with regard to this entire problem, see Kälin and Künzli, 2008, pp. 142ff.).

(iii) **Prohibition of denial of entry**   Finally, one last question should be considered in respect of obligations that may derive from human rights: whether a principle of non-denial of entry can be deduced from human rights.

It has been established by the case law of the European Court of Human Rights in Strasbourg that Art. 3 of the European Convention on Human Rights (ECHR) forbids states to extradite persons to another state in which they may be subjected to treatment contravening that article.[17] Similarly, the Court recognized that the prohibition of expulsion was also applicable if the right to life (ECHR Art. 2) might be violated in the destination country.[18] As to the scope of Pact II, it would appear that the Human Rights Committee also recognizes, similarly, such a prohibition of denial of entry, at least in cases where torture is to be feared in the destination state.[19]

If these principles are applied to ecological refugees, then two situations may be distinguished.

- If environmental deterioration were to result in a threat to the lives of persons and their flight, and if their return to their state of origin might plausibly endanger them, then the principle of non-denial of entry should arguably also apply. This is analogous to situations in which the lives of the persons concerned are threatened by the state of origin itself, through its agents; in such situations, the principle of non-denial of entry is accepted. For the persons concerned, and more generally in respect of the extent of the right-to-life obligation, the precise origin of the threat to life should have no impact on the scope of the prohibition of denial of entry which governs states' obligations

---

[17] See the European Court of Human Rights, *Soering* v *United Kingdom*, Series A 161 No. 14038/88. For the doctrine see Caroni (2005, pp. 191ff.); Caroni (2006, pp. 187ff.); Kälin and Künzli (2008, pp. 550ff.).

[18] European Court of Human Rights, *Bahaddar* v *the Netherlands*, Rep. 1998-I, No. 78.

[19] See Human Rights Committee, General Comment 20 (1992), No. 9; Human Rights Committee, *C.* v *Australia*, 900/1999 (2002), No. 8.5.

regarding the protection of life in general.[20] That being so, although
the principle of non-denial of entry has not to our knowledge yet been
applied in the context of environmental degradation likely to endanger the lives of persons, it is considered that such a prohibition may be
deduced, in the circumstances stated here, from the case law of the
European Court of Human Rights and of the Human Rights
Committee.

- If, conversely, lives are not in danger but living conditions have
  'merely' deteriorated to the point where people nonetheless leave
  their region of origin, international practice has to our knowledge
  not yet recognized explicitly an obligation to respect a principle of
  non-denial of entry. In particular, cases of danger to health, food and/
  or basic livelihood come to mind.[21] Now, the same arguments may be
  advanced in respect of the right to food or a livelihood as for the right
  to life, with the consequence that return to the state of origin is
  prohibited where a threat to the achievement of those rights is perforce entailed.

In sum, it may be concluded that an obligation of non-denial of entry
arises in all cases where return to the state of origin would jeopardize the
achievement of human rights, including cases where the jeopardy stems
from environmental degradation.

Nonetheless, the practical scope of such an obligation should not be
overestimated, as it applies only in international relations.[22] First, it
presupposes above all – as noted – that return to the country of origin
(probably) perforce entails a risk to the human right in question (right to
life, right to food, etc.). Furthermore, in many cases the persons concerned are able to settle in other regions of their state of origin.
Moreover, even where the prohibition of denial of entry applies, the
status conferred under international law is very insecure: the only claim,
for the time being is that there is a right to remain in the destination
country; on the other hand, no other obligations relating to residence

---

[20] For example, the European Court of Human Rights recognizes that the principle of non-
denial of entry also applies where private groups threaten the persons concerned, see
European Court of Human Rights, *H.L.R.* v *France*, Rep. 1997-III.

[21] Other than exceptional cases of prohibition of denial of entry due to inadequate medical
provision in the country of origin: in this regard see with other references, particularly to
the case of law, Kälin and Künzli (2008, pp. 551ff.).

[22] In cases of internal migration, the state concerned must take appropriate measures of
protection, see III.1(a)(ii) above.

may be deduced from current international law (other than respect for human rights).

(b) Obligations relating to protection of the environment and climate deriving from international treaties

States must respect obligations under international treaties in force for them (Art. 27 of the Vienna Convention on the Law of Treaties). Consequently they must – where appropriate in compliance with the procedures established under the treaty in question – account for any violation of such a contractual obligation and must reinstate the status quo ante or, if that is not possible, repair the damage to the extent possible (see Section II).

It is not possible to examine here the scope of the host of international conventions for protection of the environment and climate. However, one 'cross-cutting' question of particular importance in this context is whether the various obligations enshrined in international conventions (such as the Framework Convention on Climate Change and the Kyoto Protocol thereto, or the Geneva Convention on Long-range Transboundary Air Pollution) comprise, together with protection of the environment *per se*, the prevention of migratory flows due to environmental degradation. This question matters above all in the context of remedies for a violation of international law. If indeed the answer is affirmative, remedies will consist not only in restoring the previous situation but also in 'repairing' the consequences of migratory flows, which could lead to considerable cost.

Although it is not possible to analyse all texts that may bear on this problem, it can be stated that it is not explicitly clear from the international conventions on the environment or on climate change that they also aim to prevent persons from being forced to leave their places of residence because of environmental degradation or climate change. However, the implicit or explicit objective of these conventions is also to protect the environment and so to guarantee that the natural basis for people's survival and quality of life is sustainably protected. Consequently, it is considered that the international conventions on the environment and particularly on climate change also aim, at least implicitly, to prevent people from being compelled to flee because of environmental degradation or climate change.

Consequently, remedies for non-fulfilment of those obligations may in principle also include costs in respect of the creation of migratory flows. Nonetheless, the question of the extent to which a state's violation of its obligations may indeed be considered to have caused migratory flows will arise regularly. As this is a problem common to various primary norms that may potentially play a part in the context of ecological refugees and, moreover, touch on basic principles of the law of the responsibility of states, it is discussed below (see III.2(b)).

In any event, no international responsibility on the basis of violation of international conventions on the environment or climate change may arise if the state in question has respected its commitments arising under a particular convention, even if migratory flows occur because of environmental degradation or climate change.

(c) Obligations under customary international law: the 'Trail Smelter' principle

According to the Trail Smelter case law,[23] a state must prevent significant ecological damage from arising in other states due to activities on its territory or, should such damage arise, must 'repair' the consequences of such damage (no-harm rule). This obligation is regardless of the legality or otherwise of the activity concerned: in other words, states are under an obligation to prevent ecological damage from arising in other states even if it arises from activities that are in essence legal (*inter alia* see Voigt, 2008, pp. 8ff.; Scouvazzim, 2001, pp. 49ff.). This principle is undeniably part of customary international law,[24] although, admittedly, it does not figure very largely in international practice even in cases where the conditions for its application have undoubtedly been met.[25]

The conditions in which this obligation arises are as follows (for a detailed account, see Epiney, 2005, pp. 344ff.; Epiney, 1995, pp. 309ff.):

● Significant environmental damage must have arisen in another state (not necessarily a neighbouring state). Furthermore, states must avoid even the danger of such damage arising, although this principle does not imply the prohibition *per se* of particularly dangerous activities – *ultra-hazardous activities*. It is therefore an obligation in respect of behaviour and not of result.

● The activity causing such environmental damage may consist equally of an act or an omission of a state. In the latter case, which is

[23] See RIAA III 1905. With regard to this principle see Epiney (1995, pp. 309ff.).

[24] The International Court of Justice also refers to this principle, see cases on nuclear armaments and on the Gabcikovo-Nagymaros Project, *ICJ Rep.* 1996, p. 241, para. 29; *ICJ Rep.* 1997, p. 7, para. 41.

[25] For example, the aftermath of the Chernobyl nuclear accident in 1987.

particularly important in this regard, the state is responsible for the environmental damage if it fails to take appropriate measures, depending on circumstances (due diligence) to prevent the damage being caused by private parties (see definition of these due-diligence obligations at III.2(a) below);

- The state behaviour at issue must be the cause of the environmental harm; a causal link must therefore be shown (see III.2(b) below);
- Only significant or considerable damage must be avoided. This requirement is generally met if the environmental degradation is such that persons leave their place of residence.[26]

(d) The responsibility of states for causing environmental refugee flows

If environmental degradation is the (predominant) cause of a refugee flow towards other states, the question arises as to the extent to which a state may be held responsible for causing (in some way) that flow of refugees and to the extent to which a state (and which one) should take measures, in particular environmental protection measures, to prevent such a flow of refugees and/or to restore the situation so that the persons concerned may return home.

In general, there are solid arguments in favour of making states internationally accountable for causing migratory flows, whatever their cause:[27] the fact of triggering migratory flows may in particular constitute a violation of the sovereignty of the state to which the refugees emigrate. The latter is at least de facto compelled to receive the refugees and, as a result, triggering an outflow of refugees violates the sovereign right of each state freely to decide whether or not it wishes to admit persons of foreign nationality on its soil. The state must, however, be found to be genuinely responsible for having caused these migratory flows. In other words, an act imputable to the state must be the root cause of the migratory flows. This is notably the case when a state systematically violates human rights, so causing part of its population to seek refuge in another state. Conversely, if the migratory flows stem, for example, from acts of violence by private parties and if the state has taken action as may reasonably be expected to prevent such acts

---

[26] In the same vein, with regard to the environmental consequences of climate change, see Voigt (2008, p. 9).

[27] See in particular the detailed studies by Achermann (1997, pp. 175ff.); Ziegler (2002, pp. 401ff.).

(due diligence), no behaviour in violation of an international legal obligation may be imputed to the state.[28] Further, a state may be held responsible only if the persons concerned may no longer reasonably be expected to remain in their region of origin; it is nonetheless not necessary for them to be (physically) compelled to flee. Finally, a causal link must be established between the factor triggering the migratory flow and the violation of sovereignty. It can only be recognized if the destination of migration is clear, that is, if, reasonably, only one migration route is available to those affected. If, conversely, the refugees have the choice of several destinations, the causal link cannot be recognized.

These principles – which can only be outlined here – may also be applicable in the particular case of persons leaving their region of origin because of environmental degradation. The conditions under which such state responsibility and obligations obtain should, nonetheless, be noted.

- First, the refugees must be genuinely environmental, which means that they cannot reasonably be expected to remain in their region of origin as a result of environmental degradation.
- Second, a state act or omission must be the cause of the aforesaid environmental degradation. With regard to the state on whose territory the environmental degradation took place, this condition is fulfilled only if that state itself is the cause of the environmental degradation or, where the cause of the degradation lies with private parties or where it results from a natural disaster without (direct) human influence, if the state has neglected to take the preventive action appropriate to the circumstances (due diligence).[29] Thus the two facets of states' obligations (the obligation to refrain and the obligation to prevent) deriving from human rights and already mentioned above are to be found here (see III.1(a)). Other states (third states) may be considered to have caused migratory flows only if a causal link is established between their behaviour and the environmental degradation in the state of origin of the environmental refugees and, thus, of the flow of refugees (see III.2(b) below).
- Third, as mentioned above, there must be a causal link between the migratory flow and the violation of the other state's sovereignty,

---

[28] With regard to the imputing of an act to the state in this context, see the detailed account by Achermann (1997, pp. 147ff.).

[29] In the same vein see Michelot-Draft (2006, pp. 433, 437); Chemillier-Gendreau, (2006, p. 447). With regard to the standard of due diligence see III.2(a) below.

which can only be recognized in situations where in essence there is only one 'migratory route'.

On the whole, it may be concluded that states are obliged to refrain, under the conditions listed above, from triggering migratory flows towards other states and, therefore, should that obligation be violated, then the state in question may be held internationally accountable. In this context, such instances may include, for example a state systematically destroying a population's natural environment or failing to take measures against such action on the part of private parties, such that the population is obliged to take refuge in the neighbouring state. Nonetheless, there are many situations in which the aforesaid conditions are not met, for whatever reason (notably the lack of a causal link between the migratory flow and the violation of the other state's sovereignty or a migratory flow not caused by state behaviour). Furthermore, this obligation only creates a link between the environmental refugees' state of origin and state of destination and does not require the protection of the persons concerned *per se*.

2   Issues under the law of international responsibility

This outline of the various primary obligations that may in theory be invoked if a state causes environmental refugee movement also highlighted two issues that arise with regard to the primary standards: (a) first, the question of the standard of 'due diligence'; and (b) second, the causal link between state behaviour and outflows of environmental refugees.

(a)   The standard of 'due diligence'

This issue arises in particular with regard to the obligations of states in respect of acts by private parties. States may indeed not be held responsible for the activities of private parties as such (because the acts of private parties may not be imputed) but they are obliged (under primary norms) to take protective or preventive action as required in the circumstances (see III.1 above). That being so, the question arises as to how the standard of 'due diligence' can be defined. Although the principle of applying the standard of 'due diligence' appears to be largely recognized in practice, there is less clarity as to the exact criteria to be used in determining whether or not the standard has been respected. However, on the strength of international practice,[30] a few general principles may be identified:

- First, it should be stressed that states can in principle only be expected to take action that they can effectively take. In other words, under their international obligations they are not compelled to apply a standard if they cannot take the necessary measures to do so. In such circumstances they are unable to guarantee such a standard so that there can be no question of appropriate measures. In other words, obligations that states cannot fulfil may not be established. That being so, if a state does not have the requisite financial and technological resources to take the preventive or adaptive measures that are in principle objectively appropriate in order to avoid outflows of environmental refugees, it cannot be held accountable in international law for failing to do so. However, if the state's inability to react in specific circumstances is the outcome of its failure to maintain a state organization sufficient to discharge its various preventive obligations, the state's responsibility must be recognized. Indeed, states are generally required to be organized in such a way as to be able to fulfil their international obligations. In other words, states are obliged to maintain 'sound government' (with regard to these principles see the detailed account in Epiney, 1992, pp. 217ff.).
- If these principles are applied to state responsibility for environmental refugees, then it stands to reason that the states of the North should in general be considered capable of taking the preventive and protective measures that the circumstances objectively require. They do have adequate financial and technological resources and there is nothing to justify any mitigation of their accountability on grounds of inability to act. Conversely, in regard to the behaviour of a state of the South

---

[30] See in detail with other references Epiney (1992, pp. 211ff.). See also Verheyen (2005, p. 174), who describes the standard of due diligence as the behaviour to be expected of a 'good government'. Another author stresses that due diligence must be ascertained in the light of the resources and capabilities available to the state in an international context, see Tomuschat (1999, p. 280). For its part the ILC has defined the standard of 'due diligence' in its commentary on articles on the Prevention of Transboundary Harm from Hazardous Activities as follows: '... to take unilateral measures to prevent significant transboundary harm or at any event minimize the risk of thereof ... Such measures include, first, formulating policies designed to prevent significant transboundary harm or to minimize the risk thereof and, second implementing those policies. Such policies are expressed in legislation and administrative regulations and implemented through various enforcement mechanisms.' *ILC Rep.*, 53rd session, Commentary, 393.

(especially that of an African or Asian state), detailed analysis is required of whether it can reasonably be expected to take specific measures to protect and adapt. There is no doubt that some particularly poor states cannot be compelled to take action that lies outside their capability. Nonetheless, specific analysis is required as to whether that is or is not the case. Furthermore, states are in principle obliged to accept foreign aid where that can contribute to the necessary measures of protection and adaptation being taken. Where such aid is refused without objective reason, states are failing to respect their due-diligence obligations.

• The standard of due diligence must be ascertained in the circumstances of each individual case and obligation. All factors that may be of importance in the context of the obligation in question must be taken into account. In the context of obligations to prevent flows of ecological refugees, the following (non-exhaustive) factors seem particularly important.

• It is not necessary for the causal link between a certain behaviour and a significant environmental degradation to be beyond doubt. Under the precautionary principle, it suffices for that link to exist with a measure of scientific evidence for states to be obliged where appropriate to take the preventive measures that the circumstances require. This being so, for example, it suffices for greenhouse gases to be the cause of climate change to a high degree of probability.

• The environmental degradation should be foreseeable. In other words, it should reasonably be possible, using established scientific methods, to foresee[31] that a particular form of environmental degradation will cause persons to leave their place of residence. This condition is met with regard to climate change but also, for example, with regard to slow desertification or, in some cases, the risk of landslides or earthquakes.

• Protective or preventive action should be the more vigorous where there is a high risk of significant environmental degradation and thus of a large number of environmental refugees. In other words, the greater the risk and scale of environmental degradation, the more extensive states' protective and preventive action is expected to be.

• In general, the principle of proportionality should also be taken into account, such that the technical and economic capabilities of states should be viewed in relation to the (potential) damage caused.

---

[31] It is therefore not necessary for the state to be in practice cognizant of the risks in question, but it suffices that it has been cognizant by applying a standard of due diligence. See also in this vein and with regard to climate change Voigt (2008, pp. 11ff.).

- State action may take effect either on the state's own territory or on the territory of the injured state or the state in which questions are raised about individual rights.

(b)   Causality issues

As noted above, a state's international responsibility for (potential) outflows of environmental refugees presupposes that a causal link may be established between the state behaviour in question[32] and the significant environmental degradation[33] responsible for the outflow of environmental refugees.

A causal link is generally easy to establish where the question is one of specific state behaviour that is clearly the cause of a specific environmental degradation, as may for example be the case of industrial accidents or failure to take precautionary measures against certain specific events that may jeopardize the quality of the environment.

Causality is, conversely, far harder to establish in cases where the environmental degradation at issue stems from a host of factors and it is not possible to establish with a degree of probability that one factor or state behaviour is clearly predominant or decisive in the environmental degradation at issue. In this particular context, that is more the rule than the exception. For example, climate change stems from the emissions generated by a host of actors in many states. The lack of water in some regions or soil erosion also usually have several causes.

In such situations, it should firstly be emphasized that the fact that many states may (potentially) be responsible does not 'automatically' mean that international responsibility on the part of at least some of those states cannot be established. Indeed, in cases where one state's 'contribution' to the environmental degradation of itself suffices to cause significant environmental degradation triggering migratory flows, the causal link must in any case be recognized (see Epiney, 1995, pp. 354f.).[34]

---

[32] This differs according to the primary obligations at issue (see Section III.1).

[33] Only significant environmental degradation may give rise to environmental refugees as the concept is understood in this paper (see Section I).

[34] If such a cause can be defined, it is self-evident that a third state may also be held accountable, although the details are not yet clearly established in international law. Thus the question arises, for example, of the extent to which states should take measures to impose the standard of due diligence to prevent enterprises headquartered on their territory from developing activities that may be environmentally damaging to other states. The chemical accident at Bhopal is one example in which this question was raised but has not been answered in international practice. See Epiney (2009), Union Carbide Case.

Furthermore, the question arises as to whether a causal link may also be recognized in cases where various states' contributions to the environmental degradation at issue are themselves (severally) the factor that triggers a significant environmental degradation, such that only the cumulative effects of several states' behaviour can produce such an effect. As this problem has not yet to our knowledge been raised in international practice, it must be addressed by referring to the basic principles of the primary obligations at issue. Two aspects should then be distinguished.

- First, the current state of international law does not, on the basis of practice, warrant recognition of an international obligation on states to refrain from behaviour that in itself does not infringe any international obligation but merely has 'illegal' consequences when combined with similar behaviour by other states. Indeed the international obligations at issue lead each individual state to engage in a particular course of conduct and it would be unacceptable for one state to be held internationally accountable simply because other states behave in the same way. Furthermore, in such a case no 'direct cause' of the environmental degradation can be ascertained. The damaging effects of climate change may be adduced as an example: although the causes of the increase in mean temperature and the resultant environmental damage (and consequently, possible flows of environmental refugees) are well established, the general rules of state responsibility do not permit one or several states in particular to be designated as responsible for them because the combination of effects is decisive in the final result achieved – climate warming. Consequently, it is not possible to 'construct' the responsibility of given states inasmuch as one state's emissions, comprising large quantities of CO_2, could not suffice to cause the effects of climate change. Furthermore, in view of the current state of international environmental law, it cannot be asserted that – beyond the scope of the obligations contained in treaties and notably in the Kyoto Protocol – states are under no obligation whatsoever to restrict, to whatever level, greenhouse gas emissions from their territory. In sum, harmful activities in themselves cannot be viewed as contrary to the international obligations of the states concerned inasmuch as each state's behaviour in isolation cannot be considered to be the origin of significant environmental degradation.

- Second, the question arises as to whether states that in conjunction are the cause of significant environmental degradation may be bound under international law to 'repair' the effects of their behaviour. On first sight, this question must be answered in the negative because, as stated above, the behaviour at issue does not in general violate any primary norm. However this conclusion is considered to be somewhat hasty: in our view, if the combined contributions of various states reach the necessary threshold, an obligation of proportional redress may be construed from the obligation of states not to cause significant environmental damage in other states. Only this approach can ensure that 'victim' states are not left unprotected, and it is also consistent with the purpose of that obligation under international law. Thus states are obliged to contribute to the prevention of damage that might result from their behaviour in proportion to those behaviours and to 'repair' the consequences of that damage if it nonetheless occurs. This obligation is, in our opinion an exception to the general rule under which states are not answerable for the consequences of behaviour that complies with their international obligations. However, it must be noted that international practice has not yet to our knowledge recognized such an obligation and that its standing in international law is thus unclear. It would be desirable for this matter to be clarified by states through state practice.

IV Conclusion

It can be concluded that the prevailing international law contains a number of binding obligations which (also) relate to environmental refugees and environmental degradation leading to outflows of environmental refugees. It would be desirable for states and international organizations to remember these obligations, as to our knowledge they have never or very rarely been invoked in the case of environmental refugees. That said, it must be observed that the international obligations that are relevant in our context do not specifically refer to environmental refugees. In that they are deficient insofar as only a fraction of the problems to which the plight of environmental refugees gives rise is covered by prevailing international law. In that regard, the following aspects should particularly be emphasized:

- In many cases of environmental degradation, especially those relating to climate change, it is difficult to 'designate' a responsible state.

The mechanisms of international responsibility cannot therefore be applied against states that cause climate change through the activities of emission themselves. This aspect restricts in particular the scope of human rights obligations, as well as states' responsibility for migratory flows. However, it would arguably suffice if the various states that in conjunction cause significant environmental degradation (such as climate change) were to bear, in proportion to their behaviour, the costs arising from such behaviour on the basis of the no-harm rule. It should nonetheless be noted that there is no international practice that interprets that principle within the meaning discussed here.

- With regard to preventive or adaptive measures, the states in which environmental degradation takes place often do not possess the resources required to take adequate measures (which in principle also prevents them being internationally responsible). As regards the other states, international law contains no obligation to provide financial or technical support, other than the aforementioned obligations that may arise from the no-harm rule. Additionally, in many situations 'appropriate' preventive or adaptive measures cannot suffice to prevent the environmental degradation that gives rise to migratory flows. Furthermore, in some situations environmental degradation is not foreseeable.

- The principle of non-denial of entry can be deduced, on certain conditions, from the right to life and the right to food. However, that principle cannot generally resolve the issues arising from the plight of environmental refugees. There is reason to recall in this regard that the status of persons to whom this principle applies remains highly insecure, that its scope of application is relatively narrow (because a threat to human rights must be demonstrated in each individual instance) and that the very concept of environmental refugee is yet to be clarified.

In view of these current shortcomings in international law, we consider that it would be desirable, and is even urgently necessary to give serious and rapid thought to the development of international law, leading to a specific legal instrument resolving a number of issues raised by the plight of environmental refugees. Such a convention should settle issues relating to the 'individual protection' of persons obliged to leave their region of origin as a result of environmental degradation and provide more 'collective' machinery for planning and funding (see considerations in Epiney, 2010).

References

Achermann, A. 1997. *Die völkerrechtliche Verantwortlichkeit flucht-verursachender Staaten. Ein Beitrag zum Zusammenwirken von Flüchtlingsrecht, Menschenrechten, kollektiver Friedenssicherung und Staatenverantwortlichkeit.* Berlin, Nomos.

Biermann, F. 2001. Umweltflüchtlinge. Ursachen und Lösungsansätze. *Aus Politik und Zeitgeschichte,* Vol. 12, pp. 24–29.

Biermann, F. and Boas, I. 2007. *Preparing for a Warmer World: Towards a Global Governance System to Protect Climate Refugees.* Global Governance Project. (Working Paper No. 33.) www.glogov.org

2008. Für ein Protokoll zum Schutz von Klimaflüchtlingen. Global Governance zur Anpassung an eine wärmere Welt. *Vereinte Nationen,* Vol. 56, No. 1, pp. 10–15.

Caroni, M. 2005. Die Praxis des Europäischen Gerichtshofes für Menschenrechte im Bereich des Ausländer- und Asylrechts. In: A. Achermann, A. Epiney, W. Kälin and M. S. Nguyen (eds), *Annuaire du droit de la migration 2004/2005.* Bern/Zurich, Stämpfli/Schulthess.

2006. Die Praxis des Europäischen Gerichtshofes für Menschenrechte im Bereich des Ausländer- und Asylrechts. In: A. Achermann, A. Epiney, W. Kälin and M. S. Nguyen (eds), *Annuaire du droit de la migration 2005/2006.* Bern, Stämpfli.

Chemillier-Gendreau, M. 2006. Faut-il un statut international de réfugié écologique? *Revue européenne de droit de l'environnement,* No. 4, pp. 446–52.

Cournil, C. 2006. Vers une reconnaissance des 'réfugiés écologiques'? Quelle(s) protection(s), quel(s) statut(s)? *Revue du droit public,* No. 4, pp. 1035–66.

Cournil, C. and Mazzega, P. 2006. Catastrophes écologiques et flux migratoires: comment protéger les 'réfugiés écologiques'? *Revue européenne de droit de l'environnement,* No. 4, pp. 417–27.

El-Hinnawi, E. 1985. *Environmental Refugees.* Nairobi, United Nations Environment Programme.

Epiney, A. 1992. *Die völkerrechtliche Verantwortlichkeit von Staaten für rechtswidriges Verhalten im Zusammenhang mit Aktionen Privater.* Berlin, Nomos.

1995. Das 'Verbot erheblicher grenzüberschreitender Umweltbeeinträchtigungen': Relikt oder konkretisierungsfähige Grundnorm? *Archiv des Völkerrechts,* No. 33, pp. 309–60.

2003. Sustainable use of freshwater resources. *Zeitschrift für ausländisches öffentliches Recht und Völkerrecht,* pp. 377–96.

2005. Umweltvölkerrechtliche Rahmenbedingungen für Entwicklungsprojekte. In: *Das internationale Recht im Nord-Süd-Verhältnis.* Heidelberg, Germany, BerDGV 41, pp. 329–81.

2009. Union Carbide Case. In: *Max Planck Encyclopedia of Public International Law*. www.mpepil.com

2010. 'Réfugiés écologiques' et droit international. In: C. Tomuschat and S. Oeter (eds), *Le droit à la vie (The Right to Life)*, Leiden, Netherlands, Martinus Nijhoff.

Felder, A. 2007. *Die Beihilfe im Recht der völkerrechtlichen Staatenverantwortlichkeit.* Zürich, Schulthess.

Gouguet, J.-J. 2006. Réfugiés écologiques: un débat controversé. *Revue européenne de droit de l'environnement*, No. 4, pp. 381–99.

Greminger, T. and Jakob, M. 2008. Klimawandel verursacht Migration. Das komplexe Verhältnis zweier globaler Herausforderungen. *Neue Zürcher Zeitung*, 28.6.2008, p. 7.

Heselhaus, S. and Marauhn, T. 2005. Strassburger Springprozession zum Schutz der Umwelt: ökologische Menschenrechte nach den 'Hatton'-Entscheidungen des Europäischen Gerichtshofs für Menschenrechte. *Europäische Grundrechte-Zeitschrift*, Vol. 32, Nos. 1–3, pp. 549–57.

IPCC. 2007. Summary for Policymakers. In: *Climate Change 2007: The Physical Science Basis*. Contribution of Working Group I to the Fourth Assessment Report of the Intergovernmental Panel on Climate Change. Cambridge, UK/New York, Cambridge University Press.

Kälin, W. and Künzli, J. 2008. *Universeller Menschenrechtsschutz*, 2nd ed. Basle, Nomos.

Klein, E. 2000. The duty to protect and to ensure human rights under the international covenant on civil and political rights. In: E. Klein (ed.), *The Duty to Protect and to Ensure Human Rights*. Berlin, Berlin Verlag, pp. 295–318.

Künzli, J. 2001. *Zwischen Rigidität und Flexibilität: Der Verpflichtungsgrad internationaler Menschenrechte. Ein Beitrag zum Zusammenspiel von Menschenrechten, humanitärem Völkerrecht und dem Recht der Staatenverantwortlichkeit.* Berlin, Duncker & Humblot.

Lassailly-Jacob, V. 2006. Une nouvelle catégorie de réfugiées en débat. *Revue européenne de droit de l'environnement*, No. 4, pp. 374–80.

Leckie, S. 1998. Another step towards indivisibility: identifying the key features of violations of economic, social and cultural rights. *Human Rights Quarterly*, Vol. 20, No. 1, pp. 81–124.

Loucaides, L. 2004. Environmental protection through the jurisprudence of the European Convention on Human Rights. *British Yearbook of International Law*, Vol. 75, pp. 249–67.

Merino, M. 2006. La protection de l'individu contre les nuisances environnementales: de la jurisprudence de la Cour européenne des droits de l'homme au système juridictionnel national de protection. *Revue trimestrielle des droits de l'homme*, No. 65, pp. 55–86.

Meyer-Ladewig, J. 2007. Das Umweltrecht in der Rechtsprechung des Europäischen Gerichtshofs für Menschenrechte. *Neue Zeitschrift für Verwaltungsrecht*, Vol. 26, No. 1, pp. 25–29.

Michelot-Draft, A. 2006. Enjeux de la reconnaissance du statut de réfugié écologique pour la construction d'une nouvelle responsabilité internationale. *Revue européenne de droit de l'environnement*, No. 4, pp. 427–45.

Myers, N. 2002. Environmental refugees: a growing phenomenon of the 21st century. *Philosophical Transactions of the Royal Society of London B*, Vol. 357, No. 1420, pp. 609–13.

Nowak, M. 2005. *United Nations Covenant on Civil and Political Rights, CCPR Commentary*, 2nd ed., Art. 2, No. 3ff.

Nuscheler, F. 2004. *Internationale Migration, Flucht und Asyl*. Wiesbaden, Germany, VS Verlag für Sozialwissenschaften.

Rensmann, T. 2008. Die Humanisierung des Völkerrechts durch das ius in bello – Von der Martens'schen Klausel zur 'Responsibility to Protect'. *Zeitschrift für ausländisches öffentliches Recht und Völkerrecht*, Vol. 68, pp. 111–28.

Scouvazzim, T. 2001. State responsibility for environmental harm. *Yearbook of International Environmental Law*. Vol. 12, pp. 43–67.

Simma, B. 2007. Human rights and state responsibility. In: A. Reinisch and U. Kriebaum (eds), *The Law of International Relations – Liber Amicorum Hanspeter Neuhold*. The Hague, Eleven International Publishing.

Tomuschat, C. 1999. International law: ensuring the survival of mankind on the eve of a new century. *Recueil des cours/Collected Courses*, Tome 281. Académie de Droit International de la Haye/The Hague Academy of International Law.

Verheyen, R. 2005. *Climate Damage and International Law: Prevention Duties and State Responsibilities*. Leiden, Netherlands, Martinus Nijhoff.

Voigt, C. 2008. State Responsibility for climate change damages. *Nordic Journal of International Law*, Vol. 77, Nos. 1–2, pp. 1–22.

WBGU. 2006. *Die Zukunft der Meere – zu warm, zu hoch, zu sauer*. Berlin, German Advisory Council on Global Change.

2007. *Sicherheitsrisiko Klimawandel*. Berlin, German Advisory Council on Global Change.

Zerger, F. 2009. Klima – und umweltbedingte Migration. *Zeitschrift für Ausländerrecht und Ausländerpolitik*, Vol. 29, No. 3, pp. 84–89.

Ziegler, K. S. 2002. *Fluchtverursachung als völkerrechtliches Delikt. Die völkerrechtliche Verantwortlichkeit des Herkunftsstaates für die Verursachung von Fluchtbewegungen*. Berlin, Duncker & Humblot.

16

Concluding remarks on the climate change–migration nexus

STEPHEN CASTLES

Introduction

Climate change has become part of *high politics*. Political parties outdo each other in claiming to be 'green' while yet safeguarding 'vital national interests'; governments issue sustainability policies and regulations; and states and international agencies seek to establish international cooperation on emission control. But the contradiction between environmental protection and economic growth – especially for poorer nations seeking to catch up with the old industrial powers – has proved intractable, and the general lack of real progress in international negotiations on climate change means that serious international collaboration is probably off the agenda for the time being. This implies that the poorer and more vulnerable areas of the world are to be left alone to struggle with the consequences of the global carbon economy, which has been at the root of rapid growth and prosperity for some, but also of environmental degradation and threatening change for many, especially in poorer regions.

At the same time, one part of the climate change debate has become part of *low politics*. Headlines such as 'millions will flee degradation', coupled with the assertion that 'there will be as many as 50 million environmental refugees in the world in five years' time'[1] reinforce existing fears of uncontrollable migration flows.[2] In the UK mass-circulation tabloids have for years run almost daily anti-immigrant – and especially anti-asylum seeker – stories. Alarmist accounts by environmentalists,

---

[1] BBC News, 11 October 2005. http://news.bbc.co.uk/go/pr/fr/-/1/hi/sci/tech/4326666.stm
[2] A web search for the term 'climate refugees' in early 2010 brought up over 700,000 items, many of which sought to outbid each other with claims about the many millions of expected displacees.

warning of tens or even hundreds of millions of displaced persons in the future, support stereotypes of migration as intrinsically bad – part of the 'sedentary bias' which has become so dominant in discourses on migration (Bakewell, 2008). Moreover, such accounts put forward apocalyptic visions of third world poverty and disease swamping the rich parts of the world (Myers, 1997; Myers and Kent, 1995). All the more regrettable then, that Myers' dubious figures (based apparently on the idea that anyone living in an area affected by sea level rise would become an 'environmental refugee'), were taken over as factual by the authoritative Stern review on the economics of climate change prepared for the UK Treasury in 2006 (Stern, 2007).

Thus the international debate on the relationship between climate change and migration has been bedevilled by a rather unproductive confrontation: in the one corner have been environmentalists, aid and advocacy organizations (both governmental and non-governmental); in the other have been scholars of refugee movements and migration. The disagreements have disciplinary, methodological and political dimensions. The *disciplinary divide* is between environmentalists, who see global warming-induced climate change as a powerful new force in population displacement, and migration scholars, who regard environmental factors as just one part of a wider constellation of economic, social and political relationships that motivate people to move. The *methodological divide* is between the use of deductive methods, through which projected changes (such as sea level rise or desertification) can simply be mapped onto settlement patterns to predict future displacement; and the call for inductive micro-level research to examine the interactions between climate change and patterns of vulnerability and resilience. The *political divide* is between those who portray 'environmental refugees' as a threat to welfare and security in developed countries (and thus tend to stigmatize all refugees), and those who seek to defend international legal notions of refugee protection.

Today, we are hopefully in a position to move on from these divisions, partly because of the increased availability of credible empirical research on the theme, and partly because of the experience of increasingly frequent environmental events and their consequences. The time is ripe for interdisciplinary approaches, in which environmental scientists and social researchers should work together to map climate change-induced environmental changes, their effects on living conditions and livelihoods, and the range of responses by affected communities. This requires analysis of past and current experiences at local and regional levels, as a basis for elaborating possible future scenarios, and planning appropriate strategies, policies and legal frameworks.

This volume makes an important contribution to advancing the debate by presenting empirical evidence derived from a variety of contexts and approaches, as well as analyses of policies and normative perspectives. In this brief afterword, I first touch on the past politicization of the climate change displacement debate; second, draw attention to some facets of the growing research-based knowledge of the phenomenon; third, point to the apparent acceleration in climate change and events linked to it in recent years; and finally, talk about the possible contours of a more inclusive and cooperative approach.

History of a controversy

The old argument about the existence of 'environmental refugees' and the extent of climate change-induced displacement has been repeated *ad nauseam*. It seems a waste of time to warm it up yet again, yet unfortunately we are still at a stage of the debate where it is necessary to emphasize the politicization of issues of climate change migration. Perhaps it will soon be possible to write about the topic without retracing these steps.

The discussion goes back at least to the mid-1980s. A much-cited paper published by the United Nations Environment Programme (UNEP) argued that large numbers of people, especially in poor countries, would be displaced by environmental change. The author coined (or at least popularized) a term that was to become a core theme of contention: *environmental refugees* (El-Hinnawi, 1985). There were two reasons why this concept was so controversial.

The first problem was the use of the label 'refugees' for people who moved because of environmental factors. As refugee scholars pointed out, this was a misnomer: in international law, the term 'refugee' refers only to people who have crossed inter-state borders to seek protection from persecution based on a range of factors clearly defined in the 1951 UN Refugee Convention (see Piguet et al. in the introduction to this volume). Environmental or climate change was not included in the Convention. The background to this seeming formalism of the refugee lawyers was the fact that, at the turn of the century, some refugee-receiving states were making serious efforts to water down the Refugee Convention. There was a justified fear that any 'reform' in the Convention would actually mean a reduction in the duty of states to protect refugees. Lawyers therefore argued either for *supplementary instruments* to protect all displaced persons, including those forced from their homes by development projects such as dams and disasters as well as by climate change, or they argued for *better use of existing legal instruments* to protect these groups. These normative issues are thoroughly analysed here by McAdam, Koser, Cournil, and Epiney.

The second problem with the term 'environmental refugee' related to the environmental part. Migration scholars pointed out that environmental factors (both positive and negative) have played a part in migration throughout history. Environmental migration should not be equated with forced displacement. In any case, it is important to distinguish between the wide range of environmental factors that have always interacted with economic, social, cultural and political factors, and the specific issue of climate change brought about by the greenhouse gases resulting from the growing spread of carbon-based industries over the last half-century or so. Migration researchers argued that changes in the environmental conditions for work and life should be seen as a factor in migration, but hardly ever the only or even the predominant cause. Migration scholars emphasized the *multicausality* of migration decisions, and accused environmentalists of postulating environmental *monocausality* for political reasons.

Environmentalists countered this critique by arguing that their emphasis on climate change had a very good purpose: that of awakening the world to the dangers of global warming, by making politicians and the public think about the consequences of large-scale human displacement. Leading proponents of this approach (such as Norman Myers, cited above) warned of tens or even hundreds of millions of displaced persons in the future. Others put forward scenarios of mass displacements as a cause of future global insecurity (Homer-Dixon and Percival, 1996), while certain non-governmental organizations even escalated forecasts of future population displacements up to 1 billion by 2050 (Christian Aid, 2007). It seems that aid and development NGOs saw neo-Malthusian warnings about the 'human tide' of refugees as a way of mobilizing public support for their fundraising. Some intergovernmental organizations have also been eager to climb on the environmental refugee bandwagon, which is one explanation for the spate of reports and books on the topic.

However well intentioned, such shock tactics are risky: not only do they present questionable data, which might undermine public trust in environmental predictions. More seriously, they reinforce existing negative images of refugees as a threat to the security, prosperity and public health of rich countries in the global North. Thus the doomsday prophecies of environmentalists may have done more to stigmatize refugees and migrants and to support repressive state measures against them, than to raise environmental awareness. In response, refugee and migration scholars have argued that such neo-Malthusian visions are based on dubious assumptions and that it is virtually impossible to identify individuals or groups forced to move by environmental factors alone (Black, 2001; Wood, 2001).

In retrospect it seems clear that the politicization and polarization of the debate on migration and the environment has had quite negative consequences. Environmentalists may have been misguided in using exaggerated and threatening images of mass displacement to raise public awareness of public change, but the defensive postures adopted by refugee and migration scholars have also held back scientific analysis and thus probably the development of appropriate strategies to respond to the challenges of climate change-induced displacement. The failure so far of international negotiations on climate change means that we are entering a dangerous new phase, in which polarized positions on the causes and consequences of migration have become distinctly unhelpful. Migration scholars must recognize the potential of climate change to bring fundamental changes in the nature of human mobility, just as environmentalists need to recognize the complex factors that lead some people to adopt migration as part of their survival strategies.

The growing empirical basis

Attention was drawn above to the *methodological divide* characteristic of the early years of the climate change displacement debate: prominent environmentalists confidently asserted that expected climate change-induced developments (such as sea level rise, drought or desertification) could be mapped onto settlement patterns to predict future human displacement, whereas migration and refugee scholars called for micro-level research to examine actual experiences of how communities coped with modifications in their living conditions and economic opportunities resulting from climate change. Knowledge has indeed moved on compared with the turn of the century: researchers have begun to carry out studies at the local and regional levels, and the empirical basis for understanding the effects of climate change is being enhanced.

This does not imply that all the research questions necessary for full understanding and for appropriate policy-making have been addressed: as several contributors point out, there are still major knowledge gaps and important unresolved controversies. For example, Hunter and David draw attention to the deficiencies of gender-blind research, which obscures the ways in which climate change-shaped migratory experiences differ for men and women. Leighton points to wide gaps in the research concerning how and why drought and desertification become a primary driver of migration, as well as on scale and methodology, and on frameworks for migration management. Nonetheless, we are now in a position to go beyond some of the simplistic and often confrontational statements of the past. It is still too early to speak of scientific consensus about the causes, extent and impacts of climate change, but certain ideas seem to be gaining acceptance as pointers for further research and action.

To start with, climate change-induced migration should not be analysed in isolation from other forms of movement – especially economic migration and forced migration. The latter results from conflict, persecution and the effects of development projects (such as dams, airports, industrial areas and middle-class housing complexes). Development-induced displacement is actually the largest single form of forced migration, predominantly leading to internal displacement of 10 million to 15 million people per year, and mainly affecting disempowered groups such as indigenous peoples, other ethnic minorities and slum-dwellers (Cernea and McDowell, 2000). The causality of migration is often complex, with many factors playing a part. Moreover, climate change-induced migration is often closely linked to other aspects of environmental change. The effects of changing farming practices (e.g. mechanization, use of fertilizers and pesticides, monocultures, irrigation, concentration of land ownership) on the environment may be hard to distinguish from cyclical weather variations and long-term climate change. Rural–urban migration and the growth of cities are key social-change processes of our times. All too often this means that people leaving the land end up in urban slums (Davis, 2006), but this cannot be seen simply or even mainly as a result of climate change. Migration scholars now emphasize that environmental factors have been significant in driving migration throughout history – even though such factors have often been neglected in the past. In other words, we should generally look for *multicausality* when studying migration processes and include climate change as one of the factors to be analysed.

The recognition of multicausality in migration clearly implies the need for *multi-* or *interdisciplinarity* in research. The best research on migration includes insights from political economy, economics, geography, demography, sociology, anthropology, law, psychology and cultural studies – and to those we must consciously add environmental sciences. Most of the studies reported here are based on the methodologies of one or a few disciplines. Combining them in a reflexive way within a single work is an important step towards multidisciplinarity – but is aggregative rather than integrative. True interdisciplinarity means going further by developing and applying common conceptual frameworks and methodologies, involving a broad range of natural and social sciences. Steps in this direction are still rare (see e.g. Kniveton et al., 2008), but should be encouraged by both universities and research institutes, and by the official bodies and foundations that help to fund them.

Migration and refugee issues have become heavily politicized. This applies – as argued above – even more strongly to climate change-induced migration. This means that when assessing claims made about climate change displacement, we need to analyse the politics that may be behind them. This need is so pressing that the authors of many empirical studies on the topic feel compelled to start by summarizing the global debate on the topic. There is a clear need for a *political science analysis* of the whole discourse. François Gemenne's thesis (2009) provides an in-depth treatment of the politics of climate change migration, summarized in this volume.

As already pointed out, major advances have been made in *empirical research* on climate change-induced migration, but such studies are still at an early stage and a great deal remains to be done. The most ambitious effort so far has been the Environmental Change and Forced Migration Scenarios (EACH-FOR) project funded by the European Union under its Sixth Framework Research mechanism. Twenty-three research projects were carried out all over the world from 2007 to 2009 (EACH-FOR, 2009). Key members of the EACH-FOR team summarize their experience and findings in their contributions to this volume (Chapter 8). EACH-FOR is an important step forward in empirical research, but some concern has been expressed over the short-term nature of the research and about the rather narrow focus of some projects on perceived environmental 'push' factors (see Jónsson, 2010, for a critical discussion of EACH-FOR African research). Studies focusing specifically on displacement as a problem may be misleading. It is important to link such research with more general studies on environmental and developmental issues in various regions, and to build on the considerable existing expertise of development sociologists, human geographers, anthropologists and area studies specialists.

This volume presents a wealth of empirical studies of migration experiences, in which climate change appears to play a significant part: the chapters by Barbieri and Confalonieri, Bohra-Mishra and Massey, Findlay and Geddes, and Oliver-Smith, all provide important insights from the field. Even studies focusing on legal and normative issues, such as those of McAdam and of Leighton (among others) reflect on significant experiences from real-life cases. Several chapters (Koser, Hugo, Hunter and David) argue that lessons can be learned from other types of migration and from the normative frameworks which have evolved in response to these. Perhaps the most significant contribution in methodological terms is made by Bohra-Mishra and Massey's study on Nepal. This study uses longitudinal data on a range of social indicators (such as class, religion, gender, livelihood patterns and environmental factors) from the Chitwan Valley study. The findings demonstrate the complexity of linkages between climate change and local, internal and international migration.

Recent research indicates that there is still little evidence that climate change has so far caused large increases in migration. Despite worrying prognoses put forward in the past, it is virtually impossible to identify groups of people already displaced by climate change alone. There are certainly groups that have been affected by climatic (or broader environmental) factors, but economic, political, social and cultural factors are also at work. Even the cases portrayed in the media as most clear-cut become more complex when looked at closely. For example, Bangladesh is often seen as an 'obvious example' of mass displacement due to sea level rise, but Findlay and Geddes question this conventional view, showing that longer-term migration is related to differential patterns of poverty, access to social networks, and household and community structures. McAdam challenges another, even more obvious example: that of the 'sinking islands' of Kiribati and Tuvalu. Field research in these places indicates that movements cannot be seen as exclusively due to climate change, and needs to be analysed in the context of policy approaches and migration management models at regional and international levels.

Acceleration of climate change

Yet the absence of the displaced millions predicted by Myers and others just a few years ago should not be taken as a reason for complacency. It seems probable that the forecast acceleration of climate change over the next few decades will have major effects on production, livelihoods and human security. As Hugo discusses in this volume, it is already possible to identify 'hot spot areas which will experience the greatest impact'. Significant changes in communities' ability to earn a livelihood will lead to a range of adaptation strategies, much of which will not involve migration. However, certain families and communities are likely to adapt through temporary or permanent migration of some of their members, while in extreme cases it may become impossible to remain in current home areas, so that forced displacement will ensue.

It is customary in the climate change-induced migration field to distinguish between slow-onset processes and rapid-onset events. For example, trends to higher or lower rainfall in certain areas take place over quite long periods, giving affected groups time to adapt in various ways. Rapid-onset events include cyclones, floods and similar catastrophes that require sudden flight in order for people to survive. Their effect is similar to natural disasters such as volcanoes, earthquakes and tsunamis, which cause similar unpredictable exoduses. Yet extreme-weather events are not altogether unpredictable: global warming seems to be leading to an increased incidence of such events, presumably due to long-term impacts on weather patterns. Similarly, periodic floods in places such as Bangladesh may well be exacerbated by higher sea levels. Yet even in such cases, it is hard to disentangle various factors. Findlay and Geddes, for example, here point to the role of water management policies in neighbouring countries in causing floods in Bangladesh. The human losses caused by the 2004 tsunami in Sri Lanka may have been compounded by settlement policies that involved use of land previously seen as marginal or hazardous. Taking this even further could involve exploring links with ethnic inequality and conflict in the region concerned.

In future, then, migration will continue to be the result of multiple factors in both origin and destination areas, but *the climate change component is likely to become increasingly significant*. Migration is not an inevitable result of climate change, but one possible adaptation strategy of many. It is crucial to understand the factors that lead to differing strategies and varying degrees of vulnerability and resilience in individuals and communities. Moreover, migration should not generally be seen as negative: people have always moved in search of better livelihoods, and this can bring benefits both for origin and destination areas (UNDP, 2009). In their search for better livelihoods and opportunities, migrants should not be seen as passive victims; they have some degree of *agency*, even under the most difficult conditions. Strategies that treat them as passive victims are counterproductive, and protection of rights should also be about giving people the chance to deploy their agency. The objective of public policy should not be to prevent migration, but rather to ensure that it can take place in appropriate ways and under conditions of safety, security and legality (Zetter, 2010).

Towards new approaches?

It is time to move forward from the unproductive confrontations of the last twenty years. This book is an important contribution with its focus on the state of empirical evidence about the relationship between climate change and migration, and on policy responses and normative issues. There is clearly still much to be done in both these areas, in order to achieve appropriate strategies for addressing climate change-induced migration and – even more important – the political willingness to implement such strategies.

In this situation, it is important for researchers to take on the challenge of developing the knowledge base needed for strategic planning, and for research-funding bodies to give priority to such work. An interdisciplinary approach would require migration scholars to recognize that environmental change has often been an important (if much neglected) factor in human mobility, and that global warming is increasing the significance of this factor. Environmental scholars would have to recognize that macro-level forecasts need to be complemented by local studies, in which climate change is only one factor among many shaping livelihood strategies. Putting environmental change into migration studies and human mobility into environmental studies could lead to a new synthesis, which recognizes both the complexity of human responses and the urgent need for understanding and action.

This means addressing normative issues of historical causality and responsibility. Climate change is experienced at the local level, but it has global causes. The ability of individuals, communities and states to respond to such changes is strongly linked to political and economic factors. The causes of climate change lie in the production systems and the consumer-oriented lifestyles of rich countries of the global North – although newly industrializing economies of the South are also beginning to play a major part. It is people in the poorest parts of the world who are most affected by climate change, yet who lack the resources for effective adaptation strategies. Weak states are not a fact of nature, but a result of the inequality arising historically from colonialism and continued today through neoliberal globalization. Decrying potential climate change migration as a threat to the security of developed countries misses the point that such migration is a consequence of the human insecurity imposed on the South in the current global order. This understanding of the inequality that underpins climate change-induced migration can be an important starting point for normative debates about the responsibility of rich nations and the 'international community'.

The main purpose of such debates has to be to galvanize the willingness to act. If interdisciplinary scientific cooperation provides deeper understanding of environmental challenges, vulnerability and resilience, and normative analyses help enhance awareness of the crucial legal and political factors, then there could be a new basis for effective action. It is clear that immigration control – that is the tightening or even militarization of borders to keep out unwanted migrants – will do nothing to address the fundamental causes of displacement. Instead, a whole range of strategies is needed, at a range of socio-spatial levels.

At *local* level, better understanding of challenges to the livelihoods and habitats of specific communities, as well as enhanced knowledge of their resources and capabilities, could provide the basis for local activities, ranging from construction measures (e.g. dykes, wetland conservation, cyclone shelters), through diversification of economic activities to community preparedness measures. Migration may often be a part of local strategies: research shows that many families have used the temporary migration of one or more members to sustain and diversify their livelihoods, while permanent migration (both rural–rural and rural–urban) may be an appropriate response when certain livelihoods and habitats become unviable. Although such activities need to be shaped by local needs, they may also affect national and international policy and funding priorities.

At *national* level, practical interventions to support people affected (or likely to be affected) by climate change are needed in many areas of social action. This may require states to change their attitudes towards rural–urban and cross-border migration, by abandoning restriction and criminalization, and helping people to move in conditions of safety and dignity. Employment policies need to be reshaped to ensure economic inclusion of hitherto marginal groups, while housing strategies must be based on recognition of traditional tenure as well as informal settlements and provision of adequate services. Welfare policies should provide support for climate change-affected populations together with measures for long-term social inclusion. Provision of adequate schooling and educational opportunities is part of this process. Health is clearly a crucial area (see Barbieri and Confalonieri in this volume): climate change can lead to many types of risk, including direct impacts of climatic events, increased incidence of vector-borne diseases (such as malaria or dengue fever), illnesses resulting from malnutrition, and diseases and accidents linked to the displacement process. The stress and uncertainty caused by the anticipation and the experience of climate change threats can also have far-reaching effects for the mental health of affected populations (see health issues in McAdam, 2010).

At *international* level, the key debate of recent years has been between the advocates of mitigation (action to drastically reduce greenhouse gas emissions to slow down climate change) and of adaptation (strategies to help affected communities cope with changes to their livelihoods and habitats). Lack of substantial progress in international action means that it will be too late for mitigation strategies to prevent or even slow down imminent changes – although it remains crucial that emissions be limited as soon as possible to reduce long-term damage. This means that the major polluters – both the old industrial states and the emerging ones – need to recognize their responsibility to work together globally to provide financial, scientific and logistical support for adaptation strategies at local and national levels, especially in the poorer regions that will bear the brunt of change, while lacking the resources for effective action on their own.

In retrospect, past political confrontations about climate change migration have probably done more to hinder than to help the development of such multilevel action strategies. By creating fears of mass influxes of the 'misery of the world' to the rich countries, politicians and the media have encouraged racist and restrictionist responses which offer nothing to address the issues. The defensive reactions of migration and refugee scholars have been understandable, but have also done little to develop understanding and capabilities for action. That debate should end now, and be replaced by a new and fine-grained collaborative effort to understand the real challenges and to find solutions. The current international climate may be hardly propitious to such endeavours, but perhaps this publication will be a contribution to creating a new global public ready to campaign for action.

References

Bakewell, O. 2008. Keeping them in their place: the ambivalent relationship between development and migration in Africa. *Third World Quarterly*, Vol. 29, No. 7, pp. 1341–58.

Black, R. 2001. *Environmental Refugees: Myth or Reality?* Geneva, United Nations High Commissioner for Refugees. (*New Issues in Refugee Research*, No. 34.)

Cernea, M. M. and McDowell, C. (eds). 2000. *Risks and Reconstruction: Experiences of Resettlers and Refugees.* Washington DC, World Bank.

Christian Aid. 2007. *Human Tide: The Real Migration Crisis.* May, London. http://www.christianaid.org.uk/Images/human-tide.pdf

Davis, M. 2006. *Planet of Slums.* London/New York, Verso.

EACH-FOR. 2009. *Environmental Change and Forced Migration Scenarios (EACH-FOR).* http://www.each-for.eu/index.php?module=main

El-Hinnawi, E. 1985. *Environmental Refugees.* Nairobi, United Nations Environment Programme.

Gemenne, F. 2009. Environmental Changes and Migration Flows: Normative Frameworks and Policy Responses. *Institut d'Etudes Politiques de Paris and Doctoral School in Social Science of the French Community of Belgium.* Paris/Liège, Sciences Po Paris/University of Liège.

Homer-Dixon, T. and Percival, V. 1996. *Environmental Security and Violent Conflict: Briefing Book.* Toronto, University of Toronto and American Association for the Advancement of Science.

Jónsson, G. 2010. *The Environmental Factor in Migration Dynamics – a Review of African Case Studies.* Working Paper 21. Oxford, International Migration Institute. http://www.imi.ox.ac.uk/news/wp21-published-the-environmental-factor-in-migration-dynamics.

Kniveton, D., Schmidt-Verkerk, K., Smith, C. and Black, R. 2008. *Climate Change and Migration: Improving Methodologies to Estimate Flows.* Geneva, International Organization for Migration. (Migration Research Series No. 33.) http://www.iom.int/jahia/webdav/site/myjahiasite/shared/shared/mainsite/published_docs/serial_publications/MRS-33.pdf

McAdam, J. (ed.). 2010. *Climate Change and Displacement: Multidisciplinary Perspectives.* Oxford, UK, Hart Publishing.

Myers, N. 1997. Environmental refugees. *Population and Environment,* Vol. 19, No. 2, pp. 167–82.

Myers, N. and Kent, J. 1995. *Environmental Exodus: an Emergent Crisis in the Global Arena.* Washington DC, Climate Institute.

Stern, N. 2007. *The Economics of Climate Change: The Stern Review.* Cambridge, UK, Cambridge University Press.

UNDP. 2009. *Human Development Report 2009: Overcoming Barriers: Human Mobility and Development.* New York, United Nations Development Programme. http://hdr.undp.org/en/reports/global/hdr2009/.

Wood, W. B. 2001. Ecomigration: linkages between environmental change and migration. In: A. R. Zolberg and P. M. Benda (eds), *Global Migrants, Global Refugees.* New York/Oxford, Berghahn, pp. 42–61.

Zetter, R. 2010. The conceptual challenges: developing normative and legal frameworks for the protection of environmentally displaced people. In: McAdam (ed.), op. cit., pp. 131–150.

INDEX

Page number in bold type indicate tables and figures

Printed in the United States
By Bookmasters